Green Chemistry
An Introductory Text
2nd Edition

Green Chemistry
An Introductory Text
2nd Edition

Mike Lancaster
Chemical Industries Association, London, UK

RSC Publishing

ISBN: 978-1-84755-873-2

A catalogue record for this book is available from the British Library

Published by The Royal Society of Chemistry,
Thomas Graham House, Science Park, Milton Road,
Cambridge CB4 0WF, UK

Registered Charity Number 207890

For further information see our web site at www.rsc.org

Preface to 2nd Edition

In the eight years since the first edition was published the concept and culture has moved on considerably. First, the dangers of global warming and climate change due to excessive green house gas emissions has become almost universally accepted and governments have put agreements and legislation in place to try and curb society's emissions. Second, the oil price has soared, reaching a peak in early-2008 before the world entered recession. Both of these have been key drivers for the use of alternative feedstocks for energy, especially transport fuels, and have focused energy intensive users into development of more energy efficient processes.

The growth of the biofuels industry in Europe has been phenomenal over the last few years, driven by legislation to incorporate renewable fuels. This has brought its own issues and has been blamed by some for the rise in food prices that has forced governments to rein back plans to increase the minimum amount of biodiesel that should be incorporated, for example.

Energy has been the key driver for change but environmental regulation has also played a part with waste disposal, especially of hazardous material, becoming very expensive, and the REACH regulations starting to focus attention on substances that can cause harm. Industry, although facing financial pressures because of high energy prices, has responded positively and Green Chemistry culture and thinking is now much more widespread than at the start of this century.

These changes have been incorporated into this edition, with renewables, biofuels, and biocatalysis as well as regulation now receiving more

Green Chemistry: An Introductory Text, 2nd Edition
By Mike Lancaster
© Mike Lancaster 2010
Published by the Royal Society of Chemistry, www.rsc.org

in-depth coverage. As in the first edition this book is about changing the culture and the way we go about developing new products and processes. There is a strong focus on chemistry that has been commercialized and comparing modern synthetic paths with those previously used.

Mike Lancaster

Preface to 1st Edition

Many academic texts are available to teach chemists the fundamental tools of their trade, but few books are designed to give future industrial research and development chemists the knowledge they need to contribute, with confidence and relevance, to the development of new environmentally benign chemical technology. This book aims to be a handbook for those chemists attempting to develop new processes and products for the twenty- first century, which meet the evermore stringent demands of a society that wants new products with improved performance, and with a lower financial and environmental price tag.

The concepts discussed in this book, including waste minimization, feedstocks, green metrics and the design of safer, more efficient processes, as well as the role catalysts and solvents can play, are outlined in simple language with the aim being to educate, rather than over complicate. Industrially relevant examples have been included throughout the text and are brought together in Chapter 9 on Industrial Case Studies. Whilst these studies are taken from across various sectors of the chemical industry, wherever possible I have drawn extensively on my own research and process development experiences in various chemical companies, in order to produce a text that will be of real value to the practising chemist.

Green Chemistry means different things to different people: some purists would argue that chemists and the chemical industry have no right appropriating the term at all. At the other end of the spectrum there are individuals and companies that see the 'green' label as a route to product differentiation and higher profits, but wish to do as little as possible in terms of making the step changes needed to achieve

Green Chemistry: An Introductory Text, 2nd Edition
By Mike Lancaster
© Mike Lancaster 2010
Published by the Royal Society of Chemistry, www.rsc.org

sustainability. My own view is somewhere in the middle and can be summarized quite simply. As a society we should be using our skill and ingenuity to develop products and processes that meet our requirements in as sustainable and environmentally benign ways as possible. Green Chemistry should not be about making products with inferior performance or using end-of-pipe solutions to get an Eco-label. It should be about using our resources to produce the materials we need with as minimal negative impact on the world as possible. Sometimes there will be a price to pay, but the ingenious 'Green Chemist' will devise win-win-win products and pro- cesses, in line with the Triple Bottom Line benefits now pursued by many industry sectors.

Whilst the content of this book is broadly based around under-graduate modules and a Masters course in Clean Chemical Technology at the University of York, it should also be of interest to industrial chemists, engineers and managers wishing to learn about Green Chemistry. Since Green Chemistry essentially covers most of chemistry and chemical engineering, the in-depth background information cannot be presented in a book of this size (or, indeed, in several books of this size). The book therefore is designed to be read at two levels. First, the principles and concepts behind the subject are simply presented, enabling them to be understood and appreciated by the 'amateur'. Secondly, those with a more thorough understanding of chemistry will be able to use their knowledge to fully understand the in-depth background to the information sum-marized. In order to keep the book simple, references to the primary literature have only been given in the chapter on Industrial Case Studies. In other chapters, further reading has been suggested, which will give in- depth information on the concepts covered, as well as reviewing particular aspects of Green Chemistry in more detail. These suggestions are given in the same order as the concepts they deal with are introduced in the text. Review questions have been included at the end of each chapter; these have not been especially de- signed to test knowledge, but are intended to encourage the reader to think about, and apply the concepts covered, to new situations.

There are many people who have contributed to my enthusiasm for, and understanding of, Green Chemistry, not least the active members of the Green Chemistry Network who have been so supportive over the last three years. Special thanks are due to James Clark, who, apart from introducing me to the subject, got the Green Chemistry movement going in the UK, not least by convincing the Royal Society of Chemistry to fund the GCN and the *Journal of Green Chemistry*. Thanks are also due to colleagues from similar organizations to the GCN based outside the UK, in particular in the USA, Japan and Italy, who have contributed so much to the global understanding and development of Green Chemical Technology. Whilst it is somewhat unfair to select one person from the many who have contributed to the pursuance of the principles of Green Chemistry, it would also be unfair not to mention Paul Anastas, who has been such a superb global ambassador. Finally a very special thank-you to my wife Gill, not only for her understanding during the writing process but also for reviewing much of the text and making constructive suggestions from a critical chemical engineer's viewpoint!

Mike Lancaster
York, February 2002

Contents

Green Chemistry: An Introductory Text, 2nd Edition
By Mike Lancaster
© Mike Lancaster 2010
Published by the Royal Society of Chemistry, www.rsc.org

CHAPTER 1

Principles and Concepts of Green Chemistry

1.1 INTRODUCTION

During the twentieth century chemistry changed forever the way we live. Perhaps the greatest perceived benefits, to the general public, have come from the pharmaceuticals industry with developments of painkillers, antibiotics, heart drugs, and targeted cancer drugs. However, it is difficult to think of an important facet of modern life that has not been transformed by products of the chemical and related industries, for example:

- transportation – production of gasoline and diesel from petroleum and more recently crops, fuel additives for greater efficiency and reduced emissions, catalytic converters, plastics to reduce vehicle weight and improve energy efficiency.
- clothing – man-made fibres such as rayon and nylon, dyes, water-proofing, and other surface finishing chemicals.
- sport – advanced composite materials for tennis and squash rackets, all weather surfaces, textiles that let the body breathe and reduce wind resistance.
- safety – lightweight polycarbonate cycle helmets, fire retardant furniture, air bags.
- food – refrigerants, packaging, containers and wraps, food processing aids and preservatives.

Green Chemistry: An Introductory Text, 2nd Edition
By Mike Lancaster
© Mike Lancaster 2010
Published by the Royal Society of Chemistry, www.rsc.org

- medical – artificial joints, 'blood bags', internal stitches that dissolve, anaesthetics, disinfectants, vaccines, dental fillings, contact lenses, contraceptives.
- office – photocopying toner, inks, printed circuit boards, liquid crystal displays.
- home – material and dyes for carpets, plastics for TVs, and mobile phones, CDs, video and audio tapes, paints, detergents.
- farming – fertilizers, pesticides.

Figure 1.1 shows the value of the chemical industry. In Europe over 1.3 million people are employed by the industry (including pharmaceuticals and plastics) and it is manufacturing's number one exporter in the UK.

In many countries, however, the chemical industry is often viewed, by the general public, as causing more harm than good. There are several reasons for this, including general ignorance of the end use and value of the industry's products, since the chemical industry rarely sells to the end consumer. However, a major reason is that the industry is perceived as being polluting and causing significant environmental damage. Although a very safe industry in general, well publicised disasters such as Bhopal, causing both environmental damage and loss of life, have led to this generic view. As well as specific disasters, general pollution that came to the public's attention in the 1960s and 1970s through eutrophication, foaming rivers, the discovery of persistent organic pollutants, and the famous 'burning' Cuyahoga River have all played a part in formulating this view of the chemical industry.

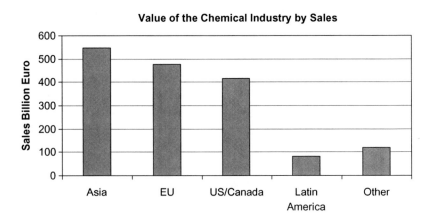

Figure 1.1   Chemical industry turnover.

Chemists and engineers engaged in development of chemical products and processes have never set out to cause damage to the environment or human health. These have occurred largely through a lack of knowledge, especially of the longer-term effects of products entering the environment and failures of procedures to ensure that hazardous operations are carried out safely. The challenge for the chemical industry in the twenty-first century is to continue to provide the benefits we have come to rely on, in an economically viable manner, but without the adverse environmental side effects. This can be achieved through development of more environmentally benign products using less hazardous processes and raw materials. With global warming being accepted as the biggest environmental challenge we face, the chemical industry must also develop more energy efficient processes and reduce its reliance on fossil fuels.

1.2 SUSTAINABLE DEVELOPMENT AND GREEN CHEMISTRY

Current thinking on sustainable development came out of a United Nations Commission on Environment and Development in 1987 (Brundtland Commission), which defined sustainable development as:

'... meeting the needs of the present without compromising the ability of future generations to meet their own needs.'

Although the ideals on which sustainable development is based are not new, indeed Thomas Jefferson made similar comments in 1789, the Brundtland Commission did catalyse the sustainability debate. Since 1987, governments, NGOs, society in general and industry sectors have considered what sustainable development really means and how best to start to achieve it from their own standpoint. Issues that will have a significant impact on how the move towards sustainability is approached include timescale, likely future technology developments, and population forecasts. Two of the key aspects of sustainable development from a chemicals and energy perspective are: (i) How fast should we use up fossil fuels? (ii) How much 'waste' or pollution can we safely release to the environment? Whilst there are no agreed answers to these questions there is general agreement to develop more renewable forms of energy and to reduce pollution.

The Natural Step, an international movement, started in Sweden, dedicated to helping society reduce its impact on the environment has developed four system conditions for sustainability:

1. Materials from the Earth's crust (*e.g.* heavy metals) must not systematically increase in nature.

2. Persistent substances produced by society (*e.g.* DDT, CFCs) must not systematically increase.
3. The physical basis for the Earth's productive natural cycles must not be systematically deteriorated.
4. There must be fair and efficient use of resources with respect to meeting human needs.

This approach recognizes that the Earth does have a natural capacity for dealing with much of the waste and pollution that society generates, it is only when that capacity is exceeded that we become unsustainable.

During the early 1990s the US Environmental Protection Agency (EPA) coined the phrase Green Chemistry:

'To promote innovative chemical technologies that reduce or eliminate the use or generation of hazardous substances in the design, manufacture and use of chemical products.'

Over the last 15 years Green Chemistry has gradually become recognized as both a culture and a methodology for achieving sustainability. Green Chemistry is not a new branch of chemistry but an approach to carrying out chemistry and engineering in a sustainable manner.

The '12 Principles of Green Chemistry' (Box 1.1) help show how this can be achieved.

Many of these 12 principles are expanded on in later chapters. When looking at Green Chemistry from an industrial perspective it is important to take costs of implementing green technology into account, so from this point of view it is helpful to look at Green Chemistry as a reduction process (Figure 1.2). From this perspective it becomes obvious that through application of Green Chemistry concepts significant savings can also be made. These come from reduced raw material use, lower capital expenditure, lower costs of waste treatment and disposal, *etc.* The fundamental challenge for the chemical industry is to continue to provide the benefits to society without overburdening or causing damage to the environment – and all this must be done at an acceptable cost.

1.2.1 Green Engineering

Of course in industry the greenness of an overall manufacturing process will depend on the input of engineers (chemical, mechanical, and electrical) as well as chemists. Like Green Chemistry the term Green Engineering was coined by the EPA. Green Engineering is the design,

Box 1.1   The 12 principles of green chemistry

(Reproduced with permission from P. C. Anastas and J. C. Warner, *Green Chemistry: Theory and Practice*, Oxford University Press, New York, 1998.)

1. Prevention
It is better to prevent waste than to treat or clean up waste after it has been created.

2. Atom Economy
Synthetic methods should be designed to maximize the incorporation of all materials used in the process into the final product.

3. Less Hazardous Chemical Synthesis
Wherever practicable, synthetic methods should be designed to use and generate substances that possess little or no toxicity to people or the environment.

4. Designing Safer Chemicals
Chemical products should be designed to effect their desired function while minimizing their toxicity.

5. Safer Solvents and Auxiliaries
The use of auxiliary substances (*e.g.* solvents or separation agents) should be made unnecessary whenever possible and innocuous when used.

6. Design for Energy Efficiency
Energy requirements of chemical processes should be recognized for their environmental and economic impacts and should be minimized. If possible, synthetic methods should be conducted at ambient temperature and pressure.

7. Use of Renewable Feedstocks
A raw material or feedstock should be renewable rather than depleting whenever technically and economically practicable.

8. Reduce Derivatives
Unnecessary derivatization (use of blocking groups, protection/deprotection, and temporary modification of physical/chemical processes) should be minimized or avoided if possible, because such steps require additional reagents and can generate waste.

9. Catalysis
Catalytic reagents (as selective as possible) are superior to stoichiometric reagents.

10. Design for Degradation
Chemical products should be designed so that at the end of their function they break down into innocuous degradation products and do not persist in the environment.

11. Real-time Analysis for Pollution Prevention
Analytical methodologies need to be further developed to allow for real-time, in-process monitoring and control prior to the formation of hazardous substances.

12. Inherently Safer Chemistry for Accident Prevention
Substances and the form of a substance used in a chemical process should be chosen to minimize the potential for chemical accidents, including releases, explosions, and fires.

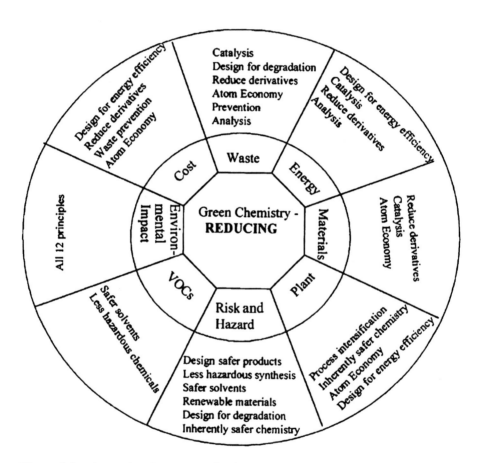

Figure 1.2   Green chemistry as a reduction process.

commercialization, and use of processes and products that are feasible and economical while minimizing (i) generation of pollution at the source and (ii) risk to human health and the environment. Green Engineering embraces the concept that decisions to protect human health and the environment can have the greatest impact and cost effectiveness when applied early to the design and development phase of a process or product. In Chapters 7–9 in particular we will see how innovative engineering has helped developed more sustainable processes. An increasingly important aspect of Green Engineering is the development of more energy efficient equipment, ranging from smaller 'process intensified' equipment to highly efficient centrifugal pumps.

1.3 ATOM ECONOMY

One of the fundamental and most important principles of Green Chemistry is that of atom economy. This essentially is a measure of how many atoms of reactants end up in the final product and how many end up in by-products or waste. The percent atom economy can be calculated as 100 times the relative molecular mass (RMM) of all atoms used to make wanted product divided by the RMM of all reactants (Box 1.2). The real benefit of atom economy is that it can be calculated at the reaction planning stage from a balanced reaction equation. Taking the theoretical reaction (1.1):

$$X + Y = P + U \tag{1.1}$$

The reaction between X and Y to give product P may proceed in 100% yield with 100% selectivity but because the reaction also produces unwanted materials U its atom economy will be less than 100%.

Traditionally chemists are taught to maximize the yield of a reaction. Whilst this is a worthy goal and is an effective measure of the efficiency of a particular reaction it is not a particularly good measure of comparing efficiencies between different reactions. Taking two of the production routes for maleic anhydride as an example, (Scheme 1.1) it is evident that the butene oxidation route is considerably more atom efficient and avoids 'wasting' two carbon atoms as CO_2. Comparison of

---

Box 1.2 Measures of reaction efficiency

$$\% \text{ yield} = 100 \times \frac{\text{actual quantity of products achieved}}{\text{theoretical quantity of products achievable}}$$

$$\% \text{ selectivity} = 100 \times \frac{\text{yield of desired product}}{\text{amount of substrate converted}}$$

$$\% \text{ atom economy} = 100 \times \frac{\text{relative molecular mass of desired products}}{\text{relative molecular mass of all reactants}}$$

Benzene Oxidation

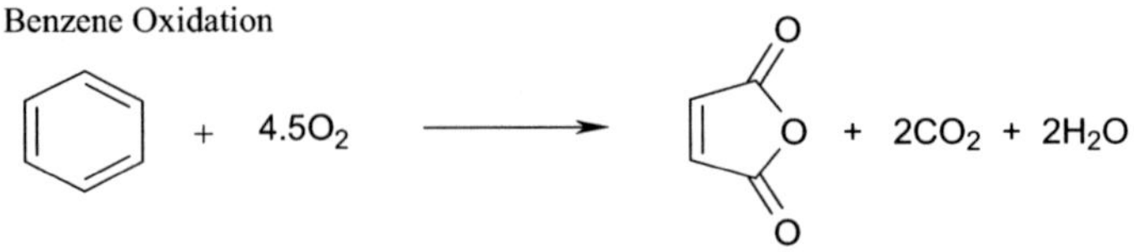

Formula weights

    78   4.5 x 32 = 144   98

    % Atom Economy = 100 x 98/(78+144) = 100 x 98/222 = 44.1%

Butene Oxidation

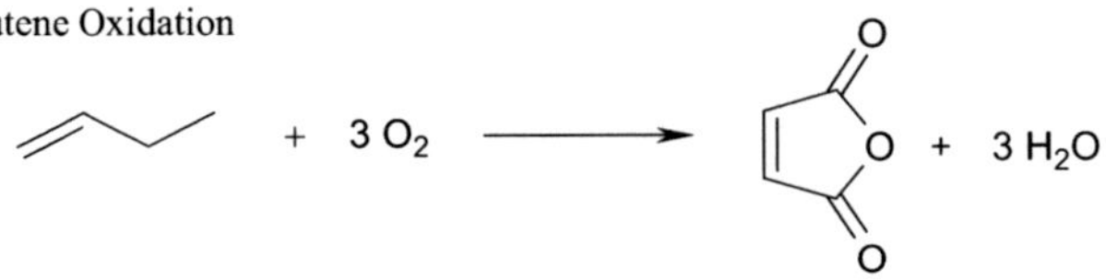

Formula weights

    56   3 x 32 = 96   98

    % Atom Economy = 100 x 98/(56+96) = 100 x 98/152 = 64.5%

Scheme 1.1   Atom economy for maleic anhydride production routes.

the two routes is interesting since both occur under similar reaction conditions of 400 °C in the presence of a promoted vanadium pentoxide catalyst. Initial processes were based on benzene but for awhile butene oxidation became the preferred route due to simpler separation technology.

Today most plants use butane as a feed stock because of the lower raw material price. Whilst, at the design stage, the choice of butene over benzene would appear obvious the two routes do have differing selectivities, negating some of the atom economy benefits of the butene route. Using benzene typical selectivities of around 65% are obtained commercially but for butene it is approximately 55%. If we multiply the theoretical atom economies by these figures we obtain practical atom economies of 28.7% for the benzene route and 35.6% for butene. This is a useful illustration of how good atom economy can compensate for poorer yields or selectivities and is a valuable additional tool in measuring overall reaction efficiency.

By taking atom economy of various synthetic routes into account at the planning stage the chosen strategy is likely to produce a greater weight of products per unit weight of reactants than may have otherwise

Table 1.1 Some atom economic and some atom un-economic reactions.

Atom economic reactions	Atom un-economic reactions
Rearrangement	Substitution
Addition	Elimination
Diels–Alder	Wittig
Other concerted reactions	Grignard

been the case. There are, however, several common reaction types that are inherently atom efficient and a number that are not (Table 1.1). Although the reactions under the atom economic heading are generally so, each specific reaction should be considered individually since, for example, unrecoverable 'catalysts' sometimes need to be used in significant amounts. On the other hand, some atom un-economic reactions may involve, for example, elimination of water, which does not significantly detract from the greenness.

1.4 ATOM ECONOMIC REACTIONS

1.4.1 Rearrangement Reactions

Rearrangements, especially those only needing heat or a small amount of catalyst to activate the reaction, display total atom economy. A classic example of this is the Claisen rearrangement, which involves the rearrangement of aromatic allyl ethers (Scheme 1.2). Although *ortho*-substituted products usually predominate some *para*-alkylated products are also formed, reducing overall yield. The reaction may, however, be both high yielding and atom economic when di-*ortho*-substituted aromatic ally ethers are used.

The Fries rearrangement of phenolic esters (Scheme 1.3) is normally 'catalysed' by stoichiometric amounts of a Lewis acid such as $AlCl_3$. The requirement for this large amount of 'catalyst' is due to it complexing with the product; work up with water hydrolyses the complex, producing copious amounts of aluminium waste. This significantly reduces the atom economy of the reaction as the $AlCl_3$ should be considered a reagent rather than a catalyst since it is not recovered in a reusable form.

A useful solution to this problem is the photo-Fries rearrangement, in which UV light is used to generate $RCOO^{\bullet}$ radicals. In this case, the reaction proceeds *via* an intermolecular free radical route rather than through nucleophilic attack as in the conventional process. Selectivity of these reactions may be improved by imposing steric control through carrying out the reactions in zeolite or cyclodextrin cages.

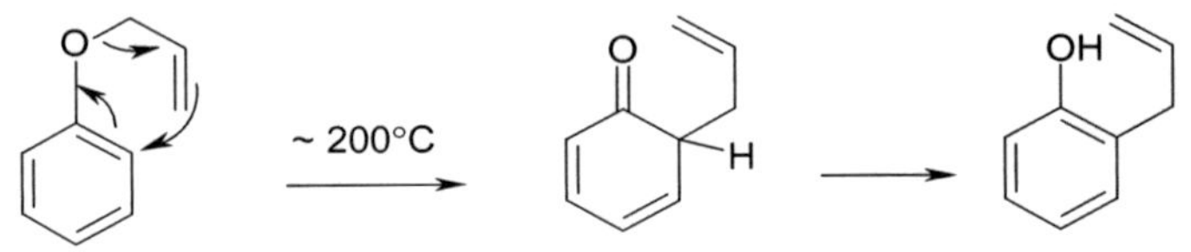

Scheme 1.2 Claisen rearrangement.

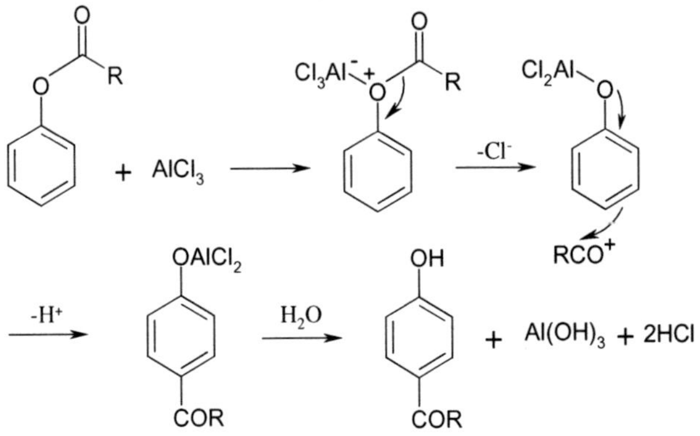

Scheme 1.3 Fries rearrangement.

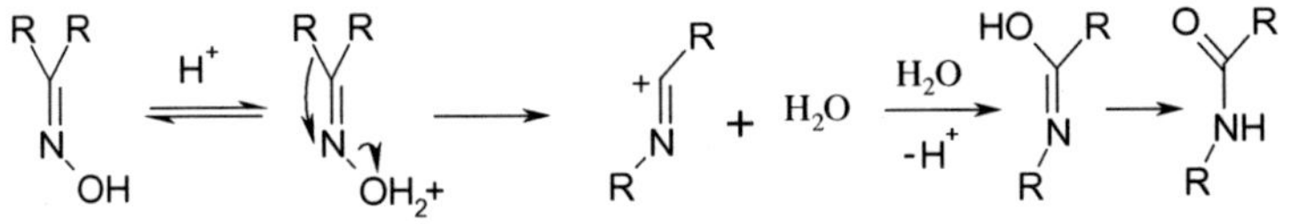

Scheme 1.4 Beckmann rearrangement.

Another valuable rearrangement reaction that is usually 'catalysed' by stoichiometric amounts of catalyst is the Beckmann rearrangement (Scheme 1.4). This reaction is used commercially for converting cyclo-hexanone oxime into caprolactam, a key intermediate for nylon 6; oleum (20%) is the usual catalyst.

A wide range of heterogeneous catalysts has been explored that avoid the need for using oleum. In particular, certain zeolites, notably [B]-MFI, have been shown to give excellent yields (up to 94% in fluidized bed reactors), and good stability and ease of regeneration. These processes often operate most efficiently in the gas phase and are relatively energy intensive, preventing widespread use.

One other 100% atom economic rearrangement is worth briefly mentioning. When vinylcyclopropanes are heated they readily undergo

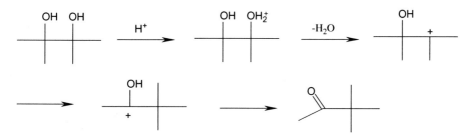

Scheme 1.5 Pinacol rearrangement.

ring expansion to cyclopentenes. The temperature required varies significantly, depending on the molecule; for example, 1-phenyl-2-vinyl-cyclopropane may be converted into phenylcyclopentene in reasonable yield at 200 °C.

Many other versatile so-called rearrangements involve elimination of water. Whilst this does reduce the atom economy these reactions are worth considering when devising a green synthesis. The pinacol rearrangement of *vic*-diols into aldehydes or ketones is catalysed by acids, frequently sulfuric acid is used, but there are also many examples of supported acid catalysts such as $FeCl_3$ on silica being used (Scheme 1.5). The mechanism involves protonation of one of the hydroxyl groups followed by loss of water, alkyl migration to give the more stable carbocation, with final regeneration of the proton. All compounds in which a carbocation can be generated alpha to one bearing a hydroxyl group may undergo similar rearrangements.

1.4.2 Addition Reactions

As the name suggest these reactions involve addition of a reagent to an unsaturated group (1.2) and as such nominally display 100% atom economy:

$$A = A + B - C \rightarrow AB - AC \qquad (1.2)$$

When the addition is initiated by attack of the π-electrons in an unsaturated bond on an electrophile to form a carbocation the reaction is an electrophilic addition – a very common class of reactions for alkenes. The reaction is governed by Markovnikov's rule, which states that in addition of HX to a substituted alkene the H will form a bond to the carbon of the alkene containing the greatest number of hydrogen atoms.

Another way of expressing this is that the most stable carbocation will be formed (Scheme 1.6).

The Michael addition reaction, involving addition to carbon–carbon double bonds containing an electron withdrawing group, is catalysed by base. Various heterogeneous bases that can be reused are known, including: alumina, KF on alumina, and phenolates supported on silica. The latter have proved particularly effective for addition of β-keto esters to enones (Scheme 1.7).

Addition reactions to carbonyl groups are also important atom economic reactions. By using chiral ligands (Chapter 4) catalytic hydrogenation can be carried out enantioselectively, adding to the overall greenness. This technology is becoming increasingly used in the pharmaceutical industry; for example, (S)-naproxen (Scheme 1.8) can be made in high enantioselectivity using a chiral ruthenium phosphine catalyst.

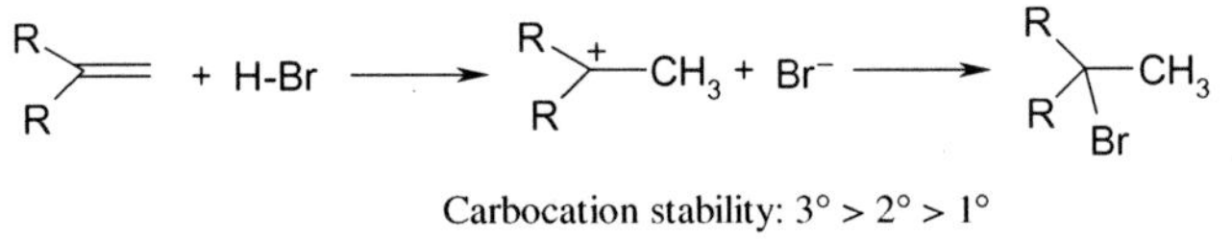

Carbocation stability: 3° > 2° > 1°

Scheme 1.6   Electrophilic addition and Markovnikov's rule.

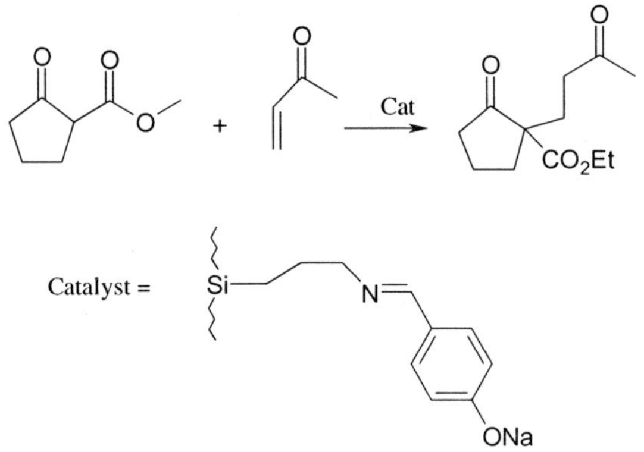

Scheme 1.7   Michael addition reaction.

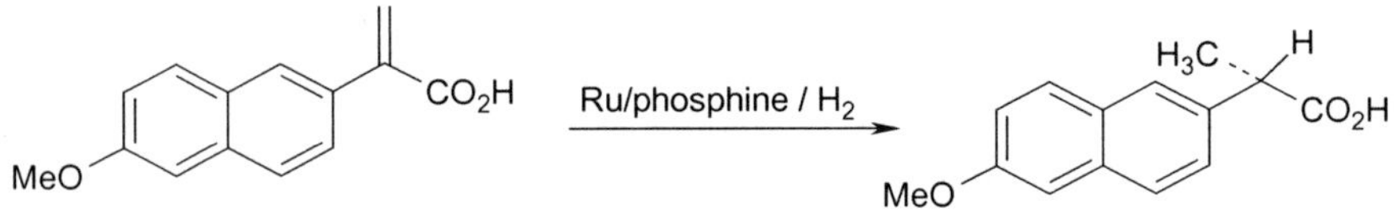

Scheme 1.8   Enantioselective hydrogenation route to (S)-naproxen.

Box 1.3   Main features of the Diels–Alder reaction

- In its simplest form the Diels–Alder reaction is a cycloaddition reaction between a conjugated diene and alkene (dienophile).
- The reaction follows a concerted mechanism and hence displays a high degree of regio- and stereoselectivity, *e.g.* cis-dienophiles give cis-substituents in the product.
- Unsubstituted alkenes are relatively poor dienophiles and dienes, with high temperatures being required to force the reaction. However, dienophiles having electron-withdrawing groups and dienes having electron-donating groups are considerably more reactive.
- The diene needs to be able to adopt the (*S*)-cis conformation in order to react, *e.g.* no reaction is observed for 2,4-hexadiene due to steric hindrance.
- Cyclic dienes give mainly endo-products due to kinetic control:

1.4.2.1   Diels–Alder Reactions.   Diels–Alder reactions provide one of the few general methods of forming two carbon–carbon bonds simultaneously. Box 1.3 describes the main features of these reactions. The reaction finds widespread industrial use; for example, hardeners for epoxy resins are made by reaction of maleic anhydride with dienes such as 2-methyl-1,4-butadiene.

An atom economic route to the insecticide aldrin (Scheme 1.9) was developed some 50 years ago. This very potent insecticide was later banned in most countries due to its toxicity to wildlife, illustrating the need to look at the whole product lifecycle, not just the synthetic route.

1.5   ATOM UN-ECONOMIC REACTIONS

1.5.1   Substitution Reactions

Substitutions are very common synthetic reactions, but by their very nature they produce at least two products, one of which is commonly not wanted. As a simple example 2-chloro-2-methylpropane can be

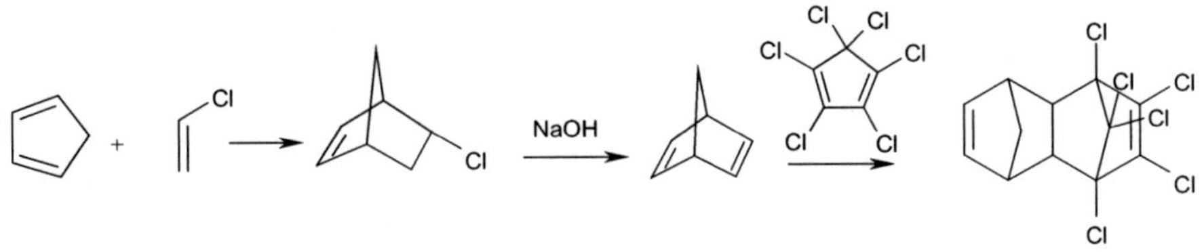

Relative Molecular Masses					
66	62.5		40	273	363
Atom economy = 100 x 363/(66 + 62.5 +40 + 273) = 100 x 363/44.5 = 82.2%					

Scheme 1.9 Manufacture of aldrin *via* Diels–Alder reactions.

Relative Molecular Masses			
74	36.5	92.5	18
Atom Economy = 100 x 92.5/(74+36.5) = 83.7%			

Scheme 1.10 2-Chloro-2-methylpropane by S_N1 substitution.

$$CH_3(CH_2)_4CH_2OH + SOCl_2 \longrightarrow CH_3(CH_2)_4CH_2Cl + SO_2 + HCl$$

Relative Molecular Masses				
102	119	120.5	64	36.5
Atom Economy = 100 x 120.5/(102+119) = 54.5%				

Scheme 1.11 An atom un-economic substitution.

prepared in high yield by simply mixing 2-methylpropan-2-ol with concentrated hydrochloric acid (Scheme 1.10). Here the hydroxyl group on the alcohol is substituted by a chloride group in a facile S_N1 reaction. The by-product in this particular reaction is only water but it does, however, reduce the atom economy to 83%.

Most substitutions have lower atom economies than this and produce more hazardous and a greater variety of by-products. Hexanol is much less reactive than 2-methylpropan-2-ol in substitution reactions. One way of converting it into the chloride involves reaction with thionyl chloride (Scheme 1.11) – here the unwanted by products are HCl and SO_2, reducing the overall atom economy to 55%. This readily illustrates how, even in simple reactions, half of the valuable atoms in a reaction can be lost as waste.

For many years phenol was made on a large industrial scale from the substitution reaction of benzene sulfonic acid with sodium hydroxide. This produced sodium sulfite as a by-product. Production and disposal of this material, contaminated with aromatic compounds, on a large scale contributed to the poor economics of the process, which has now been replaced by the much more atom economic cumene route.

1.5.2 Elimination Reactions

Elimination reactions involve loss of two substituents from adjacent atoms; as a result unsaturation is introduced. In many instances additional reagents are required to cause the elimination to occur, reducing the overall atom economy still further. A simple example of this is the E2 elimination of HBr from 2-bromopropane using potassium *t*-butoxide (Scheme 1.12). In this case unwanted potassium bromide and *t*-butanol are also produced, reducing the atom economy to a low 17.5%.

The Hofmann elimination is useful synthetically for preparing alkenes since it gives the least substituted alkene. The reaction involves thermal elimination of a tertiary amine from a quaternary ammonium hydroxide; these are often formed from alkylation of a primary amine with methyl iodide followed by reaction with silver oxide. Scheme 1.13 shows

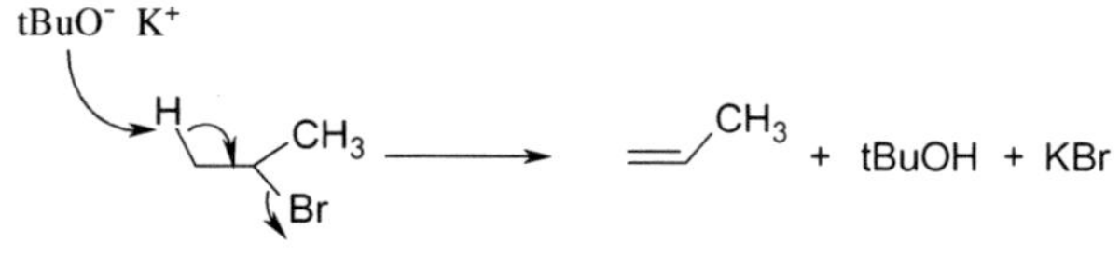

Relative Molecular Mass		
112 122		41
Atom Economy = 100 x 41/(112+122) = 17.5%		

Scheme 1.12   Base-catalysed elimination from 2-bromopropane.

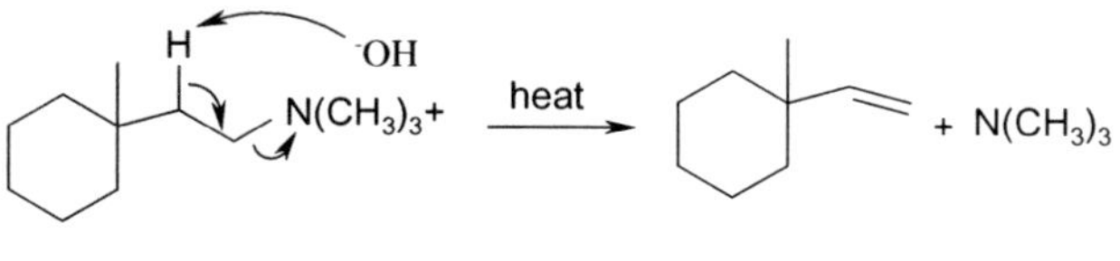

Scheme 1.13   Hofmann elimination.

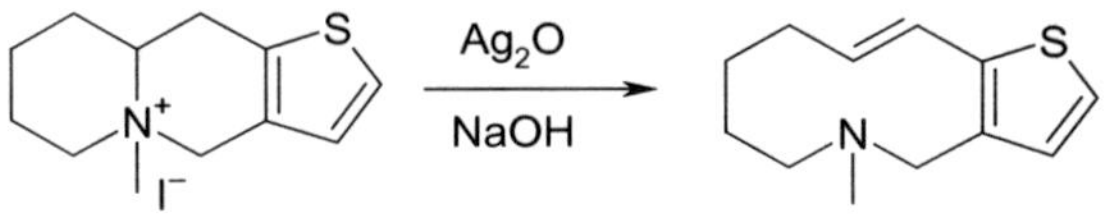

Scheme 1.14   Internal Hofmann elimination.

the mechanism of the elimination. In this synthesis of 1-methyl-1-vinylcyclohexane the atom economy is reduced to 62% through loss of trimethylamine and water.

Hofmann elimination reactions from bi- and tri-cyclic systems can, however, be used to create 'internal' unsaturation without loss of a trialkyl amine, as shown in Scheme 1.14 for the synthesis of the hexahydrothieno[b]azecine.

1.5.3   Wittig Reactions

Wittig reactions are versatile and useful for preparing alkenes, under mild conditions, where the position of the double bond is known unambiguously. The reaction involves the facile formation of a phosphonium salt from an alkyl halide and a phosphine; in the presence of base this loses HX to form a ylide (Scheme 1.15). The highly polar ylide reacts with a carbonyl compound to give an alkene and a stoichiometric amount of a phosphine oxide, usually triphenylphosphine oxide.

It is the formation of the phosphine oxide that gives the reaction a low atom economy and, due to the cost of disposal (usually by conversion into calcium phosphate and disposal as hazardous waste), has limited its commercial usefulness to high value products. Several methods have been developed to recycle $(Ph)_3PO$ into $(Ph)_3P$ but these have proved more complex than may be expected. Typically, the oxide is converted

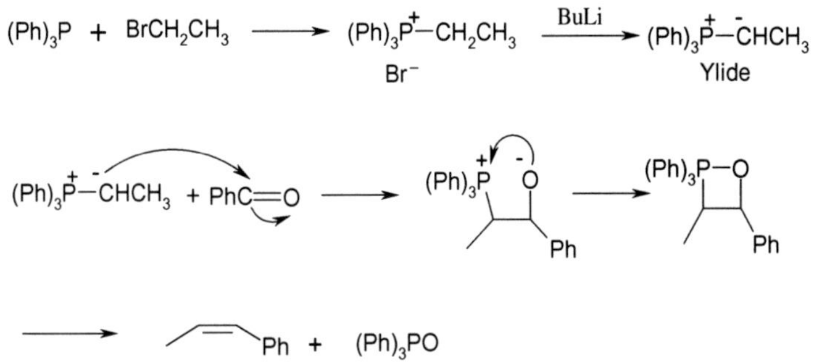

Scheme 1.15   Wittig reaction

into the chloride, which is reduced by heating with aluminium. Overall this recovery is expensive and also produces significant amounts of waste.

1.6 REDUCING TOXICITY

One of the underpinning principles of Green Chemistry is to design chemical products and processes that use and produce less hazardous materials. Here, hazardous covers several aspects, including toxicity, flammability, explosion potential, and environmental persistence. Historically, two factors have governed our approach to toxic materials. First it is only during the last 30 or 40 years that our knowledge of toxicity of chemicals has been handled scientifically. Before then we tended to only know about the toxic effects after the event, usually entailing ill health or death amongst those working with the substance. One early example of this was the early production of nickel from its carbonyl; it was only after several strange deaths amongst workers that it was finally realized that nickel carbonyl is highly toxic. This is one of many examples were we have learned through 'trial and error' of the toxic nature of chemicals. This is not a criticism of the early chemical industry since there was little knowledge regarding toxicity or how to measure it.

It may be argued that industry at the time did not take adequate precautions to prevent exposure of workers to chemicals with unknown hazards. However, this must be viewed in relation to the attitude of the pioneering society at the time to risk in all walks of life. Following the discovery that a material was toxic the 'natural' way to deal with it was to somehow prevent workers from coming into contact with the material. This philosophy led to the second factor concerning how we dealt with toxicity until recently, namely one of limiting exposure. More and more elaborate ways of limiting exposure have been developed over the years, including use of breathing apparatus and handling of chemicals remotely.

A *hazard* can be defined as a situation that may lead to harm, whilst *risk* is the probability that harm will occur. From the point of view of harm being caused by exposure to a chemical, Equation (1.3) is applicable:

$$\text{Risk} = (\text{function}) \text{ Hazard} \times \text{Exposure} \qquad (1.3)$$

As stated above, our traditional approach to reducing the risk of some harm being caused has been to limit the exposure by some physical means or by introduction of systems and working practices. Whilst this has worked relatively well, no control measure or system can be 100% perfect. The alternative way to minimize risk, which is the Green Chemistry approach, is to reduce the hazard. This approach gets to the root cause of the problem and is based on the principle of 'what you don't have can't harm you'. Recent legislation is starting to reflect this change of approach. The COSHH (Control of Substances Hazardous to Health) regulations require that assessments be made of all laboratory and production work where potentially harmful chemicals are used. As part of this assessment alternatives procedures, which avoid use of hazardous materials, must be considered and the use of personal protective equipment to prevent worker exposure should only be used when all other possibilities have been thoroughly considered.

The whole area of designing synthetic procedures using low-hazard reagents is now attracting much attention. Carbon–carbon bond formation *via* free radical chemistry is a very versatile method of producing a range of materials of interest to pharmaceutical companies. Unfortunately, the traditional way of generating free radicals is to use highly neurotoxic organotin compounds, especially tributyltin hydride (Bu_3SnH), an excellent radical chain carrier. Handling this compound poses a significant hazard and there is the slight, but real, possibility that trace amounts of organotin compounds may be left in the product at the end of the reaction and purification stages. This is completely unacceptable for any product destined for consumption by humans. Replacement of organotin compounds has therefore been the subject of much recent research. One possible alternative is to use an easily oxidized sulfide that could transfer an electron to an electrophile to give a radical anion Equation (1.4). This in turn could fragment to give an organic radical:

$$Ar_2S + RX \rightarrow RX^{-\bullet} + Ar_2S^+ \rightarrow R^\bullet + Ar_2^+X^- \qquad (1.4)$$

One of the most readily available easily oxidized sulfides is tetra-thiafulvalene (TTF) and this material readily transfers electrons to good electrophiles such as arene diazonium salts. This reaction has been used to form the tricyclic precursor to the natural product aspidospermidine, stereospecifically (Scheme 1.16). Since water is very inert towards attack by free radicals the reaction is thought to proceed *via* attack by the radical on the liberated $TTF^+BF_4^-$ complex. In this particular case the reaction is thought to go *via* the Wheland intermediate, which, due to steric hindrance of the top face, controls the stereochemistry of the hydrolysis reaction.

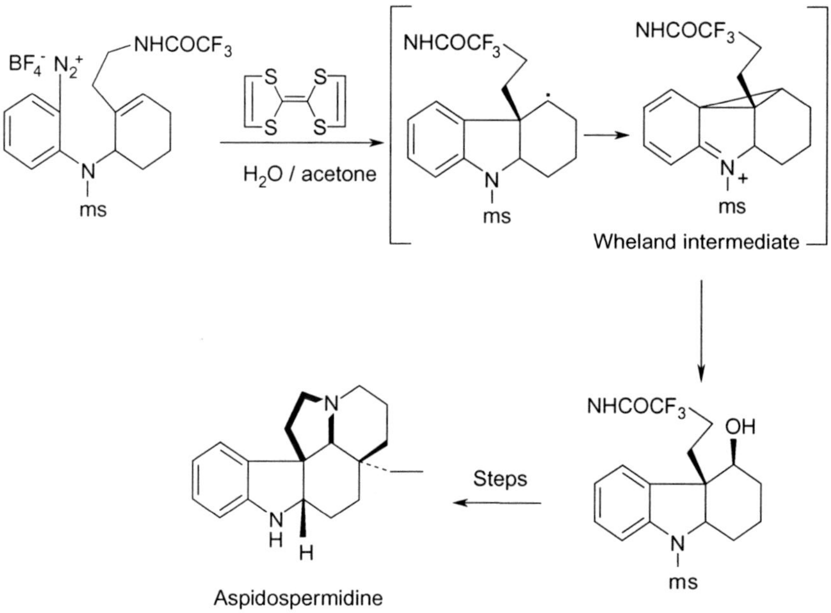

Scheme 1.16 Radical generation without using Bu$_3$SnH; ms = methanesulfonyl.

1.6.1 Measuring Toxicity

Many methods have now been developed for measuring the potential harmful effects chemicals can have. Common tests include those for irritancy, mutagenic effects, reproductive effects, and acute toxicity.

1.6.1.1 LD$_{50}$ and LC$_{50}$. LD and LC stand for lethal dose and lethal concentration, respectively. LD$_{50}$ is the dose of a chemical at which 50% of a group of animals (usually rats or mice) are killed, whilst LC$_{50}$ is the concentration in air or water of the chemical that kills 50% of test animals. These tests are the most common ways of measuring acute toxicity of chemicals. LD$_{50}$ tests are done by injecting, applying to the skin, or giving orally a known dose of pure chemical. The result is usually expressed in terms of mg chemical per kg animal, *e.g.* LD$_{50}$ (oral, rat) $= 10\,\mathrm{mg\,kg^{-1}}$ means that when given orally at the rate of $10\,\mathrm{mg\,kg^{-1}}$ animal weight the chemical will kill 50% of rats tested. Similarly, LC$_{50}$ tests are usually carried out by allowing the animal to breath a known concentration of the chemical in air, with the results being expressed in parts per million (ppm) or $\mathrm{mg\,m^{-3}}$.

Although there is much controversy over using animals in tests such as these the information is an essential part of the legal testing

Table 1.2   Hodge & Sterner toxicity scale.

Toxicity rating	Toxicity term	LD_{50} (oral, rat) ($mg\,kg^{-1}$)	Likely LD for man
1	Extremely toxic	<1	A taste
2	Highly toxic	1–50	~4 cm³
3	Moderately toxic	50–500	~30 cm³
4	Slightly toxic	500–5000	~600 cm³
5	Practically non-toxic	5000–15000	~1 L
6	Relatively harmless	>15000	≫1 L

required when new chemicals are introduced onto the market in significant quantities. These and other toxicity test results are used to help develop Material Safety Data Sheets, establish Occupational Exposure Limits and guidelines for use of appropriate safety equipment.

Whilst it is obvious that the lower the LD_{50} or LC_{50} is the more toxic the chemical it is difficult to obtain a feel for how toxic the chemical may be to man. Several scales have been developed to help compare toxicity data, a commonly used one developed by Hodge & Sterner is shown in Table 1.2.

1.6.1.2   Fixed Dose Procedure.   One problem with the above tests is that different laboratories get different results, probably because of the different genetic make up of the animals used. To reduce this difference and also reduce the number of animals used in tests the Fixed Dose Procedure is being used more frequently. This test uses about 75% fewer animals and the chemical is administered until the animals first show signs of toxicity (not death). From these figures the LD_{50} is calculated.

1.6.1.3   Ames Test.   Named after its inventor, Bruce Ames, this test has become one of the common screening tests for measuring potential carcinogen effects of chemicals, largely due to its simplicity. The test is based on observations of mutations from the bacterium *Salmonella typhimurium* that carries a defective gene, making it unable to synthesize histidine from the ingredients of a culture medium. The theory is that if a chemical is mutagenic (and therefore a possible human carcinogen) it will cause mutations in the bacterium, a certain number of which will enable it to synthesize histidine. Growth in the population of bacteria resulting from the mutation can be observed directly. Some chemicals may not be mutagenic themselves but their metabolites might be; the culture medium therefore contains liver enzymes to include this possibility. The Ames test has not only been used to identify synthetic chemicals that are possibly carcinogenic but has indicated the presence of natural mutagens in food. Two notable discoveries are aflatoxin (**1.1**) found in peanut butter (from use of mouldy peanuts) and safrole (**1.2**) present for some years in root beer. The test, however, is not perfect; for example, dioxin a known animal carcinogen gives a negative Ames test.

Aflatoxin (**1.1**)   Safrole (**1.2**)

REVIEW QUESTIONS

1. There are several past and present commercial routes to phenol using benzene as a feed stock. Outline two such processes, writing balanced equations for the reactions involved. Compare the two routes in terms of atom economy.
2. Anthraquinone is widely use in the manufacture of a range of dyes. Two possible routes for manufacturing anthraquinone are (i) from the reaction of 1,4-naphthoquinone with butadiene and (ii) reaction of benzene with phthalic anhydride. Describe mechanisms for both these reactions and identify likely reaction conditions and any other reagents required. Compare the atom economy of the two routes. Identify three factors for each route that may influence the commercial viability.
3. Give an example of a S_N1 and a S_N2 reaction, explaining the mechanism and calculating the atom economy of the reaction. Suggest alternative synthetic routes to your products that are more atom economic.
4. Show how styrene can be prepared using the following reactions somewhere in your synthetic procedure: (a) Hofmann elimination, (b) Grignard reaction, and (c) Diels–Alder reaction. Compare the atom economies of each process. Identify any issues raised by using this approach to determine the most efficient synthetic route.

FURTHER READING

P. T. Anastas and J. C. Warner, *Green Chemistry Theory & Practice*, Oxford University Press, Oxford, 1998.

M. Lancaster, in *Handbook of Green Chemistry & Technology*, ed. J. H. Clark and D. J. Macquarrie, Blackwell Publishing, Abingdon, 2002.

P. Tundo, A. Perosa and F. Aecchini, *Methods and Reagents for Green Chemistry: An Introduction*, Wiley-Blackwell, London, 2007.

CHAPTER 2

Waste: Production, Problems, and Prevention

'In an ideal chemical factory there is, strictly speaking, no waste but only products. The better a real factory makes use of its waste, the closer it gets to its ideal, the bigger is the profit.'
R. W. Hofmann, 1848 (First President of The Royal College of Chemistry, London).

2.1 INTRODUCTION

Waste is a natural consequence of all human activity, including the actual process of living; the average adult produces over 300 g of faeces and 1 L of urine per day. In the UK, however, sewerage sludge accounts for less than 1% of the total waste produced. Between them construction and mining and quarrying account for 60% of waste production (Figure 2.1) but industry, including the chemical industry, accounts for 12% (40 million tonnes per annum).

The problems posed by waste, including the inefficient use of resources and capital, together with the risks to welfare and the environment are widely recognized by most sectors of society. Many countries now have active programmes to reduce the amount of waste disposed of to land, air, and water through increased recycling and deploying waste minimization initiatives. An accepted hierarchy for waste management has been developed (Figure 2.2) with the most preferred solution being reduction of waste at source. Lower down the

Green Chemistry: An Introductory Text, 2nd Edition
By Mike Lancaster
© Mike Lancaster 2010
Published by the Royal Society of Chemistry, www.rsc.org

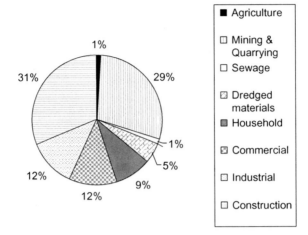

Figure 2.1   UK waste by sector (source DEFRA).

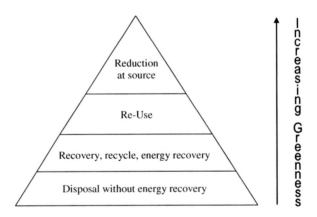

Figure 2.2   Waste or pollution prevention hierarchy.

hierarchy comes re-use followed by recycling either to recover materials and/or energy. Disposal of treated or untreated waste should only be considered as a last resort. In some instances the strict order of this hierarchy, for a particular process, may be questioned. For example, in the case where the production of a relatively small amount of dilute sulfuric acid waste cannot be avoided it may be more eco-efficient to neutralize the waste acid and send the stream directly to the sea rather than transport the acid a few hundred miles and recover it through a high-energy process. These are questions that need to be addressed through life cycle assessment and other detailed studies.

Waste reduction and increased reuse and recycling are now considered key sustainability indicators by many countries and companies.

2.2 SOME PROBLEMS CAUSED BY WASTE

The problem of chemical waste is not new; in fact it is as old as the modern chemical industry. The Leblanc process for the production of sodium carbonate was of vital importance to the development of the textile industry in the early nineteenth century. Although a fairly complex process it was based on three readily available raw materials, rock salt, coal, and limestone together with sulfuric acid (Scheme 2.1).

Owing to the geographical availability of these raw materials both the UK sodium carbonate and textile industries became centred around South Lancashire. As production volumes increased so the waste products started to accumulate, generating local environmental problems. In this process, for every mole of sodium carbonate produced, 2 moles of HCl, 4 moles of CO, and a mole of calcium sulfide are also produced.

Initial problems were caused by direct emission of HCl into the atmosphere; many people living around the factories started to develop asthma and other respiratory problems. This eventually led to a significant reduction in life expectancy for those living in South Lancashire compared to the rest of the country. The second problem was caused by the calcium sulfide which, even though foul smelling, was initially dumped inside and around the factories. Eventually, volumes became so large that factory owners persuaded local farmers to take the material to use as a pesticide. Although the material did work as a pesticide it was soon discovered that crops would not grow on the contaminated land either. The resulting alkali act of 1863 did much to ease the atmospheric pollution caused by this process, This act was the first major piece of environmental legislation and was essentially prescriptive in nature, for example, stating how HCl emissions should be controlled – through erection of stacks of a given height. It was partially due to the increased industrial costs imposed by this legislation that HCl from this process became a source of increasingly valuable hydrochloric acid. This is an excellent early example of how by-products from one process can become the raw materials for another, which is the concept behind many

$$2NaCl + H_2SO_4 \rightarrow Na_2SO_4 + 2HCl$$
$$Na_2SO_4 + 4C + CaCO_3 \rightarrow Na_2CO_3 + CaS + 4CO$$
(products in *italics* are waste)

Scheme 2.1 Leblanc process for sodium carbonate production.

current integrated chemical sites. The much more cost-effective, and less wasteful, Solvay process was introduced soon after the legislation came into force; the Leblanc process quickly became extinct.

A more recent example of chemical waste causing human death occurred in Minamata Bay in Japan in 1965. For some time people, animals, and birds living round the bay started to have problems with their eyesight and co-ordination. Eventually the cause was traced to eating fish contaminated with dimethyl mercury. The mercury had been discharged from a nearby plastics factory and microorganisms living in the mud had converted the mercury into the much more toxic dimethyl mercury. Fifty people are thought to have died due to poisoning and as a result Japan introduced very stringent discharge limits.

Although there is still concern over waste emissions from the chemical and allied industries, thanks to stringent legislation and a much more responsible attitude by industry these concerns are usually of a more general nature, to do with the cumulative effect, rather than toxic emissions from one particular factory. For example, some current areas of public concern are global warming, the ozone layer, and endocrine disruption in fish, which are all linked to pollution from industry, energy generation, transport, and the way the developed and developing world now expects to live. There is still, however, growing pressure, as previously highlighted, to minimize waste production for both economic and sustainability reasons.

2.3 SOURCES OF WASTE FROM THE CHEMICAL INDUSTRY

There is in general a lack of readily available, cumulative information on the specific types and amounts of waste produced by the chemical and allied industries, although information from individual companies can be obtained from sources such as the Environment Agency. Most governments do publish general emission data, in the UK, for example, total emissions of nitrous oxide were almost 125 000 tonnes in 2006, dropping from over 200 000 tonnes in 1990. Agriculture is responsible for two-thirds of this, with transport and power plants making up most of the rest. Emissions of N_2O from the chemical industry fell sharply in 1998 with the closure of a major adipic acid plant. The amount of volatile organic compounds (VOCs) entering the atmosphere has also fallen in the UK. Total emissions in 1990 were around 2 300 000 tonnes but 70% of this came from industrial processes and solvent use in surface coatings, *etc.* In 2006 the figure had fallen to 910 000 tonnes, largely due to reformulation to water based surface coating products and installation of abatement technology. Overall the chemical industry in the

Table 2.1 Waste produced as a proportion of product – the 'E'-factor.[a]

Industry segment	Annual production (Te)	kg By-products/kg products (E-factor)	Approx. total waste (Te)
Oil refining	10^6–10^8	~0.1	10^6
Bulk chemicals	10^4–10^6	<1–5	10^5
Fine chemicals	10^2–10^4	5 to >50	10^4
Pharmaceuticals	10–10^3	25 to >100	10^3

[a]Data from R. A. Sheldon, Chem. & Ind., 1 December 1992, p. 904.

UK produces over 2 million tonnes of hazardous waste a year, which is significantly more than any other industry sector. This figure has dropped by a factor of 3 in the last 10 years due to technical improvements to some extent driven by regulatory change.

As may be expected, relatively benign inorganic salts resulting from the widespread use of sulfuric and hydrochloric acids and sodium and potassium hydroxides and carbonates form a significant proportion of waste. Sheldon undertook one of the most quoted studies comparing waste produced by various sectors of the chemical industry. He defined the term 'E'-factor as the ratio of kg of by-product (waste) to kg of product and measured it for various industry sectors (Table 2.1). Sheldon reviewed the basis of his findings in 2007, some 15 years after development, and found that they are still broadly true, but industry is gradually moving to the lower end of the estimated ranges.

There are several ways to look at this data. For example, oil refining can be viewed as being fairly clean, with an E-factor of less than 10%. On the other hand, it could be viewed as being highly polluting if the total amount of waste (some 10^6 kg on Sheldon's figures) is taken into account. The converse argument could be applied to the pharmaceuticals sector.

Why should different sectors of industry have such widely differing E-factors? The answer involves the degree of technical development of the industry, the competitiveness of particular products within the sector, the extent of process regulation, and the cost of waste as a percentage of the products selling price. Consider the bulk chemical industry: 70 years ago it was in its infancy, competition was less severe, and sales margins high compared to production costs. Production volumes both in total and from any one plant were also low by current standards. Through a combination of these and other factors, waste production, in particular relatively benign waste, was generally not considered an issue by the industry or regulators. Taking phenol production as an example, until the 1950s the production process was based on sulfonation of benzene as outlined in Scheme 2.2. Although this process had many inefficiencies it allowed phenol production to expand significantly from the previous

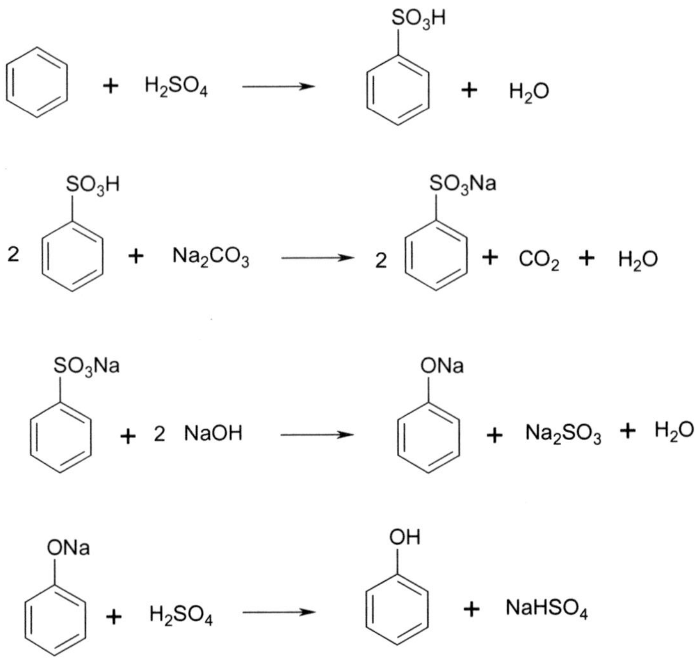

Scheme 2.2 Benzene sulfonation route to phenol.

coal tar extraction process. From a waste point of view it is the production of large amounts of inorganic salts and, to a lesser extent, carbon dioxide that are significant.

As demand for phenol started to grow due to the increased use of phenolic resins and the advent of polycarbonates more companies looked for process improvements that gave economic benefits. Although avoidance of waste was not a prime target it became a factor in the overall production costs.

With the discovery of the cumene route (Scheme 2.3), for the co-production of phenol and acetone (propanone), the benzene sulfonation route quickly became obsolete in developed countries. Today the most modern phenol plants produce very little waste, the initial alkylation step being carried out using zeolite catalysts, and overall yields based on benzene are almost 90%.

In the pharmaceutical industry the drivers for developing new processes are typically very different. The process of getting a new pharmaceutical approved for use is long and costly. Past experience has shown that small amounts of by-products present in the final drug formulation may have a profound effect on the efficacy and side effects produced. As a result many countries now not only require the drug to

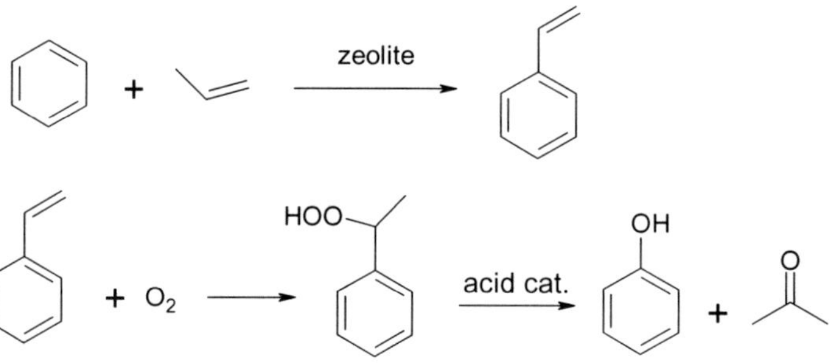

Scheme 2.3 Cumene route to phenol.

be approved but also the process by which it is manufactured. The aim of this action is to ensure a totally consistent drug composition. Once this approval process has been started it is expensive and time consuming to change. Since, for patented drugs, there is effectively no competition there is no real driver for changing the process as production costs are generally relatively small compared to the selling price. In this highly regulated industry it is therefore important to get it 'right first time', ideally at the research stage.

Ibuprofen was developed as a new analgesic by Boots in the mid-1960s using the rather inefficient route shown in Scheme 2.4. Although the synthesis is an elegant demonstration of classic (laboratory) organic chemistry, it suffers from the major flaw of using reagents rather than catalysts to carry out transformations, resulting in copious amounts of waste.

It was not until the product came out of patent, opening up competition with consequent reduction of margins, that significant process improvements were made. Scheme 2.5 shows the current manufacturing method. All three process steps are genuinely catalytic; of particular note is the use of catalytic amounts of HF in place of stoichiometric amounts of $AlCl_3$ to carry out the initial acylation. Whilst use of HF does produce less waste it poses some potential safety issues – one of many examples of the need to prioritize and choose between the various principles of green chemistry.

2.4 COST OF WASTE

The concept of the 'Triple Bottom Line' (Box 2.1) is commonly used as an indicator of business performance. Like all commercial organizations, the chemical industry exists to make profit for its shareholders, however, it is now widely recognized that successful companies must

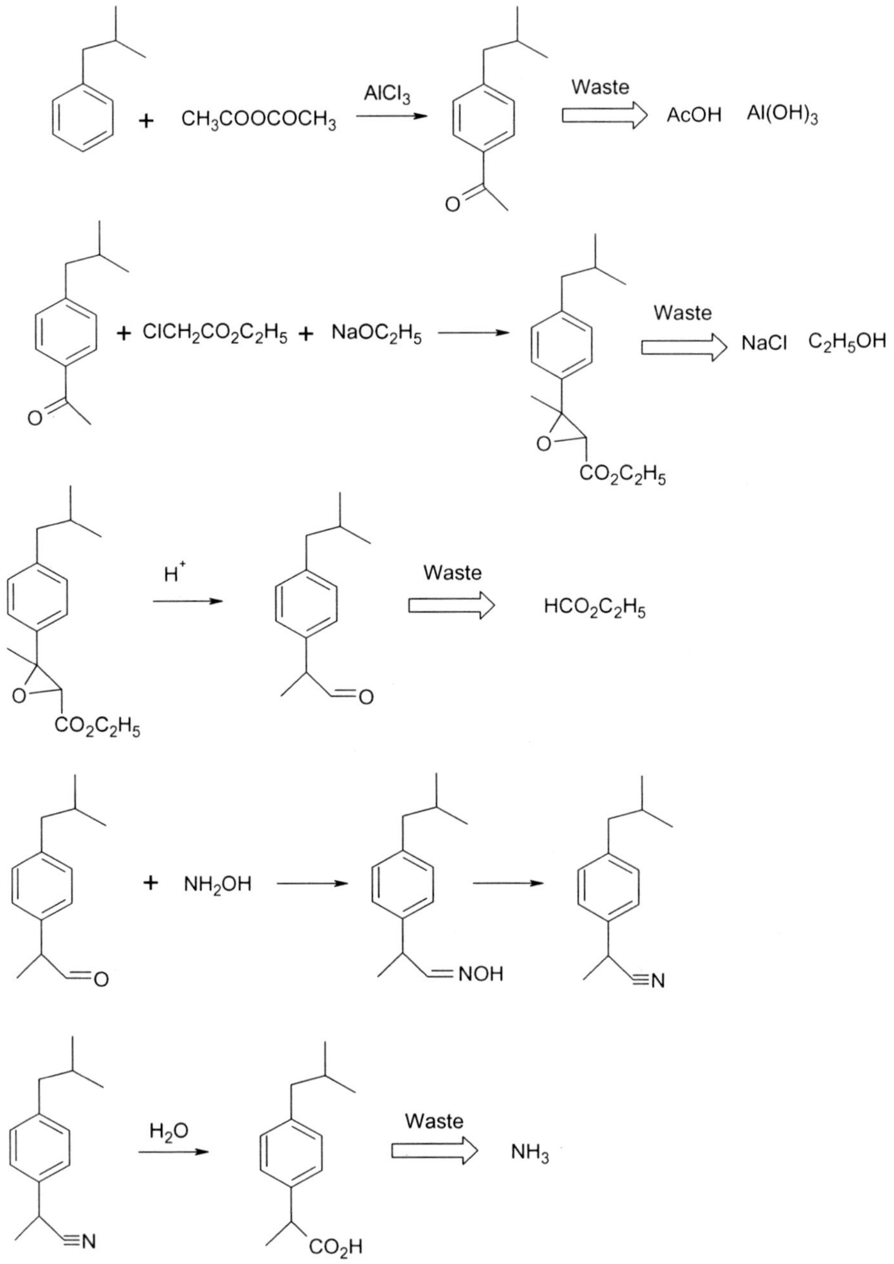

Scheme 2.4   Original Boots route to ibuprofen.

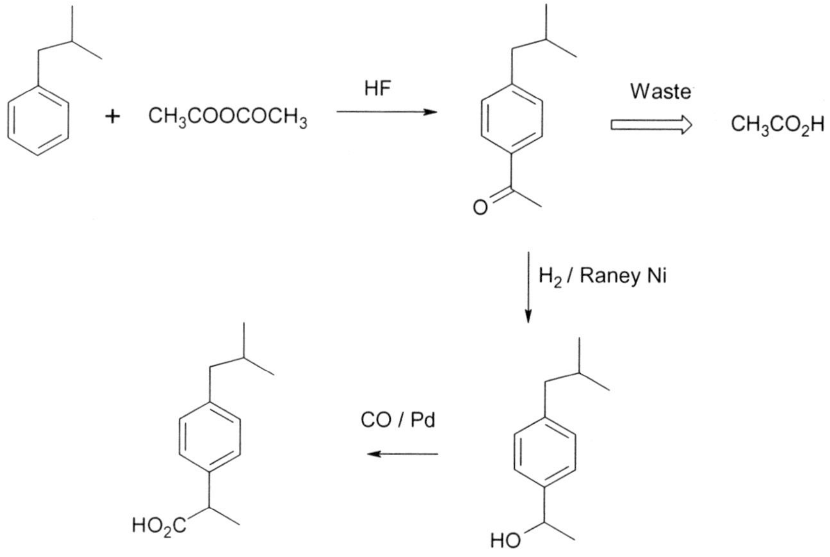

Scheme 2.5   Current manufacturing route to ibuprofen.

Box 2.1   The triple bottom line

The triple bottom line (TBL) is a term coined by management consultant John Elkington in 1997. It refers to the three interlinked strands of social, environmental, and financial accountability.

TBL is directly tied to the concept of sustainable development; if analysed properly it will provide information to enable others to assess how sustainable an organization's operations are. The premise is that to be sustainable in the long term the organization must be financially secure, it must minimize (or ideally eliminate) its negative environmental impacts, and it must act to conform to societal expectations. These three aspects are highly related.

To find a common currency for reporting all three aspects of the TBL is not simple. Currently it is widely recognized that different indicators will often need to be assessed in different ways, sometimes quantitative, sometimes qualitative.

Many governments, including in the UK, are now actively encouraging TBL indicators to be covered in annual reports and are naming companies that fall below expectations. Green investment companies are also using TBL indicators as one criterion on which their investments are made.

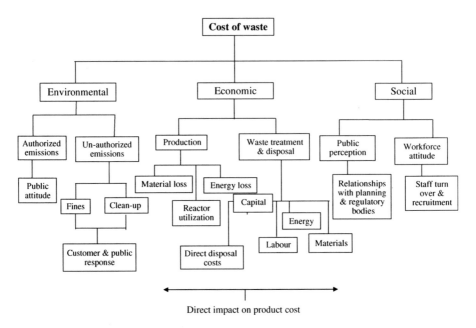

Figure 2.3   Cost of waste expressed in terms of the triple bottom line.

also have sound environmental and social policies. With increasing public interest and extensive media coverage, all aspects of business are now under constant scrutiny. Successfully merging economic and 'softer' issues is vital if a company is to be perceived as a good neighbour and its right to operate maintained.

The cost of waste to a chemical company can be expressed in terms of the triple bottom line (Figure 2.3). Evidently, waste generation will have a significant impact on direct production costs through loss of raw materials, wasted energy, and by giving a free ride to the waste products (low reactor utilization). Additionally, the direct costs of waste treatment and disposal are also significant, especially when new end-of-pipe treatment equipment is required. In the short term it is these costs that will impact on the profit margin of a particular product and may even determine whether a product can still be manufactured competitively.

There are, however, many other hidden costs to waste production, as shown under the environmental and social headings in Figure 2.3. In most cases, with the exception of fines and clean-up costs, these are very difficult to quantify in economic terms but may be very significant, and ultimately may affect the actual viability of the company itself. In many cases poor public perception and bad media exposure can quickly result in lost orders that may take many years to recover. In recent years we

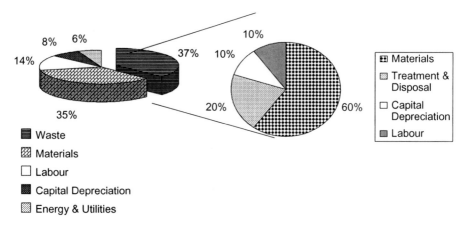

Figure 2.4 Cost of waste as percentage of manufacturing costs for speciality chemicals.

have seen concerted campaigns by non-governmental organizations (NGOs) to persuade the public from purchasing products from several well-known companies due to perceived poor environmental policies or decisions.

The direct effect of the cost of waste on production costs varies widely from product to product and industry sector to industry sector. The fact that a sector or particular process has a high E-factor does not necessarily mean that waste is a significant proportion of production costs. For example, many pharmaceutical processes use expensive chiral reagents, are made in small batches entailing expensive labour costs, and require extensive analysis and quality assurance procedures. Here the production of say 50 kg of acetic acid and sodium sulfate waste per kg of product may be a fairly insignificant cost. Looking at the speciality chemicals sector, an analysis of several different processes from different companies shows that the cost of waste (Figure 2.4) may approach 40% of production costs. For the typical relatively benign waste the actual cost of lost materials is usually the most significant component, with disposal costs being relatively small. For special cases, where treatment and disposal of highly hazardous waste is required the converse is true, with regulations continually forcing up the cost of hazardous waste disposal.

2.5 WASTE MINIMIZATION TECHNIQUES

Legislation has an important role in setting the framework for waste minimization. Apart from making the emission and disposal of certain noxious substances illegal, legislation encourages cleaner technology through environmental taxes and regulations such as Integrated Pollution & Prevention Control (IPPC). Waste minimization within this legislative framework is primarily the responsibility of corporations. If a company has a culture, set by senior management, that reducing waste is of corporate significance it will happen relatively quickly. There are several examples where a change in corporate strategy has led to beneficial (economic & environmental) waste reduction exercises. In the UK, for example, ICI/Zeneca went through such a programme in the early 1990s. Initially many waste minimization projects focus on the relatively simple, but cost-effective, aspects of reducing water, energy, and chemical use through repairing leaking valves and general 'good-housekeeping'; this is generally the best place to start and enables several quick wins. But there is a growing acceptance of the need to take 'good housekeeping' on to the next stage through a more fundamental look at the chemistry and the manufacturing process. This may also yield dramatic dividends.

One of the most widely publicized successful waste minimization programmes was developed by 3M. The initiative was called '3P' – standing for Pollution Prevention Pays. The 3M management realized the high cost of end of pipe treatments and aimed the 3P programme at preventing pollution (waste) at source, thereby reducing the need for retrofitted control measures. Typical projects under the programme included product reformulation, raw material and energy recovery and reuse, and process changes such as using water based coatings for the manufacture of tablets. Some of the successes of 3P include saving 1.5 billion gallons of wastewater containing 10 000 tonnes of chemicals per year and a reduction of chemicals emitted to atmosphere by 100 000 tonnes per year.

2.5.1 The Team Approach to Waste Minimization

Research & development leading to the manufacture of a new product (or the manufacture of an existing product *via* a new route) is usually, at least initially, carried out by a small team of chemists. The steps adopted along the way vary in sophistication, depending on the industry sector and the product, but are generally:

- literature & patent search to see how the product has been made before (or at least how the individual steps have been carried out);
- identification of options that are free from patents;
- COSHH assessment on chosen route(s);
- experimental work to assess if chosen route(s) work;
- further literature & experimental work to refine route;

- laboratory scale-up to provide material for application testing;
- discussion of production/scale-up with chemical engineers;
- refinement of route based on engineering input;
- variable & capital cost assessment/business approval;
- carry out hazard & operability studies (HAZOP) and other Safety, Health and Environmental studies (SHE);
- production of pilot plant batches (or risk analysis to assess potential for bypassing this stage);
- modification/construction of production plant;
- transfer technology to production department.

A more efficient way, now being adopted by some companies, is to establish a multi-disciplinary team at the outset of the project. Such a team would usually consist of chemists, chemical engineers, production personnel, a SHE advisor, and possibly a business representative. Mechanical and instrument engineers and quality control experts may also be co-opted onto the team as the project progresses. By involving the whole team at the route selection stage, each viewing the problem from their own perspective, many pitfalls can be avoided. By considering all the aspects identified in Table 2.2 at an early stage it is much more likely that a process is developed that takes account of all relevant concerns.

Having the right team in place may help to raise the issues of waste at an early stage but it will not necessarily provide all the answers. Both the chemist and chemical engineer should be aware of the various techniques and technologies to help avoid waste production. Many of these are encompassed in the '12 Principles of Green Chemistry' discussed in Chapter 1 and are covered in more detail in subsequent chapters.

Table 2.2 Role of teams in developing a new process.

Chemist	Chemical engineer	Production	SHE (safety, health and environmental)	Business
Yield	Flow sheet	Operability	Emissions	Production cost
Purity	Heat & mass transfer	Convenience for shifts	Waste treatment	Waste disposal cost
Selectivity	Process costs	Operator safety	Regulatory compliance	Product packaging
By-product identification	Equipment choice	Materials handling	Operator safety	Product liability
Mechanism	Product isolation			

2.5.2   Process Design for Waste Minimization

Once various practical routes, based on sound chemistry have been identified it is important to visualize the whole process through a process flow sheet (PFS) incorporating expected mass balances and atom efficiencies (Figure 2.5). For a complex multi-step synthesis this may be a significant

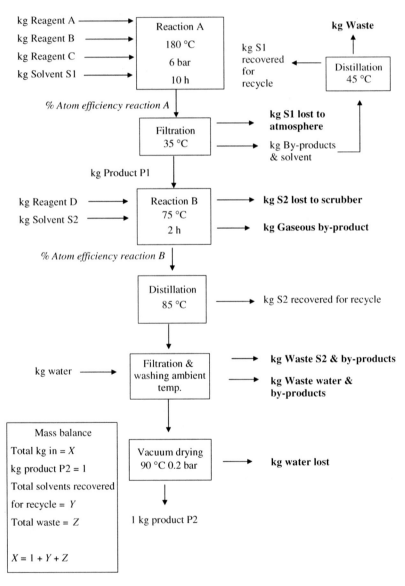

Figure 2.5   Example of a process flow sheet.

exercise, but the result will be a valuable visual guide that highlights which areas of the process produce most waste, which are the least energy efficient, *etc.* When comparing PFSs for various options it is important to compare like with like, *i.e.* they should be written for synthesis of a constant, convenient weight of product, *e.g.* 1 kg or 1 te. As in all theoretical exercises, assumptions will need to be made regarding, for example, the amount of solvent required, solvent losses, reaction yields, *etc.* Close estimates may be made from literature data, from previous experience within the company, or will need to be determined from early experiments.

From the PFS the overall mass balance can be calculated; water should be included in these since it is of vital importance in energy balance calculations, determining distillation efficiencies and in assessing waste treatment options. From evaluation of the flow sheets the various routes can be compared in terms of:

- volume of waste
- nature of waste
- mass balance
- product contaminants
- outline material costs
- complexity of processing and associated costs
- requirement of any special equipment
- energy requirements
- toxicity/handling issues

This will normally lead to a narrowing down of the possible routes and will lead to in-depth questioning of the process. Typical questions may include:

1. Can we use an alternative to solvent 1, which is volatile and is lost from the process in significant quantities?
2. Reaction 1 is fairly energy intensive, slow, and is expected to only produce a moderate yield, is there an alternative?
3. Is there an alternative to Reaction 2, which has a very low atom economy due to the generation of an unwanted gaseous by-product?
4. Do we need to wash the product with water, since this produces a very large, dilute waste stream requiring special handling?

Armed with the PFS and questions for the various process options the team can then discuss the most appropriate way forward. For example, considering question 2, production staff may comment that this particular plant only runs on the day shift and, therefore, a 10 hour reaction is not viable; the chemical engineer may conclude that the problem is likely to be one of mass transfer and other reactor design options such as a spinning disc reactor should be considered. The SHE advisor may comment that not only is solvent 1 volatile but it is also moderately harmful and would require specialist handling equipment; making it very important to find an alternative. As waste minimization starts at the reaction stage it is critical to study this area in particular detail. Questions that can be asked include:

- Do we need to use organic solvents at all?
- Is there a viable alternative to using protecting groups?
- Can a catalyst be used in place of a reagent?
- Is the proposed reactor the most efficient, from an energy efficiency and waste minimization point of view?
- Can we use a less hazardous raw material?
- Is there a viable alternative to using an elimination reaction?
- What is the reason for lack of selectivity for a given reaction? Can it be overcome?
- Can the pressure be reduced?
- Are processing aids, such as filter aids, necessary?
- Can any waste or by-products be recovered for use in another process or product?

From discussions of this type a research programme can be developed that integrates the key issues into the process development. It is important that the research programme is continuously reviewed by the whole team. As progress is made the PFS will require updating to include actual proven quantities of both raw materials and by-products; energy balances will also normally be added. One of the chemical engineer's roles at this stage is to try and minimize energy requirements using process integration techniques, *e.g.* by using a warm effluent stream from one process to warm an incoming stream for another. This team will then have the knowledge required to take the process through to the HAZOP stage, when it is important to get an independent review. By adopting this approach the overall time from conceptual idea to first production is likely to be shortened and the final process is likely to be both more cost-effective and environmentally benign.

2.5.3 Minimizing Waste from Existing Processes

The getting it 'right first time' approach to waste minimization outlined above is obviously the preferred economic and environmental option, and, in general, new processes are more environmentally benign than older ones. Approaches to waste minimization for existing processes are usually either connected to 'good housekeeping' or are part of process development work. Occasionally, however, there is a more holistic approach driven by corporate philosophy. It is worth mentioning the good housekeeping approach even though it does not involve any chemistry. If an overall mass balance is conducted for a site or operational unit there is often a significant difference between the quantity of raw materials that should have been required and that actually used. The reasons for this discrepancy are frequently process operating inefficiencies and general leakage and spillage. Dealing with these is often termed good housekeeping and is addressed by tackling aspects such as valve leakage, losses due to inadequate cooling on distillation columns, drainage and cleaning of reactors.

When undertaking a more holistic approach to waste minimization it is again beneficial to assemble a multi-disciplinary team from similar disciplines to that outlined above, bringing in additional members with expert knowledge of a particular process. The team will usually be set targets to achieve, *e.g.* a reduction of 25% in site waste over 2 years, or a saving of £400 000 pa through waste minimization, or to completely eliminate production of hazardous waste within 1 year. It is important that appropriate targets are discussed and agreed otherwise the team may end up working on aspects that provide little environmental or economic benefit. The first step (Figure 2.6) will be to gather data so that an audit of all process flow sheets can be conducted to identify the problem processes; it will often be found that approximately 80% of the waste is produced from 20% of the processes (Pareto rule). From this audit the processes to be studied to meet the teams objectives can be identified.

The priorities for the work will be dependent on the objectives. For example, if the prime objective is to minimize total waste leaving the site then it may be preferential to select a process that has a large aqueous stream containing inorganic salts rather than a process producing a very small amount of hazardous material requiring specialist off-site treatment. In all cases the action plan should be devised taking account of the waste minimization hierarchy, *i.e.* source reduction should be the first priority.

2.6 ON-SITE WASTE TREATMENT

No matter how much effort is put into cleaner production and waste minimization at source it is rarely practical to convert 100% of raw materials into product or to design a process with a 100% recoverable and reusable mass balance. Effluent from most production processes can

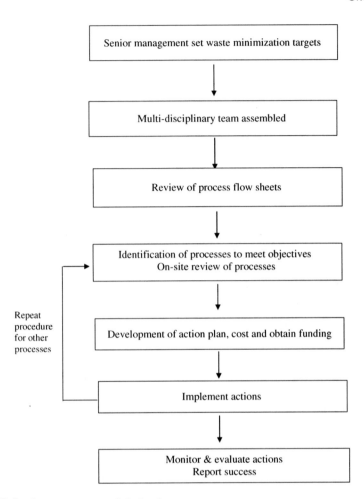

Figure 2.6   Steps to waste minimization.

be minimized but can not usually be eliminated completely; some kind of treatment is often required to render the waste less harmful – this is termed end of pipe treatment. Waste treatment falls into three broad categories: physical treatment, chemical treatment, and biological treatment. These may be used alone or frequently in combination with each other. The method chosen will depend on several factors, including toxicity, volume, and nature of the waste, together with the associated treatment costs that include:

- capital costs of new equipment;
- equipment running costs, including materials, energy, and labour;
- sampling and analysis requirements;

- waste disposal taxes and charges;
- transport costs.

Other considerations to be taken into account include the cost–benefit of treating waste from each process individually or combining all site wastes for treatment in one single unit. Although treating mixed waste is more complex, the cost of installing individual treatment units on each plant is usually prohibitively high.

An important aspect of waste treatment, particularly of wastewater streams, is concerned with controlling the oxygen demand of the waste stream. Under natural conditions many organic materials will biodegrade; this degradation process consumes oxygen. Introduction of wastewater containing trace organic materials into rivers and lakes will therefore tend to deplete the oxygen supply in the river or lake. Because of the poor rate of solubility of oxygen in water this biodegradation process may make life unsustainable for aquatic organisms. Disposal of significant amounts of wastewater is usually subject to some oxygen demand limitation. Two types of oxygen demand are normally measured:

1. Biological oxygen demand (BOD) is the amount of oxygen consumed in 5 days on treatment of the waste stream with a mixture of microbes.
2. Chemical oxygen demand (COD) is related to the amount of chromic acid consumed in oxidizing the waste stream.

In addition to these measurements the theoretical oxygen demand (THOD) is also often calculated. This is done by calculating the amount of oxygen required to convert all organic substances into carbon dioxide and water. Additionally, oxidation of certain inorganic species to nitrate, sulfate, and phosphate is also included. In practice it is often found that BOD is roughly 50% of THOD whilst COD is usually 70–80% of THOD.

2.6.1 Physical Treatment

The main purpose of the physical treatment process is to separate the waste material into like phases, usually to reduce the total waste volume or make treatment simpler. The number of physical treatment processes available is large but the most commonly used ones tend to centre on various forms of filtration or distillation. Types of filtration processes include:

- traditional cloth filters, as used in plate and frame filtration or pressure filtration; these are relatively inexpensive, simple to use, and find wide application in the removal of medium and course particles in the range 10^{-2} to over 1 mm;

- centrifuges are highly efficient at removal of a wide range of particle sizes;
- micro and ultra-filtration involve use of polymer or ceramic membranes having specific pore diameters and are especially useful for removing small amounts of solid in the range 10^{-5} to 10^{-2} mm;
- other membrane techniques, such as electrodialysis, to remove even smaller particles than micro and ultra-filtration, are less widely used but do find specialist applications; this technique relies on an electric field to move anions and cations in opposite directions through a membrane, producing two streams rich in particular ions;
- the use of resin beds, natural reed beds, or algae to adsorb relatively high levels of ionic materials, including metals and organics. This is becoming increasingly common in the electroplating and related industries.

Steam stripping is used to remove small amounts of volatile materials from aqueous waste. The steam is passed upwards through a distillation tower with the waste stream passing downwards. The volatile components are extracted into the steam, which is condensed to form a much more concentrated solution of the volatile component. Similarly air stripping is used for very low levels of contaminants where release of the volatile component to atmosphere would otherwise exceed permitted emission levels.

2.6.2   Chemical Treatment

Neutralization is probably the most common method of waste treatment. This is a simple and cost-effective way of rendering some wastes fit for disposal directly into the effluent system but it normally adds to the quantity of waste since, for example, a basic stream will normally be neutralized by addition of dilute sulfuric acid, thereby considerably adding to the mass of waste. In a fully integrated factory it may be possible to use the waste stream from one process to neutralize the waste stream from another process, reducing the overall environmental burden. Apart from using aqueous acid or base other neutralization methods that are a little more benign include passing of the waste stream through a fixed bed of limestone or acidic ion-exchange resin. Neutralization techniques are also frequently used to scrub waste acidic or basic gasses (*e.g.* SO_2 or NH_3) from reactors or distillation column vents. Here the gaseous effluent is contacted with a solution of acid or base in such a way as to reach the required exit concentration of contaminants in the gas.

Oxidative treatment of waste can be a relatively expensive process, and many smaller chemical companies favour biotreatment plants (see below) for mineralization (conversion into CO_2 and H_2O) of organic waste. Oxidation is, however, a very powerful way of dealing with low levels of toxic waste in solution. Hydrogen peroxide is one of the most common oxidizing agents used and is effective for treatment of waste containing sulfur compounds, phenols, and cyanides. Other oxidizing agents used include ozone, sodium hypochlorite, and potassium permanganate. Ozone and hydrogen peroxide are considered clean oxidants since the reaction by-products (oxygen and water, respectively) can be safely released into the environment. In contrast potassium permanganate gives rise to manganese(IV) oxide as a by-product that must be removed by filtration.

Wet air oxidation is becoming an increasingly popular technology for treating aqueous effluent streams containing species that are difficult to treat by other means, *e.g.* polyphenols, or which contain relatively high levels of organics (*ca* 5%), *e.g.* some surfactant waste streams. Typically the process is carried out at moderately high temperature, 200–300 °C, and pressures up to 100 atm, with reaction times of around half an hour. Under these conditions highly active hydroxyl radicals are generated that oxidize most organic compounds. In some cases complete degradation of organic material to CO_2 and water is achieved whilst in other cases breakdown to products for treatment by other techniques results. Wet air oxidation finds widespread use in treatment of municipal sewage sludge and landfill liquors and is gaining in popularity for treatment of toxic waste streams that would kill the bacteria used in biotreatment plants. It is also common for refineries to use this technique to treat caustic waste from ethylene crackers. Commercial systems typically use a bubble column reactor, where air is bubbled through a vertical column that is full of the hot and pressurized wastewater. Fresh wastewater enters the bottom of the column and oxidized wastewater exits the top. The heat released during the oxidation is used to maintain the operating temperature.

In some cases process efficiencies may be improved using catalysts based on, for example, cerium and ruthenium. In many cases this improves removal of COD to well over 90%. Recently, more efficient types of wet air oxidation processes based on using supercritical water have been developed (Chapter 5, Section 5.3.2).

Advanced oxidation processes (AOP) refers to a set of chemical treatment oxidation procedures carried out in parallel. Contaminants are oxidized by four different reagents: ozone, hydrogen peroxide, oxygen, and air, in precise, pre-programmed dosages, sequences, and combinations. These procedures may also be combined with UV irradiation and specific catalysts. This results in the development of hydroxyl radicals. The AOP procedure is particularly useful for cleaning biologically toxic or non-degradable materials such as aromatics, pesticides, petroleum constituents, and volatile organic compounds in waste water.

Chemical reduction is less frequently used for waste treatment but does find one important application in treatment of waste containing highly toxic Cr(VI). Commonly employed reducing agents include Fe(II) salts and sodium metabisulfite ($Na_2S_2O_5$). The reduced product, Cr(III), has a lower toxicity and can be removed from basic solution by filtration:

$$Cr^{6+} + 3Fe^{2+} \rightarrow Cr^{3+} + 3Fe^{3+} \tag{2.1}$$

The removal of trace metals from effluent streams is common. Several techniques are employed in addition to the physical ones identified above, including precipitation as the sulfide, hydroxide, or carbonate followed by filtration.

2.6.2.1 Electrochemical Waste Treatment. Industrial use of electrochemical processes for treating effluent streams is increasing due to the usually competitive running costs compared to other methods. This method is especially valuable for treatment of streams containing relatively high concentrations of metal ions where recovery of the metal is important from an economic or environmental perspective. Electrochemical reduction of metal ions to the free metal, which can be recovered for reuse, proceeds according to Equation (2.2) and can be applied to a wide range of metals, including precious metals such as gold, silver, and palladium as well as more hazardous waste containing nickel, cadmium, or cobalt:

$$M^{n+} + ne^- \rightarrow M \tag{2.2}$$

In many cases the metal ion is complexed in solution; electrochemical reduction is more difficult in these situations but still finds industrial use. Recovery of silver from photographic fixing processes is of economic importance. In simple terms the waste stream can be described as a solution of complexed silver thiosulfate from which silver metal may be recovered according to Equation (2.3):

$$Ag(S_2O_3)^{3-} + e^- \rightarrow Ag + 2(S_2O_3)^{2-} \tag{2.3}$$

The process must be carefully controlled since under certain conditions reduction of thiosulfate may occur, resulting in the release of toxic hydrogen sulfide.

Before tin cans can be recycled it is usual to strip and recover the tin from the surface of the steel. This is carried out electrochemically in a reversal of the tin plating process:

At the anode:

$$Sn + 2H_2O \rightarrow HSnO^{2-} + 3H^+ + 3e^- \tag{2.4}$$

At the cathode:

$$HSnO^{2-} + 3H^+ + 3e^- \rightarrow Sn + 2H_2O \tag{2.5}$$

Electrochemical methods may also be used to clean wastewater contaminated with organic material. Many organic materials, especially aromatic compounds containing electron-donating groups, can be completely oxidized to CO_2 using tin oxide coated anodes. In this direct anodic oxidation process the oxidizing species is thought to be active OH^- adsorbed on the anode surface. Under appropriate conditions electrochemical reduction of aqueous waste containing organochlorine compounds (*e.g.* chlorophenols) is a useful method of reducing the toxicity of the waste stream (2.6). In most cases the non-chlorinated materials have significantly lower toxicity values:

$$R–Cl + H^+ + 2e^- \rightarrow R–H + Cl^- \tag{2.6}$$

A huge range of electrochemical cells types has been developed for specific purposes and different types of waste streams. The most basic design of cell type is shown in Figure 2.7, consisting of parallel banks of vertical electrodes over which the waste solution flows at a given rate.

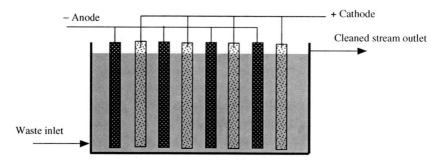

Figure 2.7   Electrochemical cell.

One of the major limitations of this basic cell is poor mass-transfer, particularly at low ion concentrations. Many cell improvements have concentrated on improving mass-transfer through, for example, injecting a fine stream of air across the surface of the cathode.

2.6.3 Biotreatment Plants

It has become increasingly unacceptable in recent years to allow even small amounts of organic waste from chemical production processes to enter public sewerage and water systems. An increasingly common solution has been the instillation of biotreatment plants. Frequently all rainwater falling onto the production site is also collected and diverted, *via* storm water drains, into the biotreatment plant as well as aqueous effluent from chemical processes. Passing all aqueous effluent (except sewerage) from the site through a biotreatment plant should ensure that no unacceptable organic chemical waste enters the water system.

Biodegradation is the breakdown of organic material by microbial activity, with the organic material acting as a food supply. The degradation process is a series of enzyme-catalysed steps that, ideally, result in complete degradation or mineralization of the product. One reason for incomplete degradation is the lack of an appropriate enzyme to carry out the required transformation. This is the same reason why some organic compounds, especially some organochlorine compound like DDT, have a very high level of environmental persistence.

2.6.3.1 Types of Biotreatment Plant. Organic compounds can degrade either aerobically or anaerobically; both processes require an adequate supply of organic substrate (food), usually together with a source of nitrogen and other essential nutrients to enable the microorganism to grow. Aerobic treatment plants also require a source of oxygen whilst anaerobic plants require an electron acceptor such as Fe^{3+}. The outputs of both processes are CO_2 and water, together with an increase in mass of the microorganism. An additional output of the anaerobic process is methane. Aerobic degradation was initially carried out in large aerated lagoons, lined to make them impermeable. This process is relatively slow due to poor mixing and complete degradation is often not achieved. Although less energy efficient, biotreatment reactors in which the microorganisms are well mixed or circulated and air (or sometimes oxygen) is well dispersed in the system are being increasingly used. This type of plant is often termed an activated sludge process, since the biomass sludge is recycled (Figure 2.8).

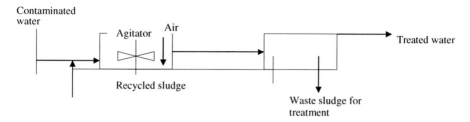

Figure 2.8   Schematic of an activated sludge reactor.

The substrate type and structure has an influence on the choice of process. However, in general terms, non-sterically hindered substrates are more easily degraded than sterically hindered ones. For aromatic substrates the presence of electron-donating or -withdrawing substituents also plays a major role in controlling the rate of degradation. The presence of a highly electron-withdrawing group such as nitro may reduce the rate of degradation several-fold compared to that achieved with a substrate containing an electron-donating group such as carboxylate. In general, substrates lacking in oxygen such as alkylbenzenes are much more readily degraded under aerobic conditions whilst substrates containing oxygen are suitable for anaerobic treatment.

The preferred aerobic degradation pathway for aromatic compounds is *via* 1,2-dihydroxybenzene (catechol), which involves the enzymes monooxygenase or dioxygenase. Scheme 2.6 shows one such oxidative degradation pathway for toluene.

The degradation path for aliphatic materials generally proceeds *via* the alcohol, ketone, and acid, to give a similar intermediate to that shown in Scheme 2.6. The final stages involving coenzyme A are then common to both substrates; this involves β-oxidation to cleave off two-carbon fragments as acetyl-CoA, which readily undergoes complete mineralization.

2.7   DESIGN FOR DEGRADATION

Although reduction in the quantity of waste produced during the manufacture of chemicals can have significant environmental and cost benefits; it is also many of the actual products of the chemical and allied industries that give rise for concern from a waste point of view. Ultimately much of the output of the chemical industry ends up in the environment as waste; three particular sources are consumer plastics, household detergents, and agrochemicals. The main mechanism for removal of such materials is through similar biodegradative pathways to those taking place in the biotreatment plants discussed above. Chemicals that do not readily degrade persist in the environment for many years and at worst

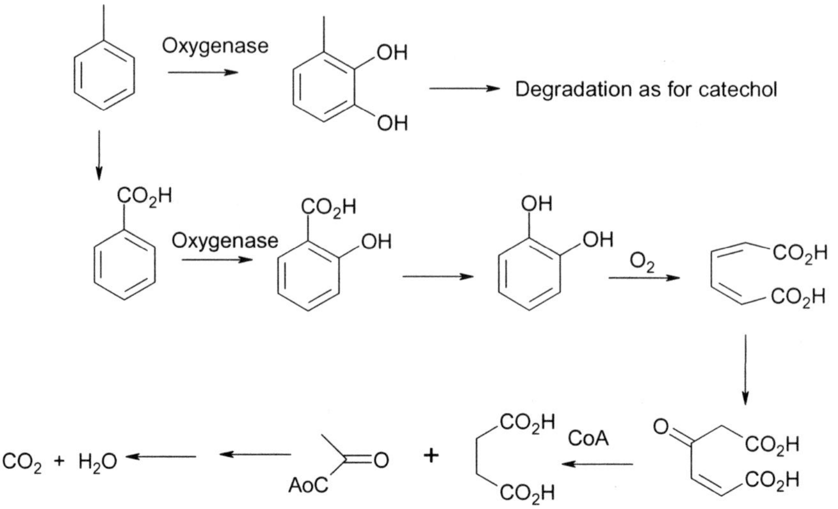

Scheme 2.6   Oxidative degradation pathway for toluene.

may have long-term toxic effects (*e.g.* current concern over persistent organic pollutants, POPs) or, at best, become an eyesore and take up valuable land space as in the case of many plastics. Increasingly, there is a recognized need for chemicals to be designed with degradation in mind.

2.7.1   Degradation and Surfactants

One of the first cases where lack of degradation came to the public attention concerned the use of alkylbenzene sulfonates in detergents. Increased use of these synthetic detergents in the late 1950s and early 1960s started to give problems with high levels of foam in sewerage plants and in turbulent parts of rivers and streams. This resulted in inefficient operation of sewerage plants, giving the possibility of harmful bacteria getting into the waterways. Initially the industry was unaware that these materials could pass through the sewerage treatment largely untouched and persist in rivers for a considerable time. Subsequent studies showed that this lack of degradation was due to the structure of the alkyl chain. Until the problem was identified alkylbenzene sulfonates had been produced by alkylation of benzene using propylene tetramer to give a mixture of products with branched chains exemplified in Scheme 2.7. It was discovered that if linear alkenes were used in place of propylene tetramer then the resulting sulfonate biodegraded much more quickly, overcoming the foam problems. This was one of the first cases of a structure–performance relationship being discovered with respect to biodegradation.

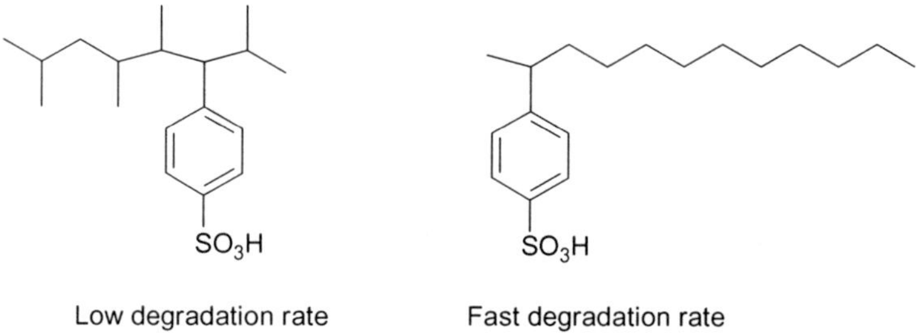

Low degradation rate   Fast degradation rate

Scheme 2.7 Structure–degradation relationship for alkylbenzene sulfonates.

A similar situation existed with alkylphenol ethoxylates, a major class of non-ionic surfactant widely used in industrial applications such as metal cleaning and in the textile industry as well as household detergents. The concern in this case was the toxicity and estrogenicity in fish. Nonylphenol ethoxylates were, until very recently, some of the most widely used materials in this class. Again the alkyl group is branched, being made from propylene trimer, which hinders degradation. This material has now been effectively banned from commercial 'down the drain' applications in Europe and is no longer present in laundry detergent products in the US.

2.7.2 DDT

When, in 1939, the insecticidal properties of DDT (dichlorodiphenyl-trichloroethane, **2.1**) became known it quickly became regarded as something of a wonder chemical, and the inventors were awarded the Nobel Prize. Not only was DDT effective against the Colorado beetle and the common housefly but it also killed malaria-carrying mosquitoes and was effective in controlling outbreaks of typhoid.

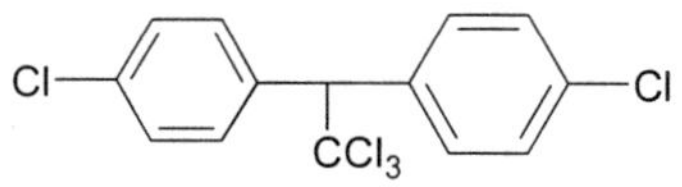

2.1 (DDT)

In many ways DDT was the ideal insecticide since it was:

- inexpensive to prepare,
- very wide spectrum of activity,
- non-toxic to humans,
- persistent, negating the requirement for re-treatment.

It was, however, this last property that caused problems. By the 1960s it was recognized that DDT was accumulating widely in the environment, since it was being produced at around 100 000 tonnes per annum. It was also found that DDT was accumulating in the fatty tissues of animals, since it is virtually insoluble in water. In some species of birds this accumulation was resulting in eggs being produced with very thin shells. Increased public concern quickly led to DDT being phased out by the developed world, but it is still used by some third world countries as a very inexpensive and effective insecticide. In the developed world DDT was replaced in the 1970s by less persistent but more expensive insecticides, such as organophosphates, and environmental persistence is now one of the criteria considered when new insecticides are developed.

2.7.3 Polymers

Plastics have transformed the way we live but are one of the most persistent and visible forms of waste. In the UK alone over 3 million tonnes of consumer plastics are produced each year, the vast majority of which are non-degradable. Recycling is becoming more common; but a high proportion of plastics eventually ends up in landfill sites, where they will probably remain for hundreds of years. There is general acceptance of the current problems with plastic waste and there is a school of thought that this is a transient issue. Bacteria are notoriously efficient at mutating to produce strains capable of using a wide variety of feedstocks, bacteria that can now digest nylon have been identified and it may only be a matter of time before bacteria capable of digesting other polymers are found. Meanwhile recycling of consumer plastic waste is on the increase and is discussed in more detail below. Attempts have also been made, with varying degrees of success, to make degradable plastics; these have fallen into four main areas:

1. manufacture of plastics from renewable resources (Chapter 6);
2. incorporation of biodegradable segments (often from renewable resources) in the polymer backbone;
3. manufacture of degradable polymers from petrochemical sources;
4. incorporation of other chemicals such as ferrocene to enhance the rate of photo- or chemical-degradation.

One concern over the production of degradable polymers is the possibility of methane production. Under laboratory conditions complete degradation to CO_2 and water may occur but under the conditions prevailing in a landfill site there may be a lack of oxygen leading to incomplete degradation and methane production. Since methane is a significantly more potent greenhouse gas than CO_2 this is a reasonable cause for concern.

Starch has been effectively incorporated into the polyethylene backbone to induce degradability and is used commercially in the production of grocery bags. There is a fundamental problem of compatibility between the hydrophobic polyethylene and hydrophilic starch. Starch may be described by the formula $(C_6H_{10}O_6)_n$, and largely consists of amylopectin, which is made up of D-glucose units. Importantly, starch contains C–O–C links that are readily broken down by enzymes such as the amylases and –OH groups that give the starch hydrophilic properties, these may be reacted with silane coupling agents to improve compatibility with polyethylene.

Typically starch is incorporated at levels of between 6% and 40% by weight; at levels over 9% the impact strength of the polymer is severely reduced, limiting its use to film-type applications. Levels of around 15% are required to bring the timescale for degradation down to below 1 year, under favourable conditions. Other methods of making polyethylene more degradable include incorporation of corn starch as an auto-oxidant (*via* peroxide formation) and ketones, which aid photo-degradation *via* radical mechanisms.

The agriculture industry is the largest user of degradable plastics. Here degradable urea-formaldehyde polymers are used in slow release fertilizer applications whilst polycaprolactone is becoming increasingly used as degradable plant containers.

Plastics produced from natural resources that are biodegradable are discussed in Chapter 6.

2.7.4 Some Rules for Degradation

Although the understanding of how to design products for degradation is not well advanced a superficial knowledge of what kinds of structural groups and molecular features are likely lead to products with a high degree of degradation has been developed. Some 'rules' for degradation, obtained from studies of various degradation pathways, that should be considered when designing a new product include:

1. natural products are all biodegradable; structures closely resembling natural materials are likely to be more degradable;
2. catechol is an intermediate in the degradation of aromatics, therefore aromatic products that do not contain two adjacent unsubstituted (or hydroxy substituted) carbons will degrade more slowly;

3. highly electronegative groups such as nitro decrease the rate of degradation of substituted aromatics;
4. many heterocyclic species are slow to biodegrade;
5. highly branched aliphatic chains degrade more slowly than linear chains;
6. aliphatic ether containing molecules degrade slowly;
7. materials containing strong C–Cl or C–F bonds generally do not undergo rapid degradation;
8. materials that are very water insoluble are unlikely to be readily biodegradable;
9. biodegradation rates normally decrease with increasing molecular weight.

2.8 POLYMER RECYCLING

Polymers do pose a waste disposal problem but they also afford many environmental advantages when compared with the alternatives. These advantages are often connected with their low weight–high strength properties. Replacement of 250 kg of metal in a car by an advanced polymeric material can save 750 litres of fuel over the lifespan of a car; extrapolating this to the whole of Western Europe would result in saving on CO_2 emissions of 30 million tonnes per annum.

Ultimately polymer degradation to CO_2 and water may be preferred from a land utilization point of view, but does add to global warming. This effect is minimized for polymers made from renewable resources, which, as raw materials, can be considered CO_2 neutral. Providing the energy involved in the collection and processing of polymers for recycle produces less CO_2 than polymer degradation then, at a simplistic level, recycling can be viewed as the environmentally preferred option. To get a more thorough understanding of the problem a full life cycle assessment (LCA) (Chapter 3) would need to be carried out.

Despite both logistical and financial barriers most governments are actively promoting post-consumer plastic waste recycling with varying degrees of success; in the UK, for example, it is currently less than 10%. There are three types of recycling, listed below in order of increasing environmentally friendliness:

1. incineration to recover energy;
2. mechanical recycling to lower grade products;
3. chemical recycling to monomers.

2.8.1 Separation and Sorting

Other than for incineration and some of the newer chemical recycling technologies, sorting of plastic waste into polymer types is of fundamental importance. To make separation easier for the consumer an international plastic recycle code mark is printed on larger items (Table 2.3). Even small amounts of a mixed plastic (sometimes as low as 1%) can have significant detrimental effects on the properties of a recycled polymer and result in it needing to be used in low value applications.

Manual recycling techniques, based on the use of the identification codes and experience, are still widely used but are becoming too costly to be economically viable and have largely been replaced in modern recycling plants. A large variety of density sorting methods are widely used; these include the well-established float–sink method in which a liquid of an intermediate density to the plastics (in the form of flakes) to be separated is used. Water is used to separate polyolefins from other plastics, but other media commonly employed include methanol–water and sodium chloride solutions. It is very difficult to separate the various

Table 2.3 Plastic identification codes.

Symbol	Abbreviation	Description & Examples
1	PET	Poly(ethylene terephthalate) Soda bottles
2	HDPE	High density polyethylene Milk & detergent bottles, plastic bags
3	PVC	Poly(vinyl chloride) Food wrap, vegetable oil bottles
4	LDPE	Low density polyethylene Shrink wrap, plastic bags
5	PP	Polypropylene Bottle tops, refrigerated containers
6	PS	Polystyrene Meat packing, protective packing
7		Other/mixed Layered articles

forms of polyolefin using this technology due to the small differences in density, *e.g.* only 0.03 g cm^{-3} between polypropylene and high-density polyethylene. A faster and more efficient form of density separation equipment is the hydrocyclone in which polymers are separated by centrifugal acceleration. By using a series of hydrocyclones virtually any polymer mix can be separated. One recently developed method involves shredding the plastic and passing through a series of pipes in suspension in water; the flow rate of the plastic depends on the density, enabling streams to be taken of at different points along the pipe.

Supercritical CO_2 (see Chapter 5 for a discussion of supercritical fluids) separators have also been developed. These are based on two important properties of supercritical fluids. First, being able to finely control the density of the medium by altering the pressure and, second, the low viscosity of such fluids which gives rise to very rapid separations. Materials with a very small density difference can be separated using this technology.

There are also several commercial separation processes based on spectroscopic techniques, two of which are briefly discussed here. Near-infrared (NIR) spectroscopy (Chapter 8) is finding increasing use in many forms of pollution control. With respect to polymer separation the common types of polymer have different absorption in the range 14300–400 cm^{-1}; this, coupled with the very short response times of the photo-detectors makes the technology suitable for the separation of bottles. The major disadvantage is that NIR is not able to accurately distinguish between black and very dark objects due to almost total absorption in the region. Figure 2.9 shows a schematic of the separation process. The bottle or article is fed into the detection chamber and after identification, based on the NIR spectrum, a controlled jet of air then blows the article into the appropriate collection chamber. Typically, around 20 articles per second can be processed.

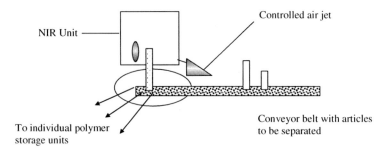

Figure 2.9   Schematic of NIR separation process.

X-Ray fluorescence (XRF) sorting is especially useful for sorting PET [poly(ethylene terephthalate)] from PVC. This is essential if the PET is to ultimately be used for food contact applications and desirable in most other applications; since under the conditions used for PET recycling the PVC would partially degrade and be seen as black specs in the final product. When bombarded with X-rays polymers containing heavy atoms such as chlorine emit a readily detectable X-ray signal. This signal can be used to control an air jet separation system similar to the one described above.

Many modern plants are starting to introduce a series of inline spectrometers to separate complex mixtures efficiently. First, an infrared array is used to detect between clear and translucent plastic. Next, a vision colour sensor programmed to ignore labels identifies various coloured plastics. X-Ray spectrometry is then used to detect the Cl atom in PVC. Finally, a near-infrared spectrometer is used to detect resin type, most importantly for the separation of HDPE and PET.

2.8.2   Incineration

Incineration of plastic waste either alone or as part of municipal waste is often perceived by the public to be an environmentally unsound process. This is due to the production of residual ash containing heavy metals and the possibility of dioxin formation from chlorine-containing waste (**2.2**). Like DDT, dioxins are persistent and accumulate in fatty tissue; unlike DDT some of the many dioxins that could be produced by incineration are highly toxic to humans. Although experts agree that waste can be incinerated safely, at high temperature, without any dioxin emission it is the potential risk that is the cause for concern to the general public.

2.2

As would be expected from the chemical structure the energy content of hydrocarbon based polymers is similar to that of heating oil and is over twice that of paper based waste. Hence plastic waste is a potentially valuable form of fuel. Incineration of plastics, as part of municipal waste, is especially suitable for small items such as thin films and small containers such as yoghurt cartons that are impractical to collect and sort for other forms of recycling. As with other forms of energy generation the heat from the combustion process is used to drive a steam turbine to generate electricity.

2.8.3 Mechanical Recycling

2.8.3.1 Poly(ethylene terephthalate) (PET). Mechanical recycling of PET is widely practised, the resulting product being used to produce PET fibres, which are in turn used to make fleece garments and carpets. The recycling process involves melting PET flake and processing it into granules using an extruder. The main problems centre on maintaining the molecular weight of the polymer (as measured by the intrinsic viscosity of the melt). The molecular weight is reduced by the presence of acidic impurities formed from trace amounts of PVC, labels, glue, *etc.*, which, at the melt temperature, cause chain scission. Several techniques have been developed to minimize this effect, including thorough drying of the flake to remove all traces of water and the introduction of chain extenders into the melt process. These are bi-functional compounds, such as diepoxides or bisoxazolines (Scheme 2.8), which react with the acid groups formed during chain scission to rebuild the chain.

2.8.3.2 High Density Polyethylene (HDPE). Along with the PET drink bottle the HDPE milk and juice containers are the most easily recognizable forms of plastic waste. Since HDPE contains no functional groups, molecular weight reduction during mechanical recycling is not an issue. A typical recycling process consists of washing HDPE flake to remove milk, juice, and glue residues and subsequent float–sink separation to remove small amounts of solid contaminants. Although expensive to carry out, some recycling companies have installed colour scanners to separate clear and coloured HDPE (*e.g.* milk bottle tops).

There are several barriers to HDPE recycling mainly connected with it not being able to be reused in food contact applications. HDPE used for milk containers is of a particularly high molecular weight, this makes it suitable for blow moulding but not for injection moulding, due to high shrinkage and long processing times. Also the HDPE used for milk containers does not have the required resistance to stress cracking required for applications such as bleach containers. For these reasons

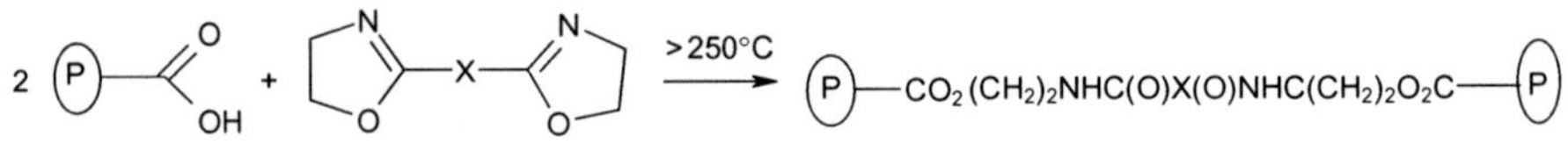

Scheme 2.8   Bisoxazoline chain extenders.

recycled HDPE is predominantly used in relatively low value applications such as drainage pipes, refuse containers and bin-liners.

2.8.3.3 Poly(vinyl chloride) (PVC). Although PVC is second only to polyethylene in terms of polymer production volumes, recycling is not widespread; this is attributable to the long-term nature of its major applications in the construction industry, *e.g.* pipes and window frames. A potentially valuable source of PVC in municipal waste is the water bottle (especially in areas of high bottled water consumption like France). PVC has found widespread use in this application since it does not 'taint' the water.

Most PVC is recycled *via* a co-extrusion process in which it becomes the inner layer of sewerage pipes and window frames (the outer layer being virgin PVC to comply with building regulations). To meet the requirements for the extrusion process it is important to remove all trace amounts of solid contaminants, hence prior to extrusion the melt is usually filtered. Further stabilizers and HCl scavengers are usually also added to prevent polymer degradation.

2.8.4 Chemical Recycling to Monomers

Chemical recycling to monomers has the advantage that the monomers can be reused to produce 'virgin' polymers, which are of higher value than their recycled counterparts. Processes have been developed to chemically recycle most functionalized polymers; however, due to the relative high associated costs, many of these processes are not widely used on a commercial scale. For polymers containing C–O bonds the process entails some form of hydrolysis or alcoholysis.

Two such processes are outlined below.

2.8.4.1 PET. There are various process options for 'depolymerizing' PET chemically; these include glycolysis, methanolysis, hydrolysis, and hybrid options. Glycolysis involves reacting PET flake with ethylene glycol at around 200 °C; the main products are bis(2-hydroxyethyl) terephthalate (BHT) and oligomers thereof. As BHT and its oligomers can not be purified by distillation it is difficult to remove soluble impurities such as dyes. This method therefore is mainly used to recycle 'production' scrap where the quality and purity are known.

Methanolysis is the major process used for chemical recycling of post-consumer PET waste. There are several variants of the process, with similar chemistry but differing energy requirements. The Dupont Pet-retec process is one of the most recently developed. Waste PET flake is

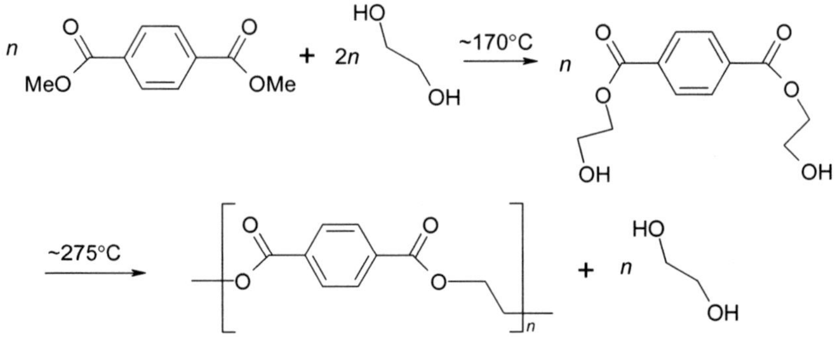

Scheme 2.9 Manufacture of PET [poly(ethylene terephthalate)] from recycled monomers.

dissolved in dimethyl terephthalate at around 230 °C and methanol is added. The mixture is heated to 300 °C under pressure, causing trans-esterification to take place. The products of this, dimethyl terephthalate and ethylene glycol, unfortunately form an azeotrope that makes separation more difficult. The azeotrope can, however, be broken by addition of methyl *p*-toluate to the reaction mixture. This forms a low boiling azeotrope with ethylene glycol, enabling dimethyl terephthalate to be separated by standard distillation techniques. Once cool the dimethyl terephthalate and methyl *p*-toluate separate into two layers. The recovered products may then be used to form PET according to Scheme 2.9.

2.8.4.2 Nylon. The largest source of nylon for recycling comes from carpets, especially in the USA. The aim of recycling is to produce high yields of high purity caprolactam and various processes based on hydrolysis, acidolysis, and ammonolysis have been developed to achieve this (Scheme 2.10).

The hydrolysis option suffers from the requirement for expensive high pressure equipment, whilst the lower energy (and capex) acidolysis process suffers from progressive poisoning of the phosphoric acid catalyst by impurities. Reaction with ammonia is quite versatile and can be used to recycle mixtures of both nylon 6 and nylon 6,6. From this, caprolactam, hexamethylene diamine, aminocapronitrile, and adiponitrile can all be recovered by distillation.

2.8.4.3 Feedstock Recycling of Mixed Waste. One of the most expensive and time consuming aspects of polymer recycling is the separation process. It would therefore make economic sense if mixed plastics waste could be taken and recycled back to suitable feedstocks. The essential

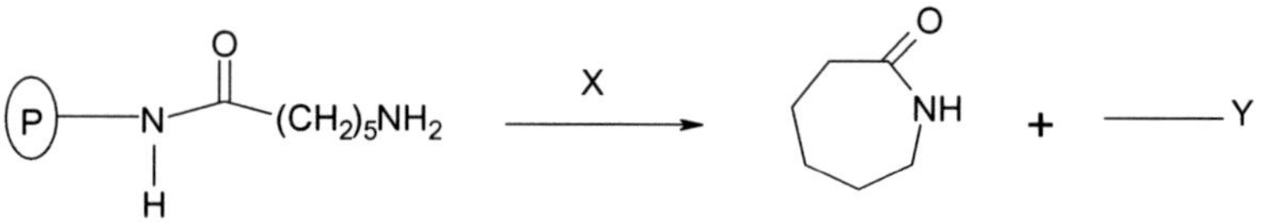

X = NH$_3$ / Phosphate catalyst,   Y = ~NH$_2$
X = H$_2$O,   Y = H$_2$O
X = H$_3$PO$_4$ / H$_2$O,   Y = ~CO$_2$H

Scheme 2.10   Chemical recycling options for nylon 6.

feature of feedstock recycling processes is the use of heat (thermolysis) to break bonds, similar to crude oil refining. The process may be carried out by heat alone (pyrolysis), in a hydrogen atmosphere, or in the presence of oxygen (gasification). In the latter case the feedstocks obtained are CO and H$_2$ (syngas). The hydrogenation process is especially beneficial when the feedstock contains relatively large amounts of Cl, S, and O atoms, these being removed as HCl, H$_2$S, and H$_2$O, respectively, maintaining the quality of the hydrocarbon products.

In the basic pyrolysis process mixed plastic waste is initially heated to around 300 °C. Under these conditions HCl is eliminated from any PVC present, for subsequent recovery as hydrochloric acid. The mixture is then heated to temperatures approaching 500 °C to produce a mixture of aromatic and aliphatic oils suitable for use as refinery feedstocks. High volumes of PET in the waste may cause some problems due to its high oxygen content. One of the problems with the pyrolysis process is that polymers have poor thermal conductivity, hence long residence times are needed to reach the high temperatures required. One way to improve heat transfer is to use a fluidized bed. In this process hot particles of sand, fluidized by hydrocarbon gasses (produced from the process), are intimately mixed with the molten polymer, thereby vastly improving heat transfer. Recently, a catalytic pyrolysis process has been developed to convert mixed plastic waste into a material suitable for use as a diesel fuel.

Gasification processes are carried out at very high temperatures, over 1300 °C, with controlled addition of oxygen. The products of the process are syngas and a 'glassy' inorganic residue that may be used in concrete. The syngas may can be converted into methanol or used as a fuel. Although a high energy process the gasification process does have some environmental benefits, since it completely avoids the possibility of forming dioxins and other toxic materials that can be produced during pyrolysis. However, process economics due to the high price of energy are preventing widespread use of this technology.

REVIEW QUESTIONS

1. Draw a process flow sheet for the production of phenol *via* benzene sulfonation. Assume all reactions proceed with 95% yield. Include mass balance data on your flow sheet.
2. Discuss the role of biotreatment plants in a modern chemical production facility.
3. Explain how mixtures of nylon 6 and nylon 6,6 can be chemically recycled using ammonia. Draw a mechanism for the reactions involved in producing caprolactam, hexamethylene diamine, and adiponitrile.
4. Discuss the environmental advantages and disadvantages of the use of polymers for packaging. Suggest ways in which the disadvantages could be minimized and discuss any add-on effects of your suggestions.

FURTHER READING

R. Carlson, *Silent Spring*, Houghton Mifflin, New York, 1962.
F. La Mantia, *Handbook of Plastics Recycling*, ChemTec Publishing, Toronto, 2002.
K. Scott, *Electrochemical Processes for Clean Technology*, Royal Society of Chemistry, Cambridge, 1995.
P. T. Williams *Waste Treatment and Disposal*, 2nd edition, Wiley-Blackwell, Chichester, 2005.

CHAPTER 3

Measuring and Controlling Environmental Performance

3.1 THE IMPORTANCE OF MEASUREMENT

Probably the most fundamental problem facing the development of greener products and processes is the measurement of progress and the development of appropriate methods for comparison of alternatives. In many instances it will be obvious that improvements have been made, *e.g.* when a toxic material is replaced by a non-toxic alternative, keeping all other process conditions essentially the same, or when the energy requirement of a process is reduced. Consider the example of dimethyl carbonate production. The traditional way of preparing dimethyl carbonate involved the reaction of highly toxic phosgene with methanol [Equation (3.1)]. As well as using phosgene the process produced two moles of unwanted HCl per mole of product that requires disposal, further increasing the environmental burden. In addition, the product was also contaminated with trace amounts of toxic chlorinated by-products, resulting in dimethyl carbonate being classed as a harmful material:

$$COCl_2 + 2CH_3OH \rightarrow (CH_3O)_2CO + 2HCl \qquad (3.1)$$

Today most dimethyl carbonate is made by methanol carbonylation (3.2) using a copper chloride catalyst that has a very long life. This

Green Chemistry: An Introductory Text, 2nd Edition
By Mike Lancaster
© Mike Lancaster 2010
Published by the Royal Society of Chemistry, www.rsc.org

process produces pure dimethyl carbonate, which is not now classified as harmful, and water as a by-product:

$$CO + CH_3OH + 0.5O_2 \rightarrow (CH_3O)2CO + H_2O \qquad (3.2)$$

Despite the toxicity of CO, and the high pressure required for the process it would be generally accepted that the carbonylation process is a significant step in the right direction, and that the resultant product is safer to use, its flammability now being the main hazard.

In other instances the 'green' improvements will be more contro-versial, *e.g.* replacement of a material used in a relatively large volume, producing large amounts of benign waste by a much lower amount of a more hazardous material or the replacement of a fossil based raw ma-terial by a renewable feedstock with an associated increase in the energy requirement of the process by 20%.

An alkene may be brominated by several methods. Traditionally the reaction has been carried out using bromine in a chlorinated solvent such as carbon tetrachloride or dichloromethane. This reaction is highly atom efficient but the use of hazardous reagents and solvents reduce its at-tractiveness from a 'green' point of view. An alternative brominating agent is pyridinium hydrobromide, typically used in an alcohol solvent. Here the hazardous nature of the raw materials has been significantly reduced but at the expense of reducing atom efficiency. The latter route is probably preferable but the choice is less clear cut (particularly when synthesis of the raw materials is taken into account) and the researcher may look to study both routes to see if either can be improved; for ex-ample, could the chlorinated solvent be avoided in the first case or can the pyridinium group could be supported and recycled in the second. In the following sections examples of process and product developments and their 'green' credentials are discussed in some detail, highlighting some of the difficulties involved in choosing one alternative over another.

3.1.1 Lactic Acid Production

There are two competing processes for the manufacture of lactic acid (2-hydroxypropanoic acid), one chemical synthesis (Figure 3.1) and a fermentation route (Figure 3.2). The synthetic route involves the reaction of HCN with acetaldehyde; whilst neither of these materials can be regarded as benign, the reaction is quite efficient in terms of atom economy. The resulting nitrile is isolated by distillation and hydrolysed by sulfuric acid. The ammonium sulfate by-product of this reaction reduces the atom economy to 60%. The resulting lactic acid can not be

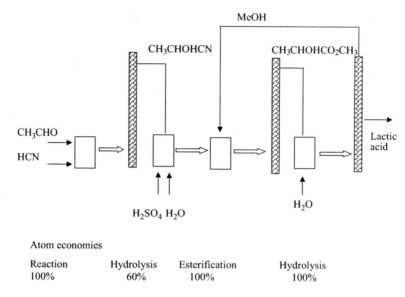

Figure 3.1   Chemical route to lactic acid.

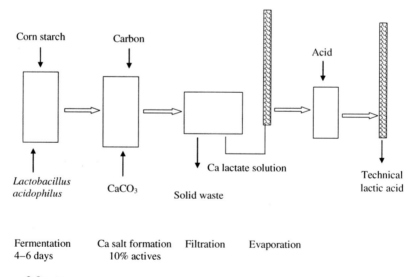

Figure 3.2   Fermentation route to lactic acid.

isolated in useable purity, however, and esterification with methanol followed by distillation, hydrolysis, and isolation by further distillation is required. The lactic acid produced is, however, of extremely high purity.

The positive features of the process are a relatively good overall atom economy, benign by-products, small volume of waste, and the high product quality. On the negative side the process uses hazardous raw materials and is relatively energy intensive, due to the three distillation steps involved. By contrast the fermentation process uses safe renewable feedstocks such as corn syrup or molasses; but the reaction time is very long, taking 4–6 days to complete. Carbon is added to the resulting fermentation broth to remove coloured impurities, as is calcium carbonate, which aids purification of the product through formation of calcium lactate. The salt solution is filtered, generating a solid waste, which requires disposal. As with many biotransformation reactions the amount of product obtained per unit of reactor volume is low. In this case the calcium lactate is only present in the resulting solution at about 10% concentration. Evaporation is therefore required to concentrate the salt, which is expensive in energy terms. Technical grade lactic acid is then produced by acidification and further distillation. The acidification process produces calcium sulfate waste, which needs to be disposed of or possibly sold as a by-product.

The lactic acid produced from this process is only technical grade and although of sufficient purity for some applications needs further purification for others. To produce the highest quality material (with a similar specification to that produced by the chemical route) the technical product needs to go through the same esterification purification process used for the chemical synthesis, thereby considerably increasing the energy requirements of the process.

The green advantages of the fermentation route are its renewable feedstock and its use of non-hazardous materials. The disadvantages are very high energy usage due to the dilute nature of the process streams – this is particularly the case for production of high purity material. Although the waste produced is non-hazardous there is a significant volume of it, especially when waste water is taken into account.

So which process is greener? Clearly more information would be needed to try and come to a rational decision. To some, simple, extent the argument could be based on a view as to the risks of using HCN compared to the additional energy requirement required in the fermentation route. To do this effectively some mechanism for prioritizing the various options is required. As a society, is it preferable to use up our increasingly scarce fossil fuel resources in providing the extra energy or accept the risk, of say, once every 50 years having an incident on the plant that will cause serious harm to one or more people? Ideally the Green Chemistry solution would be to develop a low energy, low risk process. But a way of assessing, and critically comparing, current options is also required. Chapter 6 discusses more recent developments of the process.

3.1.2 Safer Gasoline

Decisions are made to move forward to a more benign product based on good scientific evidence of the day. Later, with increased knowledge and experience, these decisions may turn out to be questionable. In the late 1970s overwhelming evidence was produced that lead emissions in car exhaust fumes were having an adverse effect on the IQ of children living in cities; the presence of lead also meant that catalytic converters could not be used to reduce noxious emissions. Most developed countries decided to ban tetraethyllead, which, until then, had been used in relatively small amounts in gasoline to increase the octane rating, thereby preventing engine 'knock', and also to provide lubrication of the pistons. Following the assumption that the world needed to keep the internal combustion engine essentially unchanged for the foreseeable future, the replacements for tetraethyllead were assessed. At the time several possible options were considered, including:

1. Altering the refinery process to put more aromatics into the gasoline pool. This would increase the crude oil requirement per litre of fuel; it would also increase exposure of the general public to higher levels of toxic benzene. This was not viewed as a significant problem.
2. Adding ethanol to the gasoline pool; this had already been done for many years in Brazil.
3. Adding methyl *t*-butyl ether (MTBE) to the gasoline pool. Of all the (cost-effective) organic fuel supplements tested MTBE had the highest octane number.

In most of Europe the first option became the preferred route whilst in the USA the third option was widely adopted. Some 20 years later, however, MTBE began to appear in drinking water in parts of the USA, most notably in California. Although observed levels were very low, at the parts per billion level, and there was no evidence that these levels were causing adverse effects the regulatory authorities rightly decided that the situation was not acceptable. The most likely cause of MTBE entering the drinking water supply is through leaks in underground gasoline storage tanks. Being somewhat water soluble MTBE can then be washed by rainwater into the watercourse and from there into drinking water. The hydrocarbon components in the gasoline pool, being less water soluble, tend to stay in the ground close to the leaking tank, causing relatively little damage. Adopting a root cause analysis approach the clear answer would be to mend all the leaks, but this would be an enormous and costly task. Instead the solution was to phase out MTBE and replace it with bioethanol. This also helps fight climate change, which had become of prime importance since the earlier decision to go the MTBE route was made.

Had it been known in the late 1970s that MTBE would be found in drinking water or that carbon dioxide emissions would have such a major effect on the climate would it have been introduced? Probably not.

Evidently, from the above examples, moving towards sustainability involves making difficult decisions. Agreement on a preferred option is often not easy to achieve, with different stakeholders finding it relatively simple to argue their case on environmental grounds. Clearly there is the need for a more standardized approach. This approach needs to cover the 'big' decisions like options for replacing tetraethyllead and also offer a set of agreed metrics for selecting options for individual processes as discussed as in the lactic acid example.

3.2 INTRODUCTION TO LIFE CYCLE ASSESSMENT

Life cycle assessment (LCA) measures or at least attempts to predict the environmental impact of a product or function over its entire life cycle. The term 'life cycle' means a holistic assessment that requires the assessment of raw material production, manufacture, distribution, use and disposal, including all intervening transportation steps necessary or caused by the product's existence. The product's life cycle is often defined as being from cradle to grave, *i.e.* from extraction of the raw materials required to make the product to its fate at the end of the product's usefulness, *i.e.* its disposal. Increasingly, however, the terminology cradle to cradle is being used to take account of the recycling of the product either into other useful material or into energy at the end of its life. LCA has been used widely to compare the environmental impacts of competing products and processes and provides quasi-quantitative date on which to make decisions as to which options are more environmentally benign. One important aspect of LCA is that it can ensure that identified pollution prevention opportunities do not result in unwanted secondary impacts on other parts of the life cycle, *i.e.* merely moving the environmental burden rather than reducing it.

Taking the manufacture of polycarbonates as an example, two manufacturing routes are shown in Scheme 3.1.

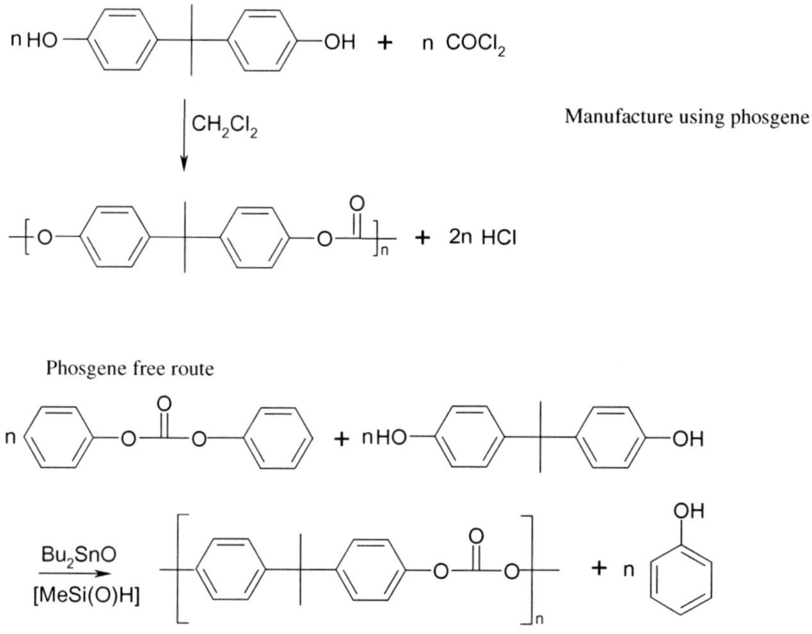

Scheme 3.1 Routes to polycarbonates.

In the first route phosgene is reacted with Bisphenol A in dichloromethane. The main environmental concern with this process is the large-scale use of phosgene. In the second route diphenyl carbonate is reacted with Bisphenol A to produce polycarbonate and phenol that can be subsequently recycled to produce Bisphenol A or diphenyl carbonate. This second route produces a lower molecular weight polymer than the phosgene route due to the presence of hydroxyl end groups; capping these by introduction of a silane ensures that material of useable molecular weight is produced.

At face value the environmental benefits of the phosgene free route are obvious. The LCA approach would, however, take into account the manufacture of all starting materials. In particular, diphenyl carbonate is normally manufactured from the reaction of phenol and phosgene. Hence this process could be viewed as simply pushing the environmental burden of phosgene use back one stage in the life cycle. However, if the diphenyl carbonate was manufactured by a trans-esterification reaction between phenol and dimethyl carbonate, as can now be done, and dimethyl carbonate was manufactured by the carbonylation route, the overall environmental burden may be less. As can be seen from this simple example, performing a full LCA, when energy use is also taken into account, is a time consuming, costly, and complex process.

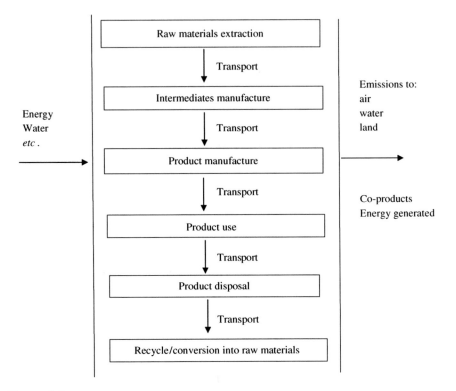

Figure 3.3   Life cycle stages.

Although used for energy analysis some 40 years ago, the LCA concept was only really developed in the early 1990s, and therefore the tools and methodologies are still being evaluated and improved. Partly as a result of its developmental history LCA results are often presented in the form of an energy balance. Whilst this is a valuable approach it can result in the different impact of various emissions being overlooked. This aspect of LCA is being more widely addressed in recent studies. There are still some variations in the approaches taken but most cover the stages shown in Figure 3.3.

Performing a full LCA is a complex process but it can be conveniently broken down into four stages.

3.2.1   Four Stages of LCA

3.2.1.1   Goal and Scope Definition (Planning).   In many respects the planning stage is the most important and will largely determine the outcome of the assessment. At this stage the objectives and goals must be clearly defined. These may well differ depending on whether the LCA is being carried out on a product with one existing manufacturing route or whether various routes to the same product are being assessed, or whether various products with the same application are being assessed. For example, the LCA would probably give different results for an assessment with the goal of establishing the lowest environmental impact process for producing flame-retardant polyurethane foam compared to one that had the goal of establishing the product with the lowest environmental impact for household seating applications.

At this stage it is also important to define the system boundaries to be studied; this will depend on the goals and the available resources. If, for example, a LCA was to be carried out on a high volume and versatile chemical like acetic acid, taking account of all the possible products would be a very costly and time consuming process and may not be relevant for the purposes of the study. It is also important at this stage to ensure that any comparative studies are carried out on the same *functional unit*. If, for example, a comparison between polystyrene and ceramic cups were being considered it would not be appropriate to compare on the basis of each being used once and disposed of. The assessment would need to take account of the ceramic cup being used hundreds of times – with washing (here an assumption needs to be made that may change the outcome). The strategy for data collection also needs to be developed – often it will not be possible to easily collect all the data required, in which case assumptions will need to be made and recorded. Goal and scope definition is an iterative process and needs to be reviewed and refined as the assessment progresses.

3.2.1.2 Inventory Analysis. The main activity of this stage is data collection. To aid this it is useful to represent the complete system as one or more flow sheets connected by material and/or energy flows. In essence these will be of a similar form to that shown in Chapter 2 (Figure 2.5) but will include energy inputs and outputs for each stage being assessed. Material and energy inputs and outputs should balance. Some of the data required such as the amounts of materials used or the energy required for a particular process will be readily obtainable, wherever possible exact data from manufacturers should be used. Other data will be less easily obtainable; lack of quantitative data is an issue for co-product manufacturing, *e.g.* co-production of phenol and acetone or production of ethene from a cracker. This is known as the problem of allocation, when some way of assessing the environmental burden of each co-product needs to be found. Wherever possible the problem of allocation should be minimized by appropriate selection of the system boundaries.

Typical methods of data collection include direct measurement, interviews with experts, literature, and database searches, theoretical calculations, and guestimates. As more assessments are carried out more databases are being established, *e.g.* by Battelle in Germany and Franklin in the USA. The data generated are typically presented in one of two ways. The simplest approach is to aggregate emissions to the medium to which they are released, *e.g.* x tonnes or litres of BOD and COD to water, y tonnes of CO_2 to atmosphere. CO_2 is usually the most significant figure since all energy use is normally quantified as CO_2 release. The other, more complex, but possibly more valuable method of presenting results is as a series of potential environmental impacts. Eight such impacts are normally reported, which are discussed in the following section. Various useful software packages are now available to help with this task.

3.2.1.3 Impact Assessment. One of the more common methods of assessing the data is to put a numerical value on various potential environmental impact criteria, namely:

1. *Abiotic depletion:* This takes account of depletion of all non-renewable resources. The impact is calculated as the sum of the resource used for each functional unit divided by the estimated reserves of that resource.
2. *Acidification potential:* The acidification potential of acid releases is expressed in terms of their potential to form H^+ relative to SO_2. The total impact is defined as the sum of the acidification potential for each released component multiplied by the quantity released.
3. *Aquatic toxicity:* This is the sum of the toxicity factor of a particular emission multiplied by the amount. Since the factors have only been agreed for a small number of materials, this measurement is currently of limited applicability.
4. *Eutrophication potential:* This is the potential to cause over-fertilization of water and soil, which can lead to uncontrolled growth of algae *etc.* This value is calculated in a similar way to acidification potential and is expressed relative to PO_4^{3-}. Potentials have been established for several common emissions, including NH_4^+ and NO_x.
5. *Global warming potential:* This value is based on known global warming factors for gasses such as N_2O, CH_4, and various organic solvents, expressed relative to CO_2. A mass weighted summation is used as for the acidification potential.

6. *Human toxicity potential:* As for aquatic toxicity, the database for human toxicity potential is still being established but is based on acceptable daily doses. The total potential is the sum of potentials released to different media.
7. *Ozone depletion potential:* This is calculated in a similar manner to global warming potential and is expressed relative to CFC-11. Factors for all common gasses having significant effects on the ozone layer have been calculated.
8. *Photochemical oxidants creation potential:* This is a measure of the potential to generate smog and is expressed relative to ethene.

3.2.1.4 Interpretation. The final stage of the LCA process is connected with identifying which parts of the life cycle have the most significant impacts and at identifying possibilities for improving the total environmental impact of the system under study.

Whilst this approach to quantifying the environmental impact portrays the LCA approach as being highly scientific, the results obtained are currently open to debate, due to the lack of information on many of the emissions and products entering the environment. The other issue is one of ranking the relative importance of the various criteria. This ranking or *valuation*, process is probably the most subjective part of the impact assessment process. Although several valuation techniques have been proposed there is no agreed hierarchy and difficult conclusions such as the relative importance of increased ozone depletion *versus* increased smog forming potential or human toxicity potential *versus* aquatic toxicity potential need to be made on a case by case basis. This of course does leave the whole process open to some degree of manipulation, particularly were LCAs are used to market one product over another.

Both the value of, and problems associated with LCA are evident from a study of the many assessments carried out on the environmental impact of disposable *versus* reusable nappies (diapers). The general public perception is that cotton nappies, reusable and made from a renewable resource, are much more environmentally friendly than disposable nappies made from 'chemicals' that need to be disposed of and hence contribute to an unpleasant form of landfill. Actual LCA studies suggest that the differences are less pronounced and depend on the assumptions made. In fact a 2008 study by the Environment Agency suggests that due to weight reduction disposable nappies have a marginally better global warming impact than reusable ones. With the former most of the environmental impact is in production, which has been quantified with reasonable accuracy, for reusables most impact is in their use, which is harder to quantify and is based on survey information. In fact, for people who use full washing loads at lower temperatures and air dry the nappies they will be causing a lower environmental impact than by using disposables. Although much of the same basic technical information such as amounts of agrochemicals used to grow cotton, energy usage, emissions, *etc.* for the two products has been used in the various assessments the conclusions have been different because of assumptions made, including:

- the number of times a cotton nappy can be used before it needs replacing;
- the number of nappy changes each day; since disposable nappies are more absorbent;
- the temperature of the wash process for cotton nappies;
- the drying method used following nappy washing (natural or tumble-drying).

All these assumptions can make a significant difference to the outcome of the study and, perhaps not surprisingly, assumptions made by organizations with a vested interest tended to support their product. Perhaps the overall conclusion from these studies is that the difference in overall impact is relatively small. How these products are used in practice may in fact be the most significant determinant of their relative environmental impact.

LCA is a powerful tool, which will become increasingly useful, as it is refined and becomes more objective. But its complexity, cost, and the length of time taken to carry out a full analysis make it an impractical tool to use on a day to day basis for research and development chemists and chemical engineers. What is really required for most practising technologists is a simple set of metrics to aid the decision making process involved in choosing one synthetic route or product over another.

3.2.2 Carbon Footprinting

With global warming now considered to be one of the major challenges facing mankind much more emphasis is being put on carbon footprinting rather than the full LCA. The carbon footprint of a product can be defined as 'The total set of greenhouse gas emissions caused directly and indirectly by an [individual, event, organization, product] expressed as CO_2 equivalents.' As well as being applied to specific products it is increasingly becoming common for organizations to report on their total carbon footprint.

The process framework is similar in many ways and typically focuses of four major aspects:

1. *Define the methodology*: One commonly used methodology is the GHG Protocol produced by the World Resources Institute and the World Business Council for Sustainable Development. This methodology provides detailed guidance on corporate emissions reporting and is available free of charge online. A more recent standard from the International Organization for Standardization, ISO 14064, also provides guidance on corporate footprint calculation and emissions reporting.

2. *Specify the boundary and scope of coverage*: It is important to be clear about which set of emissions will be quantified. This is commonly referred to as defining your 'boundary'. Established methodologies such as the GHG Protocol provide rules for allocation of the emissions to the organization. Having defined the boundary, consider what types of emissions will be included. Ask the following questions:

   I. CO_2 only or all greenhouse gases?
   II. Direct emissions from fuel use onsite and from transport?
   III. Direct emissions from manufacturing processes onsite?
   IV. Emissions from the electricity the organization purchased?
   V. Emissions from the organization's supply chain and other activities for which the operation is indirectly responsible, such as outsourced activities or manufacture and transport of raw materials, by another company, which your organization then uses?

3. *Collect emissions data and calculate the footprint*: The accuracy of the footprint relies on the ability to collate and analyse data relating to as many of the defined emissions as possible. For gas and electricity, collect consumption data in MWh or kWh. Data for other fuels can be collected in various units, for example, MJ, litres, and so on. For transport emissions it may be necessary to estimate the total fuel consumption based on the mileage of the vehicles and fuel economy assumptions. GHG emission data can generally be translated into equivalent CO_2 emissions data using standard emissions factors. For some emissions sources, more complex calculations may be required.

4. *Verify results*: often to gain credibility with stake holders it required that the results be verified by a third party.

3.3 GREEN PROCESS METRICS

In Chapter 1 the concept of atom economy was discussed as a design tool. Similarly, in Chapter 2 the term 'E'-factor was introduced as a measure of the amount of by-products formed per unit weight of product. Unlike atom economy the 'E'-factor is determined from an actual process or can be extrapolated from laboratory work. As a valuable extension to the E-factor concept Sheldon has proposed an 'Environmental Quotient', which is the product of the E-factor and a by-product 'unfriendliness' factor Q, *e.g.* Q may be 100 for a heavy metal and 1 for sodium chloride, for example. This concept has not been expanded upon and it is unlikely that a consensus amongst various stakeholders, regarding appropriate Q-factors, could easily be reached. The E-factor, however, takes account of the mass of all materials used in a process, including water; production of significant amounts of benign waste such as this can make the environmental impact appear much worse than it actually is. The concept of effective mass yield (EMY) has been proposed to overcome this. EMY approximates to the reciprocal of the E-factor expressed as a percentage but does not take into account benign materials such as water, dilute ethanol, or acetic acid or low concentrations of benign inorganic salts.

By way of illustration, taking the simple esterification of butanol with acetic acid, the balanced equation for the reaction is (3.3):

$$CH_3(CH_2)_3OH + CH_3CO_2H \rightarrow CH_3(CH_2)_3CO_2CH_3 + H_2O \qquad (3.3)$$

The atom economy for this process is 85% ($100 \times 102/120$), which is reasonable. To calculate the E-factor and EMY further information is needed. From published literature (Vogel's *Practical Organic Chemistry*), a standard procedure is to mix butanol (37 g) with glacial acetic acid (60 g), and a small amount of sulfuric acid catalyst (ignored in all calculations). Following completion of the reaction the mixture is added to water (250 g). The crude ester is washed further with water (100 g), then saturated sodium bicarbonate solution (25 g) and finally water (25 g). After drying over 5 g of anhydrous sodium sulfate the crude ester is distilled to give the product (40 g) in a yield of 69%.

Calculating the E-factor as kg waste/kg product it can be seen that 40 g of product has been made from a total input of 527 g, giving an E-factor of 487/40 or 12.2. This could be considered relatively high and indicative of a not particularly efficient process. If the environmental quotient was considered a Q-factor of somewhere between 1 and 2.5 might be assigned, since the waste is relatively benign. Finally, in calculating the EMY as 100 × (mass of product)/(mass of non-benign material used), with the water, inorganic salts, and acetic acid being ignored, the EMY is 100 × 40/37 (weight of butanol) or 108%. Hence using EMY the reaction is exceptionally environmentally benign. The EMY concept is again prone to misuse since there is no agreed consensus as to what constitutes an environmentally benign material – some of the most difficult to treat waste is large volumes of aqueous waste containing trace amounts of hazardous material. Table 3.1 summarizes the process measures. This highlights the difficulty in reaching a conclusion as to the 'greenness' of even such a simple process.

Some of these measures attempt to take account of the hazardous nature of by-products and effluent, but they do not take account of the hazardous nature of any starting materials. The other major omission of these measures is in measuring the energy involved.

Evidently, there is no simple but comprehensive method for selecting the 'greenest' route to a particular product, but by assessing certain criteria at the planning stage and then refining this analysis after some experimental work, moves can be made in the right direction. Table 3.2

Table 3.1 Measuring the environmental efficiency of butyl acetate synthesis.

Measure	Value	'Greenness'
Yield	69%	OK
Atom efficiency	85%	Quite good
E-factor	12.2	Poor
Environmental quotient	Approx. 12.2–30	Fairly Good
EMY	108%	Very Good

Table 3.2 Pre-experimental route selection pro-forma.

Parameter	Measure
Atom economy	
Expected overall yield	
Atom economy × expected yield	
Number of individual stages	
Number of separation/purification steps	
List of VOCs to be used	
List of toxic or other environmentally hazardous raw materials	
List of toxic or other environmentally hazardous waste products	
Significant energy requirements,	
i.e. estimated reaction time at $> 150\,^{\circ}C$ or $< -15\,^{\circ}C$	
% Raw materials from renewable resources	
List of specialist equipment required	
Estimated raw material cost/tonne product	

shows a suggested pro-forma to be completed at the planning stage when various routes are being considered. Ideally this should be done as part of the team approach to waste minimization discussed in Chapter 2.

This qualitative approach will provide information for the team to select several routes for experimental evaluation. Many routes may be ruled out, relatively simply, on the basis of cost and/or the need for specialist equipment. Also by considering the hazardous nature of materials and by-products as well as their likely quantity it will be easier to make reasoned assessments that will include the impact of environmental issues. By looking at the detail behind these findings it may become evident that the poor atom economy or use of a particular hazardous reagent is associated with only one step of a multi-step process, *e.g.* production of large amounts of aluminium waste from a Friedel–Crafts acylation. If this is the case it could be worth revisiting the literature to see if this part of the process could be modified. Although this may seem a bureaucratic process the information gathered will be valuable in completing COSHH assessments, HAZOP studies, and in making cases for capital release.

Following an assessment of the criteria in Table 3.2 it may become evident that there is only one viable route, but more likely two or three options will be difficult to choose between. In this case the alternatives should be screened experimentally to refine the data in Table 3.2 and to complete a pro-forma similar to that shown in Table 3.3.

From this further analysis, the actual amount of waste (and its nature) per kg of product will become evident. At this stage it is also important to look forward and assess options for recycling or reusing the 'waste' on site, for example if a solvent can be efficiently recovered then this should be taken into account in calculating the E-factor. Although the choice of which route to fully optimize may not be obvious, even from this further analysis, it will facilitate a reasoned discussion of the issues.

Table 3.3   Additional metrics from experimental work.

Measure	*Value*
E-factor	
Effective mass yield (excluding water)	
kg VOC/kg product	
kg Waste/kg product to be treated on-site	
kg Waste/kg product to be treated off-site	
Additional hazardous by-products identified	
Identified options for recycling solvent/by-products on site	
Estimated E-factor after on-site recycling	
Refined estimate of energy requirements, or ideally direct measurement	

3.4   ENVIRONMENTAL MANAGEMENT SYSTEMS (EMS)

The need for effective rules or systems governing the actions of all parts of an organization is essential to the efficient working of that organization and helps ensure it meets its objectives. Most companies have a plethora of such systems covering health & safety, production, purchasing, quality, *etc.*; ideally such systems will be coherent and complementary. In response to growing international concern over environmental issues, as well to the demands of shareholders, the inclusion of Environmental Management Systems (EMS) is widespread in most manufacturing industries. As some chief executives find to their cost at annual general meetings, companies can not afford to ignore environmental concerns. Many companies realise that environmental indicators have become essential tools for decision making and that taking these indicators into account can lead to cost savings and new business opportunities.

An effective EMS will:

- define environmental responsibilities for all staff;
- identify opportunities to reduce waste, including raw materials, utility use and waste disposal costs;
- increase profits;
- reduce the risk of fines for non-compliance with environmental legislation;
- ensure all operations have procedures to minimize their environmental impacts;
- record environmental performance against set targets;
- provide a clear audit trail;
- attract shareholders and investors.

To have external value any auditing system must have the approval of, and be certified by, a recognized independent body. Several such bodies exist, including the International Organization for Standardization (ISO).

3.4.1   ISO 14001

In 1996 ISO 14001 was introduced as one of the first widely recognized environmental management systems. The intent of the standard is for the organization to develop a systematic approach to dealing with environmental concerns and to follow a philosophy of continual improvement in its chosen environmental indicators. Figure 3.4 shows the five key elements of ISO 14001. For each element there is a standard that the organization must meet to get accreditation.

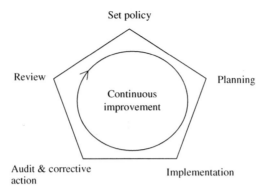

Figure 3.4   Key elements to ISO 14001.

3.4.1.1   Environmental Policy.   Before an environmental policy can be put in place the *status quo* of all the company's operations and systems must be established for consideration. Following this an analysis of how to meet the Standard can be undertaken (Box 3.1). Ideally the policy should provide a unifying vision, stating the overall goals of the organization. A key part of the standard is concerned with communicating goals and actual performance, both internally to employees, and externally to the general public.

3.4.1.2   Planning.   The planning stage is one of the more time consuming aspects of ISO 14001 and includes several elements.

3.4.1.2.1   Environmental Aspects.   This is concerned with the establishment of procedures to identify all aspects of activities, including products and services that may have an environmental impact and over which the company has some degree of control. Some important activities that a chemical company would normally consider include emissions, chemical storage, energy use, and raw material supply. Some aspects of these activities may come under regulatory control, but the organization must be seen to consider all aspects. Significant aspects should then form the basis for setting targets for minimizing environmental impact.

3.4.1.2.2   Legal & Other Requirements.   This covers the rather obvious obligation of the organization to be aware of and adhere to any legal regulations covering its business. There is also the requirement to adhere to any voluntary codes of practice, customer agreements, Responsible Care, *etc.* that the organization signs up to.

Box 3.1   Environmental policy statement

Top management shall define the organization's environmental policy and ensure that it:

1. *is appropriate to the nature, scale, and environmental impact of its activities, products, or services;*
2. *includes a commitment to continual improvement and pollution prevention;*
3. *includes a commitment to comply with related environmental legislation and regulations, and with other requirements to which the organization subscribes;*
4. *provides the framework for setting and reviewing environmental objectives and targets;*
5. *is documented, implemented, and maintained and communicated to all employees;*
6. *is available to the public.*

3.4.1.2.3   Objectives & Targets.   These are the essential features of the standard against which performance is measured. Objectives should be in line with the policy statement and take into account identified environmental issues and legal requirements; usually, most departments within the organization will have at least one associated objective. Quantifiable performance improvement targets should accompany the objectives. Typically, some of the targets will be associated with employee training and awareness programmes whilst others may deal with reducing emissions and/or energy use and/or increasing the percentage of on specification material.

3.4.1.2.4   Environmental Management Programmes.   Once the objectives have been set a programme of how to achieve them needs to be set up. An auditing system to monitor performance also needs to be established. These procedures may not necessarily be new, for instance many companies will already have a system in place for monitoring emissions and this could be used for auditing progress towards emission reduction target.

3.4.1.3   Implementation and Operation.   This element is again broken down into several sub-sections.

3.4.1.3.1   Structure and Responsibility.   This is concerned with ensuring that there is a senior manager with overall responsibility for the programme and that the responsibilities of other staff are clearly identified and communicated. These individual responsibilities should fit with the overall policy and objectives.

3.4.1.3.2 Training, Awareness, and Competence. Training of employees is key to the success of most programmes. A training plan needs to be developed for all employees who work in an area that may produce an environmental impact (usually most employees). This training programme should highlight the importance of the policy and plan, and explain the employee's role in helping to achieve the objectives. Additionally the consequences of adverse environmental impacts of the organization's activities should be clearly explained. In any programme of this type success will only be achieved if most employees buy into it – appropriate training is key to achieving this buy-in.

3.4.1.3.3 Communication. Communication is also an essential feature of achieving employee buy-in. It is important to ensure that mechanisms are in place for two-way communication, *i.e.* that employees also get an opportunity to comment, provide feedback, and make suggestions to senior management. In addition to top down communication it is also essential that procedures are in place for communication to the wider population, which includes stakeholders, and the public regarding the organization's environmental policy and performance against stated objectives.

3.4.1.3.4 Systems Documentation and Control. For any system to be auditable there must be adequate documentation covering procedures, actions taken, and results obtained. Ideally most of this documentation should be included in an Environmental Manual that each employee has access to.

3.4.1.3.5 Operational Control. Specific documented procedures and instructions must be written for all activities where absence of instructions may lead to an event that has a significant environmental impact. For these cases the instructions should include the corrective actions be taken in the event of a problem arising.

3.4.1.3.6 Emergency Preparedness. The purpose of this element is to reduce the environmental damage from any unplanned event. By establishing and publicising clear lines of command and control, as well as the actions to be taken in particular circumstances, the time taken to get an emergency under control can be significantly reduced. Emergency procedures should also detail where specific information, *e.g.* regarding the hazards of a particular chemical, can be found.

3.4.1.4 Checking and Corrective Action. Procedures should be established to periodically monitor all operations that may have an environmental impact, likewise scheduled periodic checks should be made to ensure procedures and operations are still in line with regulatory and other requirements. If equipment is involved in the monitoring process then procedures should also be in place to ensure correct calibration. Procedures are also needed to record any preventative or corrective actions identified and to ensure that they are completed to schedule. The whole EMS should be audited periodically to check that it is consistent with the planned policy and has been properly implemented. The details of audits should be discussed and reviewed by management at the highest level with a view to continuous improvement through the modification of both policy and procedures.

3.4.2 The European Eco-Management and Audit Scheme (EMAS)

The European Commission, as part of its policy to encourage industry to adopt more sustainable practices, created the EMAS system in 1993. EMAS was also seen as a move towards stated the EU goal of encouraging industry to adopt more self-regulatory practices and moving away from command and control type central legislation.

The EMAS regulations were revised in 2001 to include:

- extension of the scope of EMAS to all sectors of economic activity, including local authorities;
- integration of ISO 14001 as the environmental management system required by EMAS;
- adoption of a visible and recognizable EMAS logo (Figure 3.5) to allow registered organizations to publicise their participation in EMAS more effectively;
- strengthening of the role of the environmental statement to improve the transparency of communication of environmental performance between registered organizations and their stakeholders and the public;
- a more thorough consideration of indirect effects, including capital investment and planning strategies.

At most levels EMAS is very similar to ISO 14001, and, as implied above, organizations must have ISO 14001 to apply for EMAS. EMAS is a little more prescriptive than ISO 14001 in some ways; for example,

Figure 3.5 EMAS logo.

the frequency of external audits must be at least every three years. In addition, whereas under the ISO the organization is left to identify its own environmental aspects, under EMAS several prescribed aspects must be considered where relevant, these include emissions to air and water, contamination of land, use of natural resources, raw materials and energy, transport, and effects on biodiversity. Many companies believe EMAS to be bureaucratic and costly and it has not been taken up as widely has hoped. Taking these criticisms into account changes were proposed in 2008 to make the brand more widely recognized and encourage uptake, particularly amongst SMEs. One major proposed change is the requirement for EMAS registered organizations to take account of environmental considerations when selecting their suppliers.

3.5 ECO-LABELS

Although EMSs help to ensure that systems are in place to aid the development of more sustainable processes at specific sites, amongst the public at large there is a high degree of confusion over what are and are not 'green' products. In an attempt to improve public awareness the EU (and other bodies) has developed a system of 'approval labels' (Figure 3.6) for specific groups of products that comply with specific ecological criteria for that product group. Companies applying to the scheme may then use the label as a marketing aid on products that comply with the criteria.

Figure 3.6   Eco-label.

The range of products for which Eco-labels are available is continuously being updated and currently includes items such as washing machines, computers, and bed mattresses. Labels are also available for chemical products such as paints & varnishes, cleaners, and all kinds of detergent. A LCA approach is at the heart of this system, which sets out to compare the relative environmental impact of products with a specific function.

Consider the example of Eco-labels for dishwasher detergents. One of the criteria for awarding an Eco-label for dishwashing detergents is that the product must achieve a certain environmental impact score calculated from the various components in the product on a per wash basis. The impact score is based on the total weight of chemicals, a measure of the total toxicity based on long-term effects, the amount of phosphates in the product, and measures of the aerobic and anaerobic biodegradability of the material. Each factor has a 'cut-off' limit above which the product fails and in the calculation of the total impact figure the individual factors are weighted to reflect their relative importance. This approach ensures that total impacts are regulated but allows the formulator freedom to ensure product differentiation. For example, a formulator may choose to add a small amount of a highly active surfactant, which has a relatively high toxic score, rather than a larger amount of material with a lower toxicity.

3.6   LEGISLATION

As stated in Chapter 2 the first major piece of environmental legislation was the 1863 alkali act, which set out to limit HCl pollution from the

Leblanc process. For the next 100 years there was very little additional legislation; for example, in 1950 in the US there were only 20 environmental laws. However, the subsequent 50 years saw the number increase seven-fold. This recent growth in legislation has gone hand in hand with extensive growth in manufacturing and in particular widespread use of products from the petrochemical industry, in applications as diverse as intensive farming, detergent washing powders, and baby toys. This increased activity and general improvement in our quality of life has been accompanied, during the last 50 years, by a few well-publicised health or environmental disasters such as Bhopal, Flixborough, Seveso, Love Canal, and the burning Cuyahoga River in Ohio; accompanied by increasing public concern over the effect of chemicals in the environment. These concerns have resulted in governments in most parts of the world introducing a plethora of environmental legislation.

Until recently the philosophical approach to environmental legislation had changed little since 1863, with legislation being largely introduced after a problem had been discovered, there was little emphasis on prevention of potential problems. Also much of the legislation and control of environmental impact was more directly concerned with heath and safety and was generally prescriptive in nature. Modern environmental legislation is becoming much more internationally coherent and less prescriptive and focussed on prevention of pollution through control hazardous materials and processes as well as protection of ecosystems.

3.6.1 Integrated Pollution Prevention and Control (IPPC)

The control of emissions to air, land, and water in the UK had, until 1990, been regulated by different authorities. In some instances it was possible to divert pollution from one medium to another to avoid regulation. In 1990 the UK Government brought in the concept of Integrated Pollution Control (IPC) as part of the Environmental Protection Act. Under IPC, emissions to land, air, and water for certain (more polluting) categories of industrial processes were regulated and a permit to operate the process was required from the Environment Agency. Authorization to operate the process was only granted if the Agency was satisfied that it could safely be operated within the emission limits set and that Best Available Technology Not Entailing Excessive Cost (BATNEEC) has been applied. IPC legislation has now been superseded by the related IPPC.

IPPC is the successor to the UK Integrated Pollution Control Act (IPC), the important difference being the prevention aspect. In England

and Wales the Environment Agency (EA) is responsible for the implementation of IPPC (SEPA in Scotland). In summary, the aims of IPPC may be expressed as:

- to protect the environment as a whole;
- to promote use of clean technology to minimize waste at source;
- to encourage innovation by leaving responsibility for developing satisfactory solutions with industry.

IPPC regulates a wide range of industry sectors, including most of the chemical and allied industries. To operate a production process within one of these sectors a permit must be obtained from the EA. To obtain a permit a detailed application must be submitted in which all aspects of the production process as well as waste treatment and management systems are included. One of the underlying requirements is to use the Best Available Technique (BAT) and guidelines on what are considered various appropriate techniques for certain sectors are available, *e.g.* of chemical pulping technologies for the pulp & paper sector. There is an important difference here to IPC were the onus was to use BATNEEC. In theory the cost of best available technology should not be an excuse for not employing it, in practice regulators have freedom to compromise based on the geographical location, local environmental conditions, and technical characteristics of the installation. Other differences to IPC include regulation of a sites activities rather than a specific process. It includes broader aspects such as energy efficiency and raw material selection, and it also regulates site closure to prevent pollution being left behind. Typically the following main areas need to be covered in an application:

1. *Management systems:* The operator needs to have an effective management system in place to control the production process. Registration under EMAS or ISO 14001 would meet this criterion although this is not a specified requirement.
2. *Materials input:* This covers all raw materials including water. The operator is required to show that unnecessarily hazardous materials are not being used and that adequate precautions are in place to prevent release or operator harm. In some cases an improvement programme may be required to reduce either the quantity or hazardous nature of materials being used.
3. *Main activities:* A detailed description of the process, including process flow sheet is required. All emissions to air, water, and land need to be quantified and control and emergency measures identified.

Table 3.4   List I and II substances under groundwater directive.

List I[a]	List II[b]
Organohalogens	20 metals, including Zn, CU, Ni, Mo & Co
Organophosphorus compounds	Biocides
Organotin compounds	Substances affecting taste/odour of groundwater
Mercury and its compounds	Toxic or persistent organic compounds of Si
Cadmium and its compounds	Inorganic compounds of P
Cyanides	Fluorides
Mineral oils	Ammonia
	Nitrates

[a]List I substances may not normally be discharged.
[b]List II substances may only be discharged under strictly controlled conditions.

4. *Emissions to groundwater:* This is a separate section where all hazardous materials on List I or II (Table 3.4) and which may enter the groundwater need to be identified. Surveillance and control measures should be specified. An improvement programme to limit their use may be required.

5. *Waste handling recovery and disposal:* All aspects of waste management and compliance need to be identified to ensure safe handling, appropriate treatment, and reuse where possible. Unless waste is minimal and of a benign nature an improvement plan is likely to be needed.

6. *Energy:* An energy balance for the operation is required together with the associated emissions from each source of energy. Best practice, in terms of reusing and minimizing energy requirements across the whole production site, is the target against which progress is judged.

Although the application may look complex and time consuming the overall approach being taken by the regulating bodies is one of partnership, and constructive help is offered with the aim of reaching a mutually acceptable agreement.

3.6.1.1 Environmental Permitting. The Environmental Permitting Regulations (EPR) came into force on 6 April 2008. As a first step, they combine IPPC and Waste Management Licensing regulations. Operators having an IPPC permit or a Waste Management Licence automatically transferred to having an Environmental Permit from April 2008. The Regulations provide a single, common, risk-based framework for permitting and compliance. They enable the Environment Agency to maintain environmental protection standards whilst reducing bureaucracy.

A review of IPPS and related regulations by the EU has resulted in proposals to introduce the Industrial Emissions Directive in 2011. IED will combine several pieces of regulation, including IPPC, waste incineration directive, large combustion plant directive, and solvent emissions directive. The new directive is designed to cut bureaucracy for operators who needed to comply with several regulations and to address the finding that there was poor implementation of the BAT element of IPPC.

Under the new directive no operator would be given approval for emissions above those set in the BAT. It is also proposed that some elements of the directive become more prescriptive, reversing the recent trend. These include regular monitoring of soil and ground-water as well as an annual operation comparison against BAT and reporting on compliance.

3.6.2 REACH

The REACH (Registration, Evaluation, Authorization of Chemicals) regulations were introduced into Europe in 2007 and will be phased in over a period of 11 years. The aim of REACH is to improve the protection of human health and the environment through the better and earlier identification of the intrinsic properties of chemical substances. At the same time the innovative capability and competitiveness of the EU chemicals industry should be enhanced. They give greater responsibility to industry to manage the risks from chemicals and to provide safety information on the substances.

There are around 100 000 different chemical substances registered in the EU, of which 10 000 are sold at over 10 tpa and a further 20 000 sold at between 1 and 10 tpa. Since most of these substances were registered prior to 1981, when significant testing was introduced for new substances, there has been concern about the lack of knowledge of the impact chemicals have on the environment and human health. The REACH regulations set out to address these concerns through ensuring a 'high level of protection of human health and the environment' as enshrined in the EU treaty, through regulations covering new and existing chemicals.

REACH, the most wide reaching piece of legislation ever introduced in Europe, affects manufacturers, distributors, and importers of chemicals and requires registration of all chemicals manufactured or imported into the EU in more than 1 tonne per year. Whilst maintaining the free movement of goods the regulations aim to ensure that organizations placing chemicals onto the market or using them understand the associated risks. A further stated aim is to enhance innovation and competitiveness, although this is often disputed by industry.

Registration involves preparing a Technical Dossier and Chemical Safety Report covering information on properties, uses, hazards, exposure scenarios, and risk control measures. If a substance is not registered it can not be manufactured or placed on the market within the EU.

Downstream users also have obligations under REACH. Whilst they are not required to register substances themselves, they may be asked to provide information on typical exposures and how they use substances to support substance registrations. They are responsible for implementing any hazard and risk management measures identified by their suppliers, and for ensuring this information is communicated downstream. Manufacturers and distributors of specific chemicals are encouraged to get together to share data, collaborate on any further tests, and produce one dossier per chemical. It is compulsory to share all animal test data to minimize the quantity of animals required.

Evaluation of the Technical Dossier – which may take companies many months to complete, as they have to find the data and carry out toxicity and other tests – by the European Chemicals Agency in Helsinki is designed to highlight substances of very high concern to health or the environment. The Agency acts as the central point in the REACH system: it manages the databases necessary to operate the system, coordinates the in-depth evaluation of chemicals, and runs a public database in which consumers and professionals can find hazard information. For substances of high concern, such as carcinogens and very persistent or accumulative chemicals, authorization is required for the substance to be used in a particular application. This information also needs to include possible substitutes and why they are not appropriate as well as what R&D is being undertaken to find a solution. If it is deemed that the risks are too high or the benefits do not outweigh the risk the substance may be restricted from use in a given application.

REVIEW QUESTIONS

1. The world requirement for lactic acid is increasing rapidly. Your company has decided to build three new plants each with a capacity of 50 000 tonnes per year. The locations for the new plants have been decided as (a) Grangemouth, Scotland on an existing petrochemical manufacturing site, (b) on the outskirts of London on a new green field site, and (c) in Nebraska USA on a new green field site. You are

responsible for deciding which technology will be adopted at each site. Write a two-page report to your board justifying your choices.

2. Construct a LCA process flow sheet for a PET drinks bottle, indicating what data you would wish to collect. Compare this with a similar LCA for an aluminium can. Discuss the meaning of the term 'functional unit' in this context.

3. Review a recent synthetic reaction you have carried out in the laboratory. Write a balanced equation for the reaction(s) and calculate the atom economy. From your experimental results calculate the yield, E-factor, and effective mass yield (ignoring any water used). Identify ways in which this reaction could be made greener.

4. You are the production manager for a plant producing adipic acid by the nitric acid oxidation of a mixture of cyclohexanone and cyclohexanol. Your company is preparing for ISO 14001 registration.

   (a) Draw a process flow sheet for the operation and identify three elements that may lead to environmental damage.

   (b) Set an environmental improvement objective for each of these elements, outlining how it may be achieved and what resources you require.

   (c) For one of these elements predict an unplanned event that may occur and outline an emergency plan to deal with it.

FURTHER READING

T. E. Graedel, *Streamlined Life Cycle Assessment*, Prentice Hall, New Jersey, 1998.

E. Rowland and B. Day, *Health Safety and Environment Legislation*, RSC Publishing, Cambridge, 2003.

P. Smith, in *Clean Technology for the Manufacture of Speciality Chemicals*, eds. W. Hoyle and M. Lancaster, Royal Society of Chemistry, Cambridge, 2001, p. 25.

CHAPTER 4

Catalysis and Green Chemistry

4.1 INTRODUCTION TO CATALYSIS

Nitrogen and hydrogen will sit happily together in a sealed vessel without reacting to form ammonia, with the equilibrium for the reaction being completely over to the left-hand side of the equation under ambient conditions (4.1):

$$N_2 + 3H_2 \rightleftharpoons 2NH_3 \quad \Delta H = -92 \text{ kJ} \tag{4.1}$$

According to Le Chatelier's principle the equilibrium will be shifted to the right-hand side by high pressures and, since the reaction is exothermic, by low temperatures. Indeed early work by Haber showed that at 200 °C and 300 atmospheres pressure the equilibrium mix would contain 90% ammonia, whilst at the same pressure but at 700 °C the percentage of ammonia at equilibrium would be less than 5%. Unfortunately, the activation energy is such that temperatures well in excess of 1000 °C are needed to overcome this energy barrier. The conclusion from this is that direct reaction is not a commercially viable option (Figure 4.1).

In the early 1900s Haber and, later, Bosch discovered that the reaction did, however, proceed at reasonable temperatures (around 500 °C) in the presence of osmium and subsequently iron based materials. These catalysts acted by lowering the activation energy of the reaction, in other words by interacting with the starting materials they altered the reaction pathway to one of lower energy. Catalysts do not, however, alter the equilibrium position of a reaction, which is under thermodynamic

Green Chemistry: An Introductory Text, 2nd Edition
By Mike Lancaster
© Mike Lancaster 2010
Published by the Royal Society of Chemistry, www.rsc.org

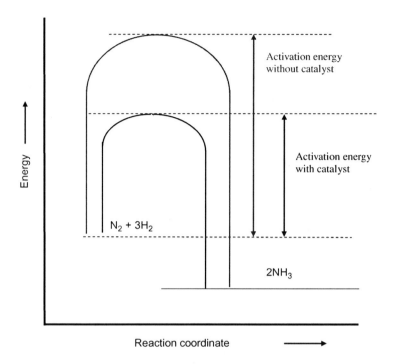

Figure 4.1 Activation energy for catalysed and uncatalysed ammonia synthesis.

control; therefore high pressures are still needed to force the reaction. Hence a catalyst is commonly defined as:

A material which changes (usually increases) the rate of attainment of chemical equilibrium without itself being changed or consumed in the process.

By increasing the rate of attainment of equilibrium through lowering the activation energy catalysts reduce the energy requirement of a process and therefore can be considered to be inherently green. Many catalysts are also highly selective, either preferring one synthetic pathway over an alternative or preferring one reagent in a mixture over another. Often catalysts can be used in place of stoichiometric reagents. In all of these cases waste is generally reduced compared to the non-catalytic alternative (if any), highlighting the green credentials of catalyst technology. As discussed later in this chapter there are shades of greenness; for example, some catalysts, particularly those based on heavy metals, are highly toxic and unless they can be totally recovered at the end of the process pose a significant environmental threat.

Today it is estimated that some 90% of the chemicals used have, at some stage in their manufacture, come into contact with a catalyst. The range is truly broad, from bulk chemicals such as acetic acid and ammonia to consumer products like detergents and vitamins. Virtually all major bulk chemical and refining processes employ catalysts. The number of fine, speciality, and pharmaceutical processes currently using catalysts is still relatively small by comparison, but a combination of economic and environmental factors is focussing much research on this area. The great economic benefit of catalysts lies with their incredible activity, sometimes converting tens of millions times their own weight of chemicals. This results in a catalyst market worth $<1\%$ of the value of products they create. Three important parameters impact on both the commercial viability and inherent greenness of a particular catalyst:

1. Selectivity – the amount of substrate converted into the desired product as a percentage of total consumed substrate (a catalyst will be of limited benefit if it also enhances the rate of by-product formation).
2. Turnover frequency – the number of moles of product produced per mole of catalyst per second (low turnover frequencies will mean large amounts of catalyst are required, resulting in higher cost & potentially more waste).
3. Turnover number – the amount of product per mole of catalyst (this is related to catalyst lifetime and hence to cost and waste).

4.1.1   Comparison of Catalyst Types

Catalysts are commonly divided into two basic types, heterogeneous and homogeneous, depending on their state relative to the reaction medium. Heterogeneous catalysts, sometimes referred to as surface catalysts or contact catalysts due to their mode of action, are in a different phase to the reaction medium. Heterogeneous catalysts are widely used industrially, in most cases the catalyst is a solid with reactants being in the gaseous phase. The actual reaction takes place on the surface of the catalyst; this may be the external surface or, more effectively, at a surface within internal pores of the solid. Homogeneous catalysts are in the same phase as the substrate and are uniformly distributed. In almost all cases the reaction takes place within the liquid phase, the catalyst being dissolved in the reaction medium.

Overview comparisons are often difficult to make and there are always exceptions to any generalizations. That said, there are some differences

Table 4.1 Comparison of heterogeneous and homogeneous catalysts.

Heterogeneous	*Homogeneous*
Usually distinct solid phase	Same phase as reaction medium
Readily separated	Often difficult to separate
Readily regenerated and recycled	Expensive/difficult to recycle
Rates not usually as fast as homogeneous	Often very high rates
May be diffusion limited	Not diffusion controlled
Quite sensitive to poisons	Usually robust to poisons
Lower selectivity	High selectivity
Long service life	Short service life
Often high energy process	Often takes place under mild conditions
Poor mechanistic understanding	Often mechanism well understood

between heterogeneous and homogeneous catalysts that have a significant impact on their greenness (Table 4.1). This table is not meant to be interpreted in a way that would lead to a general view that one type of catalyst is greener than another. For any given situation there may be specific environmental objectives that can be better achieved by one type of catalyst or another. It is, however, true that the ultimate goal of many researchers working in this field is to combine the best characteristics of both types of catalyst. One of the main aims of this work is to combine the fast rates and high selectivities of homogeneous catalysts with the ease of recovery and recycle of heterogeneous catalysts. In most (but not all) cases this results in attempts to hetrogenize a homogeneous catalyst.

By definition all catalysts must be heterogeneous or homogeneous, within and across these, though, there are other classifications that are important to green chemistry. Most important among these are:

- Asymmetric catalysts: these are still relatively rare in industrial processes but they are playing an increasingly important role in the development of pharmaceuticals. This is because they offer one of the most efficient, low waste methods for producing enantiomerically pure compounds.
- Biocatalysts: these are essential for life, and play a vital role in most processes occurring within the body as well as in plants. In the laboratory biocatalysts are usually natural enzymes or enzymes produced *in situ* from whole cells. They offer the possibility to carry out many difficult transformations under mild conditions and are especially valuable for producing enantiomerically pure materials. Their huge potential is currently largely untapped, partially due to the time and expense of isolating and screening enzymes. Research and use of these catalysts is rapidly expanding as chemicals from

renewable resources become a more viable commercial option. This area is discussed more fully in Chapter 6.

- Phase transfer catalysts: these have been around for about 50 years and were developed as a means of increasing the rates and yields of reactions in which the reactants are in two separate phases. In these cases poor mass transport often limits the reaction. Phase transfer catalysts act by transporting the reactants from one phase into another, thus overcoming mass transport limitations.
- Photocatalysts: these harness energy from the sun for carrying out chemical transformation. These energy efficient catalysts are proving especially beneficial in destroying harmful waste and water clean-up.

4.2   HETEROGENEOUS CATALYSTS

Heterogeneous catalysts have been used industrially for well over 100 years. Amongst the first processes was the catalytic hydrogenation of oils and fats to produce margarine using finely divided nickel. It is quite likely that when this process was first operated in the late 19th century unhealthy amounts of nickel remained in the product. The issue of leaching and the avoidance of trace catalyst residues are still an important aspect of research from both economic and environmental points of view.

4.2.1   Basics of Heterogeneous Catalysis

There is a whole spectrum of heterogeneous catalysts, but the most common types consist of an inorganic or polymeric support, which may be inert of have acid or basic functionality, together with a bound metal, often Pd, Pt, Ni, or Co. Even if the support is inert its structure is of vital importance to the efficiency of the catalytic reaction. Since the reactants are in a different phase to the catalyst, both diffusion and adsorption influence the overall rate, factors that, to some extent, depend on the nature and structure of the support.

Surface area is one of the most important factors in determining throughput (amount of reactant converted per unit time per unit mass of catalyst) – many modern inorganic supports have surface areas of 100 to >1000 $m^2 g^{-1}$. The vast majority of this area is due to the presence of internal pores; these pores may be of very narrow size distribution to allow specific molecular sized species to enter or leave; or be of a much broader size distribution. Materials with an average pore size of less than 1.5–2 nm are termed microporous whilst those with pore sizes above this are called mesoporous materials. Materials with very large pore sizes (>50 nm) are said to be macroporous (see Box 4.1 for methods of determining surface area and pore size).

The catalytic reaction can be conveniently divided into several sequential steps, all of which impact on the overall efficiency of the reaction. First the reactants must diffuse to the catalyst surface; the rate of diffusion depends on several factors, including fluid density, viscosity, and fluid flow rate. Some reaction will take place at the external surface but the majority of reactants will need to diffuse into the internal pores. For a given substrate this is largely determined by the pore radius but collisions with other molecules also hinder the overall rate of diffusion. The diffusion rate may be a particular problem when using microporous catalysts; for this reason many reactions using these materials are carried out in the gas phase at relatively high temperatures to minimize effects of viscosity, density, and intermolecular collisions as well as increasing molecular velocities.

The next stage in the catalytic cycle is adsorption of the reactants onto the catalyst surface. There are two types of adsorption process:

1. Physisorption, originating from van der Waals interaction between reactant & surface. This weakly exothermic process is reversible and does not result in any new chemical bonds being formed. In general, physisorption does not lead to catalytic activity but may be a precursor to chemisorption.
2. Chemisorption, which results in new chemical bonds being formed between reactant and catalyst and is usually more exothermic than physisorption. Understanding orbital interactions between reactant and catalyst during chemisorption can enhance the development of efficient catalysts for specific reactions. Bonds formed during chemisorption should be strong enough to prevent desorption yet not too strong to prevent reaction with other reactant molecules.

In many cases surface diffusion of the adsorbed species is required before chemical reaction can take place. Following reaction [through a low activation energy process (Figure 4.1)] the product should be adsorbed weakly enough to diffuse readily from the surface and into the bulk fluid. Again pore size is important in determining the rate of diffusion of (often larger) product molecules. Figure 4.2 summarizes the generic catalytic cycle.

Box 4.1   Measurement of 'texture' properties

Surface Area.   The basic technique involves physical adsorption of N_2, which has a cross sectional area of $0.162 \, nm^2$, on the surface. The problem is that multilayers of gas start to build up on the catalyst surface before a monolayer is completely formed. The BET (Brunauer–Emmett–Teller) equation describes these phenomena:

$$P/(P_o - P) = 1/V_m C = [(C - 1)/V_m C](P/P_o)$$

where V_m is the volume required to form a monolayer, V is the volume uptake of nitrogen, P_o is the vapour pressure of N_2 at the adsorption temperature, P is the equilibrium pressure, and C is constant for a given class of materials and includes heats of adsorption and liquefaction. The amount of N_2 adsorbed at 77 K as a function of the N_2 pressure is measured using one of various techniques, including the dynamic method (see below). A plot of $P/V(1-P)$ against P then gives a straight line; V_m may then be calculated from the intercept ($= V_m C^{-1}$) and the slope ($= C{-}1/V_m C P_o$). From V_m the surface area (SA) of the sample may be calculated using the equation:

$$SA = 0.162 \, V_m N_A / V_A$$

where N_A/V_A is Avogadro constant per unit volume of gas. By dividing by the weight of the sample the surface area in $m^2 \, g^{-1}$ can be found.

Dynamic Method.   Here a flow of He is passed over the sample at 77 K. A small amount of N_2 is introduced into the He stream. The gas stream coming from the sample is monitored using mass spectroscopy. N_2 is only detected after a monolayer is formed. The N_2 supply is then switched off and the desorption curve plotted. Integration of this curve gives the information required for the BET equation.

Pore Volume.   Pore volumes are determined by forcing N_2 (for micro- and mesoporous materials) or Hg (macroporous materials) under pressure into the pores. The quantity of N_2 or Hg entering the catalyst is directly related to the pressure and the radius of the pores. The Kelvin equation describes this:

$$\text{Radius} = 2SV_o/RT \ln (p/P_o)$$

where S is the surface tension and V_o is the molar pore volume. This information can also be used to determine average pore size.

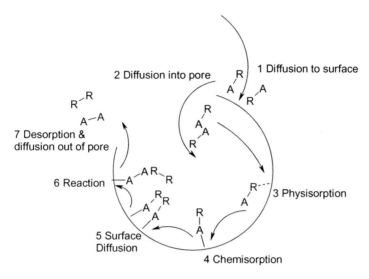

Figure 4.2 The catalytic cycle.

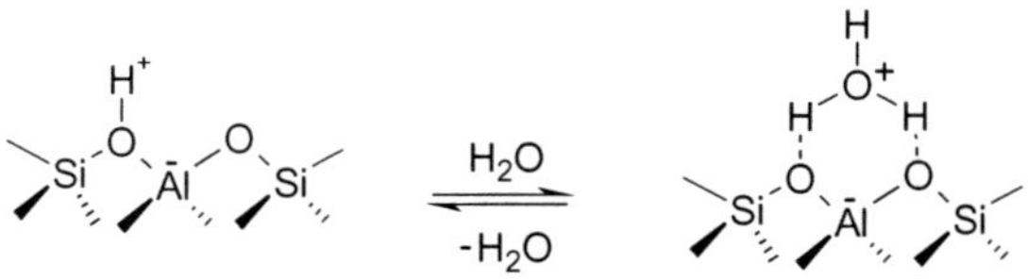

Scheme 4.1 Brønsted acidity of zeolites.

4.2.2 Zeolites and the Bulk Chemical Industry

Zeolites are made of three-dimensional frameworks of crystalline hydrated aluminosilicates consisting of TO_4 tetrahedra (T is Si or Al in most zeolites). The tetrahedra are arranged in several different ways to give well-defined microporous structures, the pores often being interconnected. Around 45 naturally occurring zeolites have been discovered, some of which have well-known names like mordenite, clinoptilolite, and chabazite. The hydrated nature of zeolites imparts significant Brønsted acidity (Scheme 4.1); it is this property, together with the opportunity for carrying out selective catalysis in the pores, that has resulted in the commercialization of many zeolite-based chemical processes.

Although natural zeolites are widely used (around 4 million tpa) they are not particularly valuable as commercial catalysts. This is due to several factors, including natural variations in crystal size and porosity as well as the actual small pore size limiting their synthetic usefulness. Natural zeolites do, however, find widespread use in applications such as removal of heavy metals from water, odour removal, and building materials.

To overcome the limitations of natural zeolites a whole range of synthetic zeolites have been manufactured in the last 50 years; these have tailored pore sizes and tuned acidities as well as often incorporating other metal species. The basic synthesis involves mixing a source of silica, usually sodium silicate or colloidal SiO_2, a source of alumina, often sodium aluminate, and a base such as sodium hydroxide. The mixture is heated at up to 200 °C under autogenous pressure for a period from a few days to a few weeks to allow crystallization of the zeolite. The exact nature of the zeolite is determined by the reaction conditions, the silica to alumina ratio, and the base used. For example, zeolite beta, a class of zeolites with relatively large pores, in the range of 0.7 nm, of which mordenite is an example, are usually made using tetra-ethylammonium hydroxide as the base. This acts as a template for the formation of 12-membered ring apertures (Figure 4.3).

Zeolite A is by far the most widely produced synthetic zeolite, with an annual production of some 1.3 million tonnes. As may be expected from this large volume its main use is not as a catalyst but as a detergent builder (Box 4.2).

Over the last 30 years the use of zeolite catalysts has provided huge economic and environmental benefits to the bulk chemical and petroleum refining industries.

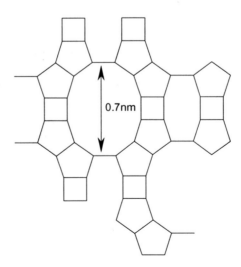

0.7nm

Figure 4.3 Mordenite, showing a 12-membered ring aperture.

Box 4.2   Zeolite A and detergent builders

Builders are added to detergents to soften the water by removal of calcium and magnesium ions. Although modern detergents do not form scum in the presence of hard water, as soap does, calcium and magnesium ions can form hard agglomerates and cause loss of anionic components in the detergent. Apart from reducing the efficiency of the detergent the agglomerates cause wear both to the washing machine and the clothes being washed. For many years the solution to this was to add sodium tripolyphosphate, an inexpensive and efficient way to ensure the calcium and magnesium ions remained in solution. During the late 1950s and 1960s, a period that coincided with mass use of household detergents, eutrophication of many freshwater lakes and rivers was noticed. This excess growth of algae, caused by high levels of phosphate nutrient, was starving fish of oxygen and light.

Two different solutions were developed. In some Scandinavian countries, notably Sweden, the sewerage treatment plants were altered to remove troublesome phosphate whilst in most of the rest of the developed world phosphate was replaced with zeolite A. The zeolite is of course a solid, which is insoluble in the wash, and potentially could also cause wear damage. The zeolite is relatively soft and production techniques have been developed that produce small spherical particles with no rough edges that could damage clothes or washing machines. The overall merits of phosphate and zeolite builders are still debated; a recent Finish study has shown that zeolite-containing detergents are less efficient due to the gradual build up of zeolites on clothes, this being more pronounced with modern low-water consuming machines. It is argued that inferior washing will lead to faster replacement of clothes, which has an obvious environmental burden, and hence will have an adverse impact on the LCA (life cycle assessment) of these detergent formulations.

4.2.2.1 Catalytic Cracking.   Cracking of heavy oil fractions is required for efficient production of gasoline; there are over 350 cat cracking units worldwide, each processing up to 150 000 tonnes of feedstock per day. Originally, the process was carried out thermally in a high energy, inefficient process. Eventually this process was replaced by a cat-cracking process using aluminium chloride; however, although energy and materials efficiency improved the process produced large amounts of aluminous waste, since the catalyst can not be effectively recovered. Following the use of recyclable clay based catalysts, today's

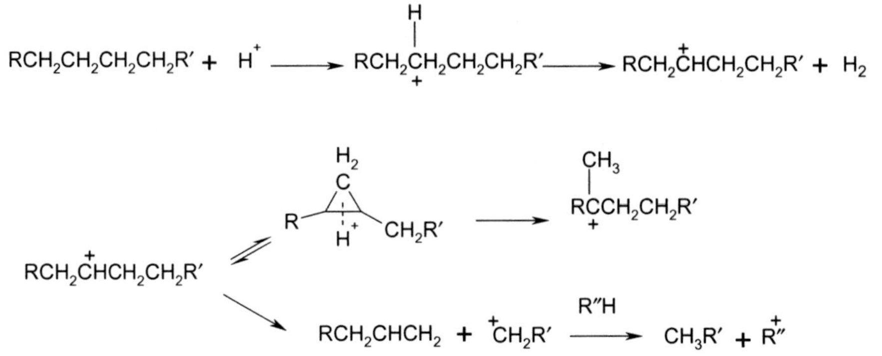

Scheme 4.2 Some basic reactions involved in FCC (fluid catalytic cracking).

processes are based on zeolites. They are commonly referred to as FCC or fluid catalytic cracking processes because of the original use of fluidized bed reactors. A wide variety of reactions occur during FCC but the two main ones are carbon–carbon bond cleavage and isomerization, hence the overall process may be viewed as a conversion of linear alkanes into shorter chain branched alkanes, terminal olefins, and aromatics (Scheme 4.2). A major benefit of FCC compared to thermal cracking is that bond breaking *via* carbocation ion intermediates (tertiary carbocations ions being the more stable) is much more selective than that encountered in free radical thermal processes. Also, fortunately, hydrocarbons containing less than seven carbons are generally not cracked during the FCC process – this limits the volume of light ends and gives a high yield of gasoline fraction (typically around 50% of feed).

Zeolite Y (synthetic faujasite), which is commonly employed in FCC, has a relatively large pore size of just under 0.8 nm and an open structure, allowing quite large hydrocarbon molecules to enter the pores containing the highly acidic sites. Small molecules rapidly diffuse out of the pores compared with larger ones, whilst highly branched hydrocarbons will not be able to enter the pores at all. As may be expected from such a complex sequence of reactions, coke formation through polymerization is a significant problem. Frequent regeneration of the catalyst is required, which is achieved by burning off the coke in air. The energy derived from catalyst regeneration is used to partially offset the energy required for the endothermic cracking process. Catalyst regeneration does, however, lead to significant pollution, arising from the sulfur in the feed becoming incorporated into the coke which when burned gives SO_x. The main control method for removing SO_x is through conversion into a metal sulfate and subsequent generation of

$$2SO_2 + O_2 \rightarrow 2SO_3$$

$$SO_3 + MO \rightarrow MSO_4$$

$$MSO_4 + 4H_2 \rightarrow MS + 4H_2O$$

$$MS + H_2O \rightarrow MO + H_2S$$

Scheme 4.3 Control of SO_x emissions from a FCC.

hydrogen sulfide for sale; this uses by-product hydrogen from the cracker (Scheme 4.3).

As with many zeolite-based processes, the zeolite is not used alone since it is highly active, resulting in rapid coke deactivation and poisoning by vanadium and nickel present in the feedstock. The zeolite is supported on a matrix such as alumina, which also has some cracking activity. The matrix pre-cracks some feedstock molecules too large for the pores as well as removing some of the metal residues. The method of synthesis initially produces zeolite Y in its sodium form. Complete removal of sodium by proton exchange can only be achieved under fairly harsh conditions, resulting in an expensive and not very stable product. Normally around 85% of the sodium ions are replaced by protons, often *via* replacement with ammonium ions followed by heat treatment. Catalytic activity can be improved if some of the residual sodium ions are replaced by rare earth ions such as La^{3+}. Partial fluorination gives enhanced acidity and reactivity.

Other methods of improving both the selectivity and yield of FCC processes are continuously being sought. The C_3 and C_4 fractions are becoming more valuable chemical feedstocks; by incorporating the smaller pore H-ZSM-5 zeolite into the catalyst, linear hydrocarbons can be cracked into these products. An added advantage of this is that, although the gasoline yield is decreased, the octane rating is improved by increasing the relative amount of branched and aromatic components.

The most modern FCC units are highly efficient and can run for 2–3 years without a shutdown. Modern FCC catalysts are fine powders with a bulk density of 0.80–$0.96\,g\,cm^{-3}$ and have a particle size distribution ranging from 10 to $150\,\mu m$ and an average particle size of 60–$100\,\mu m$. The heated fine catalyst is mixed with the hydrocarbon feed at 400–$500\,^{\circ}C$ and up to 2-bar pressure, with the hydrocarbon fluidizing the catalyst. After reaction the gaseous hydrocarbons are separated from the catalyst, which is sent to a regenerator for treatment with air at $700\,^{\circ}C$ to remove the coke before being mixed with fresh hydrocarbon feed.

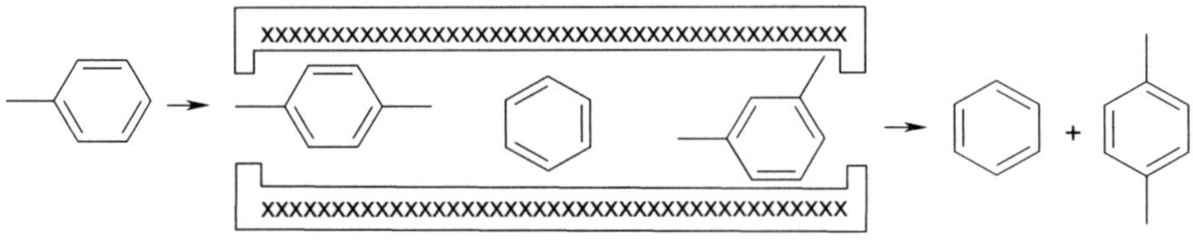

Figure 4.4 Disproportionation of toluene using H-ZSM-5.

4.2.2.2 Commercial Uses of ZSM-5. Since their development in 1974, ZSM-5 zeolites have had considerable commercial success. ZSM-5 is an aluminosilicate mineral belonging to the pentasil family of zeolites. Its chemical formula is $Na_nAl_nSi_{96-n}O_{192} \cdot 16H_2O$ ($0 < n < 27$), it has a ten-membered ring pore aperture of 0.55 nm (hence the 5 in ZSM-5), which is an ideal dimension for carrying out selective transformations on small aromatic substrates. Being the feedstock for PET, *p*-xylene is the most useful of the xylene isomers. The Brønsted acid form of ZSM-5, H-ZSM-5, is used to selectively produce *p*-xylene through toluene alkylation with methanol, xylene isomerization, and toluene disproportionation (Figure 4.4). This is an example of a product selective reaction in which the reactant (toluene) is small enough to enter the pore but some of the initial products formed (*o*- and *m*-xylene) are too large to rapidly diffuse out of the pore. *p*-Xylene can, however, rapidly diffuse out, and is produced selectively, along with some benzene. Other xylene isomers initially formed undergo isomerization within the pore, driving the reaction to produce further *para* isomer. Without processes such as these the economics of PET manufacture would be very different, with a high proportion of *ortho*- and *meta*-xylene having little value except as a fuel. This would be very wasteful of energy expended during refinery operations.

ZSM-5 is also the catalyst used in the Mobil MTG process (methanol to gasoline). Methanol (from synthesis gas) has huge potential as a green feedstock, both as a primary chemical building block and as a fuel since it can be produced from synthesis gas, which in turn can be produced from renewable resources (Chapter 6). Until commercialization of fuel cell technology (which may use methanol directly) for mass transportation becomes a reality (Chapter 6), one of the growing uses for methanol will continue to be its conversion into gasoline. This process is again highly complex but involves initial formation of dimethyl ether *via* acid-catalysed dehydration of two methanol molecules. Subsequent loss of water from the ether initially gives low alkenes that are oligomerized by the catalyst to give a gasoline grade product containing aromatics (BTX) as well as C_{6-8} alkanes and alkenes. Catalyst deactivation (sintering) is a major problem and is caused by the water formed during the high temperature reaction.

There are several other examples of ZSM-5 being used commercially to reduce waste and give high product selectivity. One of these is the alkylation of benzene with ethene to produce ethylbenzene selectively. The pore size of ZSM-5 successfully minimizes dialkylation reactions whilst the ability to regenerate the catalyst avoids waste issues associated with older catalysts like aluminium chloride.

With recent growth in bioethanol production there has been considerable interest in conversion of this into aromatic and aliphatic hydrocarbons using ZSM-5. At temperatures of 400 °C bioethanol conversion is complete and is largely independent of water concentration. The main products are aromatics (over 50%) and longer chain alkanes, but by incorporating iron into the zeolite the mechanism can be altered to give mainly C_2 and C_3 hydrocarbons. Commercialization of such processes is likely to be a few years away and will be dependant on the relative price of oil and bioethanol.

4.2.2.3 High Silica Zeolites. As mentioned above, many zeolites are not stable at high temperatures in the presence of water. This is due to a dealumination process that can cause complete collapse of the framework (Scheme 4.4). This effect will obviously be more pronounced for zeolites with low Si/Al ratios. The acidity of a zeolite is also dependent on the Si/Al ratio. It has been shown that the acid strength of a particular proton is related to the number of nearest neighbour and next nearest neighbour Al atoms, with the maximum acidity being when this number is zero. For reactions requiring very strong acidity, therefore, it would be preferential to have frameworks with high Si/Al ratios, these materials having the added benefit of increased stability to water.

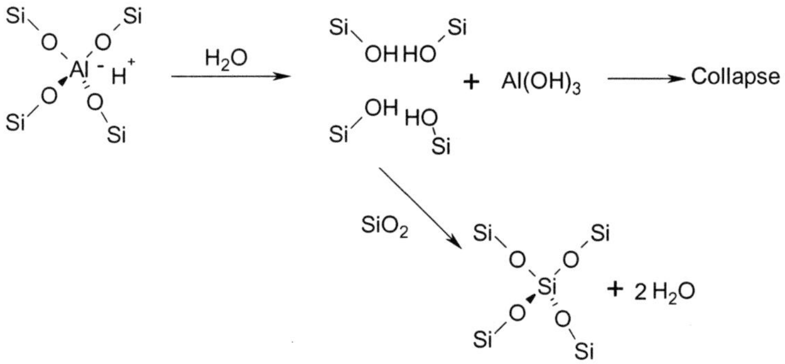

Scheme 4.4   High silica zeolites.

Some high silica zeolites can be prepared by adjusting the silicate/aluminate ratio at the synthesis stage, another method positively uses the dealumination process to cause partial framework collapse followed by stabilization with silica (Scheme 4.4).

High silica zeolites are increasingly being used in the bulk chemical industry to reduce waste and improve process economics. Potentially one of the greenest developments is the Asahi process for hydration of cyclohexene to cyclohexanol using a high silica ZSM-5 catalyst (SiO_2/Al_2O_3 ratio of 25). Cyclohexanol (mixed with cyclohexanone) is produced at over 6 million tpa and is a key intermediate in the manufacture of nylon 6,6 *via* adipic acid and nylon 6 *via* caprolactam. The most common method of manufacture (Scheme 4.5) entails catalytic hydrogenation of benzene to cyclohexane followed by air oxidation using a homogeneous cobalt catalyst. The process is energy intensive, being operated at 225 °C and 10 atm pressure and, to achieve reasonable selectivities, the process is operated at low conversions (around 6% per pass). Hence large inventories of highly flammable cyclohexane are continuously being circulated; this led to the Flixborough accident in 1974. The Asahi process is more energy efficient, operating at around 100 °C, and avoids the inherently hazardous combination of oxygen and hydrocarbon. Conversions per pass are still relatively low at 15% but a high selectivity of 98% is obtained; overall the process offers affordable eco-efficiency improvements.

Sticking with nylon production, high silica pentasil zeolites are used by Sumitomo to overcome environmental issues associated with the conversion of cyclohexanone oxime into caprolactam (Chapter 1, Scheme 1.4).

High silica zeolites are also used in the Mobil–Badger cumene process (alkylation of benzene with propene). This process is gradually replacing older ones using supported phosphoric acid since plant corrosion is

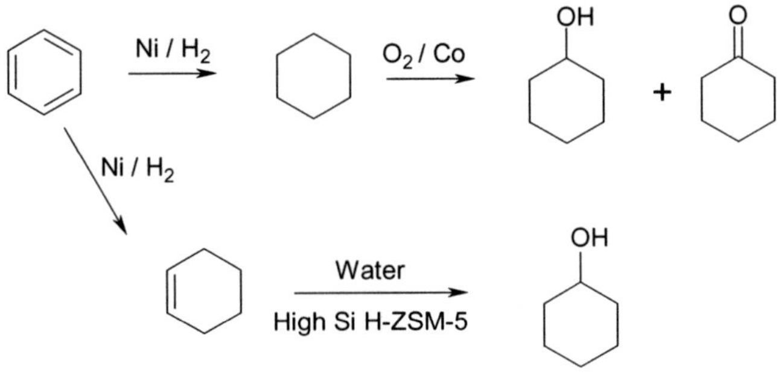

Scheme 4.5   Production of cyclohexanol.

much less of a problem. In tailoring the catalyst, problems with propene oligomerization and dialkylation needed to be overcome as well as catalyst stability. A further 'green' advantage of the zeolite process is that yield and selectivity are virtually quantitative; this reduces the complexity of down-stream distillation equipment, reducing energy requirements and capital costs.

4.2.3   Heterogeneous Catalysis in the Fine Chemical and Pharmaceutical Industries

In contrast to refinery and bulk chemical operations heterogeneous catalysts are used relatively little in the manufacture of fine chemicals and pharmaceuticals. To some extent this has been due to much of the research effort in the period 1950 to the mid-1980s going into development of zeolites for the large tonnage processes. These catalysts in general have too small a pore size and have too high an acidity for the larger more functional molecules required for these sectors. With increasing concern over waste by-products from these industries, as well as the increasing cost of waste treatment and disposal, considerable effort is now being put into using catalysts to improve efficiency and reduce the environmental impact of these sectors.

4.2.3.1   Heterogeneous Base Catalysis.   Base-catalysed reactions are not often required in the production of bulk chemicals but are highly useful in the synthesis of many fine chemicals since N- and O-containing functional groups are usually stable in basic conditions. A whole variety of solid base catalysts are available, ranging from simple metals, metal oxides, and hydroxides impregnated onto supports like gamma alumina and silica to ion (*e.g.* Cs^{+}) exchanged zeolites and highly tailored catalysts having specific organic base functionality tethered to an inert backbone.

One of the few industrial processes using a heterogeneous base catalyst – the production of ethylidene norbornene – is discussed more fully in Chapter 9. This process is an excellent illustration of the value of heterogenization, with highly hazardous sodium/potassium amalgam in liquid ammonia being replaced by relatively benign sodium/sodium hydroxide on alumina.

A highly unusual synthesis developed by Merck involves reaction of the imine formed from acetone and methyl amine with sulfur dioxide over Cs-doped ZSM-5 (Scheme 4.6). The product, obtained in 70% yield at 470 °C, is 4-methylthiazole, a key intermediate for the fungicide thiabendazole (used to control mould and blight on fruit). Whilst the

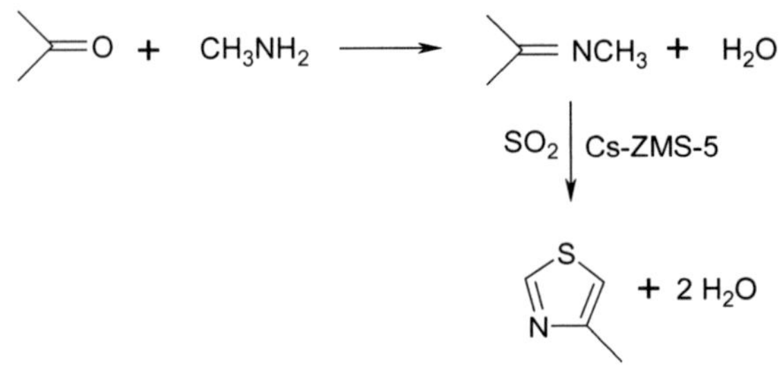

Scheme 4.6 Greener route to 4-methylthiazole.

moderate yield, high temperature, and less than ideal starting materials detract from the overall greenness of this atom efficient route, it is much more environmentally benign than the route currently used. The latter involves a five-step reaction sequence involving chlorine, carbon disulfide, and ammonia as starting materials and produces stoichiometric amounts of hydrogen chloride, sodium bisulfite, and ammonium chloride waste.

Many basic transformations such as esterifications, Michael additions and the Knoevagel reaction are carried out under basic conditions, some using aqueous bases others requiring strong soluble organic bases such as guanidines. The base is not usually recovered, producing either salt or organic waste. In the case of expensive organic bases difficulty in recovery often prevents commercialization of the process. Several methods have now been developed to heterogenize organic bases on supports such as silica or polystyrene whilst development of large pore silicate materials (mesoporous molecular sieves) has opened up wider opportunities for heterogeneous base (and acid) catalysis. For acceptable catalytic activity it is important to maximize the amount of basic sites on the catalyst whilst maintaining an open structure for access of bulky reagents.

An area of current research is the use of heterogeneous base catalysts for biodiesel production (Chapter 6). Transesterification of fatty acids with methanol is traditionally carried out with sodium hydroxide or sodium methoxide, but solid base catalysts such as sodium aluminate offer many advantages for very large-scale production. High catalytic activity, with yields of 94%, have been achieved.

Amorphous silica, with pore sizes in the range 1–10 nm, is a common support for base catalysts, while more structured pore sizes can be made by what is known as the sol–gel method. In this method a micelle is formed in water, often by using a C_8–C_{12} amine or a block glycol ether copolymer; tetraethoxysilane is then copolymerized around the micelle

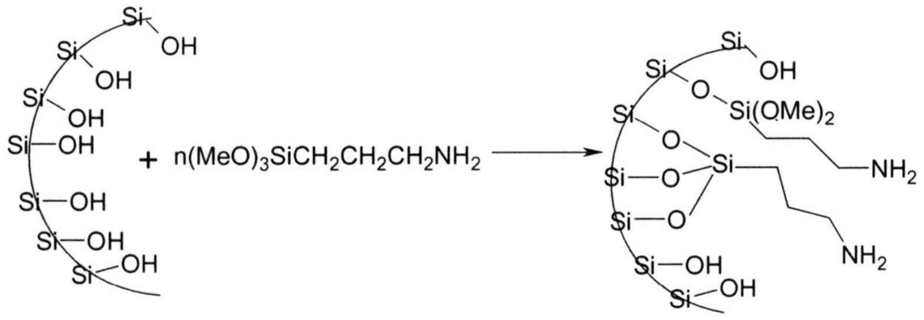

Scheme 4.7   Synthesis of solid base catalysis.

to give silica with a definite pore size. The micelle-forming material may be subsequently removed by washing in a suitable solvent or in some cases by calcination. Such materials are often referred to as hexagonal meso-porous silicas or HMS. One of the simplest methods of preparing solid base catalysts is by grafting suitable groups onto the silica. For example, aminopropylsilane (AMPS) can be grafted onto silica using trimethox-y(aminopropyl)silane by heating to around 100 °C in an inert solvent like toluene (Scheme 4.7). The actual number of bonds formed between the grafted molecule and the support (1, 2, or 3) will vary according to the support and the grafted group. Typically, a loading of around 1 mmol g^{-1} is obtained with grafting. Alternatively, in the sol–gel method the silane containing the base functionality can be directly incorporated into the polymerization process, which often results in higher loading.

Such catalyst systems have proved effective for carrying out the Knoevenagel reaction, although catalyst turnover numbers and poisoning have so far prevented commercial use. In general, HMS substituted materials have proved more active than catalysts based on amorphous silica for this reaction, but the process is highly complex with some substrates exhibiting greater reactivity on amorphous silica. Scheme 4.8 shows an example of the Knoevenagel reaction between benzaldehyde and ethyl cyanoacetate. Alternative mechanisms that involve activation of the aldehyde group by reaction with the amine have also been postulated to account for the lower rates observed with tri-substituted amine catalysts.

The Knoevenagel reaction has many similarities to the Michael addition, in which a base is required to form a carbanion from an activated methylene precursor that subsequently undergoes nucleophilic addition to an alkene containing a group such as an ester capable of stabilizing the resulting anion by delocalization. These reactions are widely used for production of pharmaceutical, perfumery, and agro-chemical intermediates (Scheme 4.9).

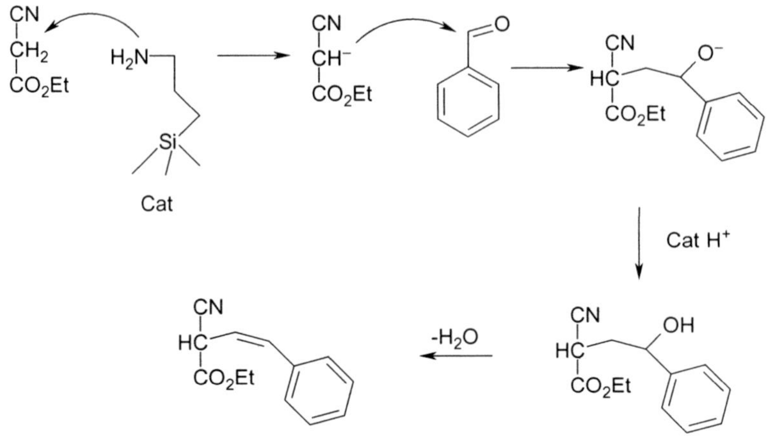

Scheme 4.8   Solid base catalysed Knoevenagel reaction.

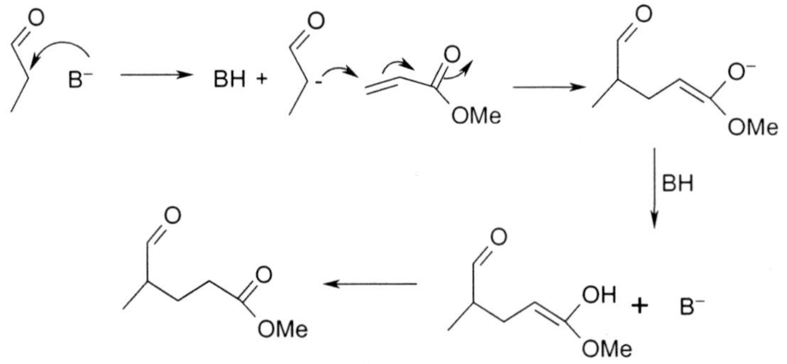

Scheme 4.9   Michael addition – used in synthesis of intermediates for methylpyridine-based herbicides.

Stronger bases are often required for Michael additions, which has been achieved by reaction of AMPS catalysts with hydroxybenz-aldehydes to form the imine. Subsequent reaction with sodium hydrogen carbonate gives the sodium phenolate which is the active catalyst. Other synthetically useful groups that have been grafted onto silica include trimethoxy(chloropropyl)silane and 3-(glycidyloxypropyl)trimethoxy-silane. Both of these have proved valuable in preparing solid catalysts from the strong bases guanidines and amidines.

The other main support used for solid base catalysts is polystyrene. Whilst polystyrene does not have a well-defined porous structure it does swell in solvents, providing an accessible high surface area on which to carry out reactions. One common method of chemically attaching groups to polystyrene involves incorporation of specific amounts of styrene-containing functional groups as shown in **4.1**; these may then be reacted to give a wide variety of catalysts (both acid and basic).

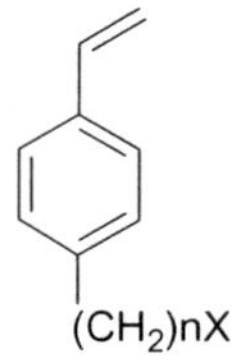

n = 1-8
X = Cl, Br, SO$_3$H, epoxide, amino, etc

(CH$_2$)nX

4.1

4.2.3.2 Oxidation Catalysts. Since chemical production is largely based on hydrocarbon feedstocks but many of the products required contain oxygen, oxidation technology is vitally important. Many oxidation reactions frequently produce large volumes of waste containing heavy metals such as chromium, cobalt, and manganese. Such waste streams are becoming increasingly costly to treat and dispose of. Recovery of the metals (often in a non-catalytically active form) is widely practised but recent research activities have centred on the development of stable, active, and selective heterogeneous oxidation catalysts that can, ideally, be used with green oxidants such as air or dilute hydrogen peroxide. Many oxidation reactions require the metal component to be present in stoichiometric amounts (*e.g.* manganese dioxide and chromium trioxide oxidations with sulfuric acid) since, although the metal is not consumed during the process, it is an integral part of the oxidant and is converted into a non-catalytic form. This is exemplified in Scheme 4.10 for an old commercial process for preparing benzoquinone, a valuable fine chemical that is also an intermediate for the production of hydroquinone. The value of heterogenizing such reagents is low compared to finding a true catalytic process.

EniChem made one of the most important steps forward in the development of general heterogeneous oxidation catalysts in the early 1990s with the commercialization of titanium silicate (TS-1) catalysts. TS-1 has a structure similar to ZSM-5 in which the aluminium has been replaced by titanium; it is prepared by reaction of tetraethylorthosilicate

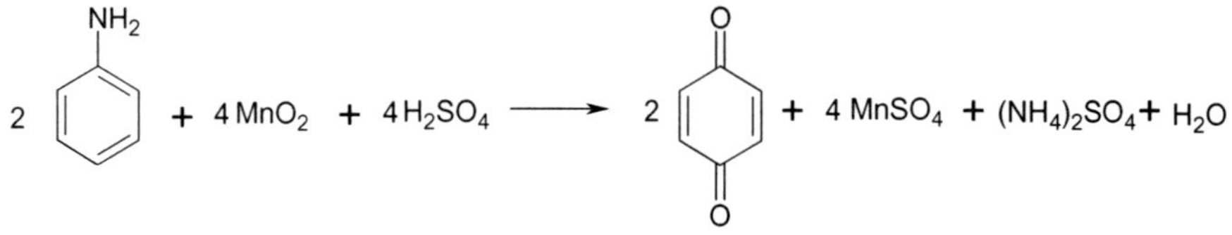

$$2 \; \underset{}{\overset{NH_2}{\bigcirc}} + 4\,MnO_2 + 4\,H_2SO_4 \longrightarrow 2 \; \underset{O}{\overset{O}{\bigcirc}} + 4\,MnSO_4 + (NH_4)_2SO_4 + H_2O$$

Scheme 4.10 Old commercial route to benzoquinone.

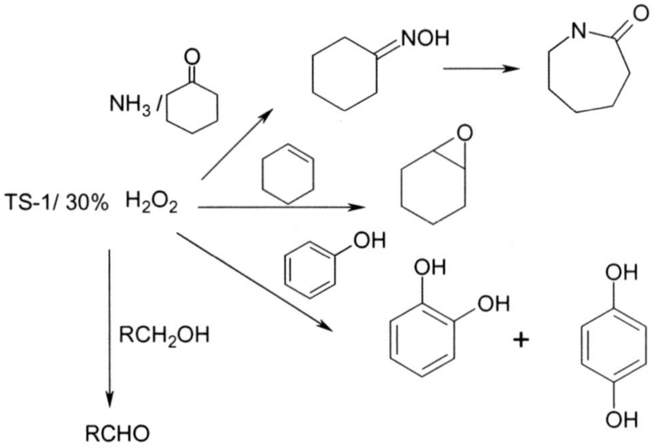

Scheme 4.11   TS-1 as a green oxidation catalyst.

and tetraethylorthotitanate in the presence of an organic base such as tetrapropylammonium hydroxide. This catalyst is especially useful for oxidation reactions using hydrogen peroxide (Scheme 4.11), from which the only by-product is water; clean production of hydroquinone being one of the possibilities.

Although alkene epoxidation using hydrogen peroxide is well known it is often carried out with concentrated hydrogen peroxide (70%) to avoid excessive glycol formation. This process is highly dangerous and costly, preventing wide-spread commercialization. TS-1, however, is an efficient epoxidation catalyst that uses 35% hydrogen peroxide, which can be handled much more safely. A wide range of alkenes can be efficiently epoxidized, including propene and cyclohexene. In the case of propene selectivities as high as 93% can be achieved, with propane diol being the main by-product. Commercialization of the process has been considered by several companies but has been limited by the cost of hydrogen peroxide. A 300 000 tonne per year plant built in Antwerp by BASF and Dow Chemical came on-stream in 2008 and a further plant is being built in Thailand. The Antwerp plant is thought to be commercially viable due to the presence of a nearby hydrogen peroxide plant, thereby avoiding expensive transport costs. The process is thought to operate in methanol solvent at 60–80 °C, in a continuous stirred tank reactor. Benefits of the process include:

- reducing wastewater by 70–80%, compared with existing PO technology;
- reducing energy usage by approximately 35%, compared with existing PO technology;

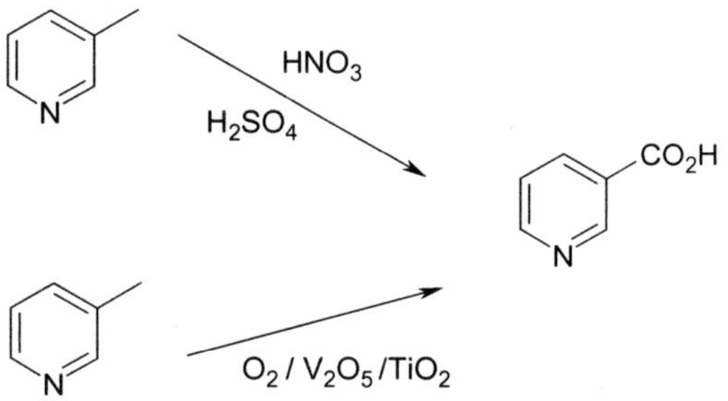

Scheme 4.12   Routes to nicotinic acid.

• reducing infrastructure and physical footprint with simpler raw material integration and avoidance of co-products.

A further interesting example of the versatility of TS-1 is in the synthesis of caprolactam, precursor to nylon 6. In this process cyclohexanone, 30% hydrogen peroxide, and ammonia are reacted at 100 °C in the presence of TS-1. Overall waste from this process is around 70% less than for the conventional process using hydroxylamine hydrosulfate.

Nicotinic acid synthesis (Scheme 4.12) provides an interesting insight into the complexities of catalyst development required to achieve efficient, commercially viable processes. Nicotinic acid is a valuable vitamin and pharmaceutical intermediate; initial production methods included oxidation of 3-methyl- or 3-methyl-5-ethylpyridine with a nitric acid/sulfuric acid mixture or with potassium permanganate, both process leading to high levels of waste and low yields (around 75%). Later synthesis methods included catalytic oxidative ammonolysis using a vanadium catalyst. Overall this process is much more efficient, producing less waste. It does, however, require purification of the nitrile intermediate and is relatively expensive to operate. The obvious goal is direct oxidation of 3-methylpyridine using an inexpensive oxidant (air) over a suitable catalyst. Several such gas-phase processes have been patented based around use of vanadium pentoxide/titanium dioxide catalysts. Early attempts, using a catalyst formed by fusing the two components together at high temperature, gave yield of less than 50%, with the majority of the reactant being converted into HCN and CO_2 at a reaction temperature of over 400 °C. The main problem with the process was the low surface area (1 $m^2 g^{-1}$) and unstructured nature of the catalyst. Preparation of a higher surface area catalyst (up to 100 $m^2 g^{-1}$) by impregnating titanium dioxide with vanadyl oxalate followed by heat treatment together with incorporation of a small amount of an alkali metal promoter enabled the reaction temperature to be reduced to below 300 °C, improving selectivity and yield (around 85%). In the latest patented processes by Degussa, conversion and selectivity have been improved to over 95%. Whilst some of this is due to engineering improvements, much of it is due to the finding that using titanium dioxide from the so-called sulfate process gives better results than that from the chloride process. This is likely to be due to the presence of trace amounts of titanium sulfate.

A further example of the importance of the support and the way the catalyst is attached to it is the use of perruthenates for selective oxidation of alcohols to aldehydes and ketones under mild conditions. Tetra(n-propyl)ammonium perruthenate has become a popular laboratory reagent widely used, for example, in producing libraries of materials for pharmaceutical screening. Heterogenization of the system through reaction of potassium perruthenate ($KRuO_4$) with ammonium chloride functionalized polystyrene enabled the catalyst to be removed by filtration. True catalytic activity was obtained when the reactions were carried out with addition of oxygen, typically at 75 °C. Unfortunately, reuse of the catalyst was limited due to loss of activity attributed to Hoffman elimination from the quaternary salt. When similar catalysts were prepared on a mesoporous silica support, MCM-41, the catalysts were completely stable and active. However, when doped onto unmodified highly hydrophilic silica reactivity was poor, presumably through poor oxygen transport. Although the exact nature of these catalysts has not been determined it does demonstrate the importance of having both an active catalyst species and a suitable support.

4.2.3.4 *Some Other Catalytic Reactions Useful for Fine Chemical Synthesis.*
Friedel–Crafts acylation reactions are still widely used in the production of fine chemicals and, despite considerable research efforts, greater than stoichiometric amounts of Lewis acids ($AlCl_3$, $FeCl_3$, and $ZnCl_2$) continue to be required, due to complexation with the product. These reactions are characterized by difficult separation procedures and large waste streams. Many heterogeneous acids have been assessed for these reactions, including a wide variety of zeolites, clays, and heteropolyacids such as $H_2PW_6Mo_6O_{40}$; however, no general catalyst type has emerged for the acylation of unactivated aromatics. Commercial success has been achieved for acylation of small molecules such as anisole using partially dealuminated zeolite Y doped with sodium or potassium (Scheme 4.13). A further advantage of this process is that it can be carried out continuously in a fixed bed reactor without solvent. Acetic acid by-product may be recovered for use or conversion back into anhydride. Screening studies clearly showed that although very high

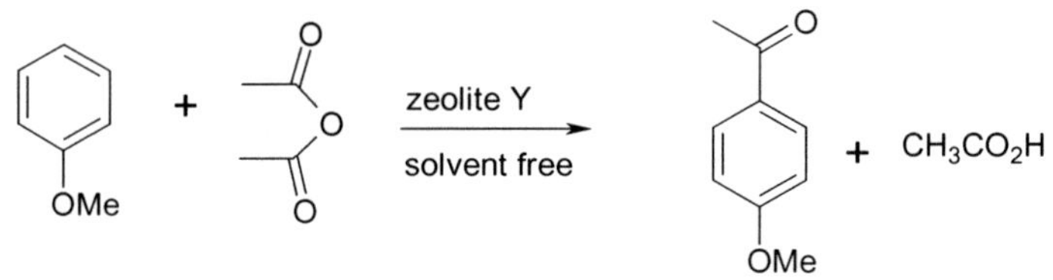

Scheme 4.13 Catalysed acylation of anisole.

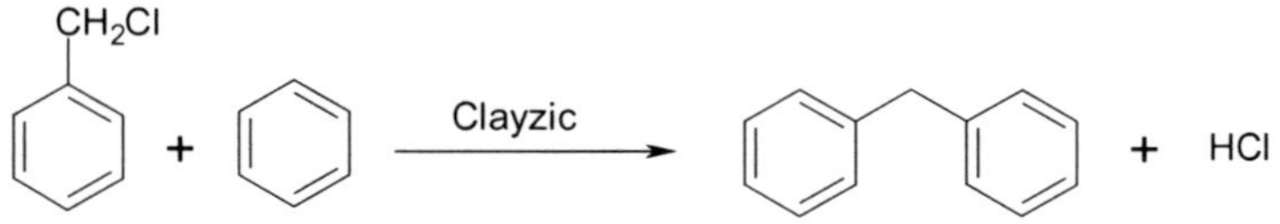

Scheme 4.14 Benzylation catalyzed by clayzic.

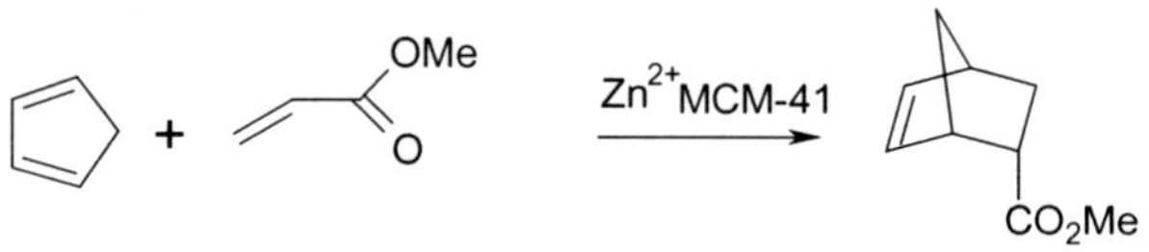

Scheme 4.15 Catalysed Diels–Alder reaction.

para selectivity could be obtained using a wide variety of solid acid catalysts only very few gave yields over 30% even for this activated substrate.

Although alkylations do not give rise to the same waste problems as acylation, replacement of soluble non-recoverable Lewis acids by re-cyclable catalysts is still beneficial. One such commercial catalyst, clayzic, is especially efficient at catalysing benzylation reactions (Scheme 4.14). Clayzic is based on an acid-treated clay containing zinc chloride, with the Brønsted effects of the clay producing a synergistic effect with the Lewis acid properties of zinc chloride. Lewis acid containing clays are also effective sulfonation catalysts, *e.g.* high yields of diphenylsulfone can be obtained from reaction of benzene and benzene sulfonyl chloride.

As noted in Chapter 1, Diels–Alder reactions are highly versatile atom efficient reactions, which can be carried out in the absence of ancillary reagents. In some instances conversions at moderate temperatures and selectivities can be improved using Lewis acid catalysts. The reaction between cyclopentadiene and methyl acrylate (Scheme 4.15) is relatively slow and gives an endo:exo ratio of 79:21 but the reaction can be catalysed by a wide range of Lewis acids. Improved yields and endo

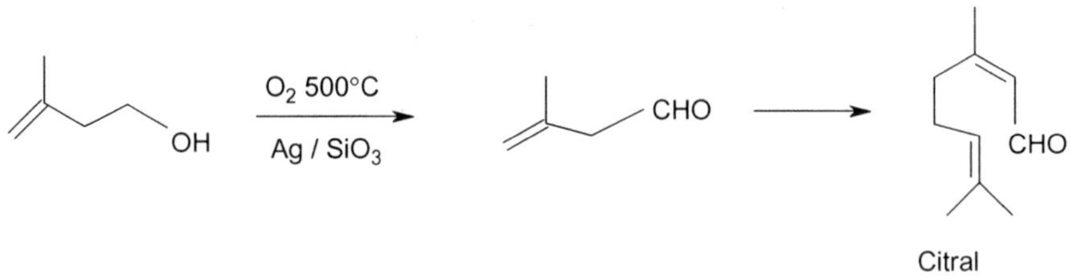

Scheme 4.16 Synthesis of citral.

selectivities (around 95:5) were obtained in all cases. Interestingly, even with highly structured solid catalysts such as a Zn^{2+}-containing MCM-41 with a pore size of 2.5-nm endo:exo selectivity was similar to that obtained using non-structured or homogeneous catalysts.

Citral is a key intermediate for fragrances and in the synthesis of vitamins A and E. Traditionally it has been made by oxidation of an alcohol to the aldehyde using stoichiometric amounts on MnO_2. BASF have developed a catalytic process using silver supported on silica with air as the oxidant at 500 °C (Scheme 4.16).

Nitration of aromatic compounds is a common reaction in the synthesis of dyes and pesticides. The most basic of these reactions, nitration of benzene, is carried out with a mixture of nitric and sulfuric acids; the sulfuric acid donates a proton, generating a nitronium ion that is the nitrating agent. Over 50% of the reaction mix is actually sulfuric acid, and at the end of the reaction this is recovered as dilute sulfuric acid that needs go through an energy intensive recycling process before it can be reused. It has now been found that a silica catalyst containing 20% molybdenum oxide can replace all the sulfuric acid. The nitration is carried out in the gas phase at 140 °C with conversions of over 90% being achieved.

4.2.4 Catalytic Converters

Control of car exhaust emissions using three-way catalytic converters is the single largest use of a catalyst and has had a significant impact on the price of the platinum group metals from which they are made, with around 150 000 kg of these precious metals being used annually. The term 'three-way catalyst converter' arises from the three environmental pollutants that they are designed to remove: carbon monoxide, nitrogen oxides, and hydrocarbons. Even though engine efficiencies are continually improving, in the absence of any 'end of pipe' clean up technology an average family car would emit 10 g of CO, 2 g of NO_x, and 1 g of hydrocarbon for every kilometre driven. These emissions have been the major cause of poor air quality in cities throughout the world for the past 40 years. With the advent of lead-free gasoline catalytic converters became the method of choice for controlling emissions and are now fitted to most new cars produced throughout the world.

The catalytic converter consists of several parts:

- A ceramic monolith catalyst support, cordierite, consisting of silica, alumina, and magnesium oxides. The purpose of this is to provide support, strength, and stability over a wide temperature range.
- A washcoat that provides a high surface area onto which the active catalyst is impregnated. The washcoat typically consists of a mixture of zirconium, cerium, and aluminium oxides. Apart from providing a high surface area the washcoat also acts as an oxygen storage system (see below).
- The active catalyst platinum, palladium, and smaller amounts of rhodium, the total weight of these metals used per converter is less than 2 g.
- A high quality steel housing to provide additional support and protection.

Scheme 4.17 depicts the main catalytic reactions occurring in the converter. Platinum and palladium are mainly responsible for removal of hydrocarbons and CO whilst the rhodium is mainly responsible for removal of NO_x.

The process runs best just above the stoichiometric point, *i.e.* about 14.8 parts air to 1 part gasoline (if no oxygenates are present). It is carried out in two stages, the first being NO_x reduction. This reduction is inefficient in the oxygen-rich atmosphere that occurs in specific parts of the combustion cycle. Hence in this lean (with respect to fuel) part of the cycle oxygen is taken up by the washcoat and released when oxygen

$$2NO_x \rightarrow N_2 + XO_2$$

$$2NO + 2CO \rightarrow N_2 + 4\,CO_2$$

$$Hydrocarbon + NO \rightarrow N_2 + CO_2 + H_2O$$

$$2NO_2 + 4CO \rightarrow N_2 + 4CO_2$$

$$Hydrocarbon + O_2 \rightarrow CO_2 + H_2O$$

$$2CO + O_2 \rightarrow 2CO_2$$

Scheme 4.17 Some reactions occurring in a catalyst converter.

concentrations coming from the engine are lower. Hydrocarbon and CO oxidations are carried out in the second stage of the process. Operating efficiently, a converter will reduce hydrocarbon and CO emissions by 95% and NO_x emissions by 90%. Efficiencies are much lower though on cold starting the engine, with optimum performance for most systems not being achieved until the temperature inside the converter reaches 180–200 °C. Improving the cold-start performance is one of the major areas of current catalytic research.

Catalyst converters have a design life of approximately 160 000 km. Although relatively robust the catalyst is slowly poisoned by low level contaminants that are in, or find their way into, the fuel system, such as phosphorous, zinc, and lead; this, combined with mechanical damage are the main limiting lifetime factors. Even though the amount of precious metals (average 1 kg per tonne of spent catalyst) is quite low they are worth recovering. One of the main processes used involves pulverizing the support followed by heating in a furnace at around 1600 °C. This temperature is below the melting point of the precious metals and a collector metal such as copper or iron is added to 'dissolve' them and remove them from the slag. The precious metals are then recovered from the collector metal using conventional refining techniques.

The local, and possibly global, advantages to air quality of using catalyst converters are obvious, but like all 'end of pipe' technologies the improvements come at a price. Full LCA studies have questioned the overall global environmental advantages of the technology. Valuable reduction in the emissions discussed above have been achieved, but significant increases in fossil fuel consumption (due to running at the stoichiometric point rather than fuel lean), carbon dioxide emission, and solid waste production have resulted.

Catalyst converters for diesel vehicles are termed two-way since they are unable to effectively remove NO_x, due to excess O_2 in the exhaust gases preventing the catalytic process. NO_x traps are starting to be installed on some diesel engines. These use zeolites to trap the NO_x molecules, acting as a molecular sponge. Once the trap is full no more NO_x can be adsorbed, and it is passed out of the exhaust system. Various schemes have been designed to 'purge' or 'regenerate' the adsorber. Injection of diesel fuel before the adsorber can cause the NO_2 in particular to react with the hydrocarbons to give water and nitrogen. Selective catalytic reduction is also employed for larger engines to reduce NO_x. These catalysts consist of titanium dioxide impregnated with vanadium, tungsten, or iron/copper. In the presence of a reductant such as urea or ammonia the NO_x is converted into water and nitrogen. The main issue is that the system needs a regular top up with reductant.

4.3 HOMOGENEOUS CATALYSTS

In contrast to heterogeneous catalysis, industrial applications of homogeneous catalysis in the manufacture of bulk chemicals are rare, use largely being restricted to the speciality and pharmaceutical sectors. Homogeneous catalysts have been well researched, since their catalytic centres can be relatively easily defined and understood, but difficulties in separation and catalyst regeneration remain major challenges for many applications.

The most widely used homogeneous catalysts are simple acids and bases that catalyse well-known reactions such as ester and amide hydrolysis and esterification. Such catalysts are inexpensive enough that they can be neutralized, easily separated from organic materials, and disposed of. This, of course, is not a good example of green chemistry, and contributes to the huge quantity of aqueous salt waste generated by industry.

4.3.1 Transition Metal Catalysts with Phosphine or Carbonyl Ligands

Many of the green benefits of homogeneous catalysis, especially that of high selectivity, arise from 'designer' catalysts made from transition metals and appropriate ligands. It is the partially filled d-orbitals that make transition metals attractive catalysts. These orbitals are of relatively high energy, enabling electrons to be readily transferred in or out. Ligands bond with transition metals *via* these partially filled d-orbitals. The maximum number of metal–ligand bonds is determined by the 18-electron rule (4.2), were n is the number of d-electrons for the particular oxidation state of the metal and CN is the metal co-ordination number, or the number of metal–ligand sigma bonds. Complexes that are co-ordinatively unsaturated, *i.e.* have less than the maximum number of ligands often make good catalysts:

$$n + 2(CN)_{max} = 18 \qquad (4.2)$$

Phosphines, for example triphenylphosphine, are probably the most common type of ligand encountered. There are several reasons for this, first phosphines are very good electron donors and hence form quite stable bonds with many transition metals, enabling isolation and characterization of complexes. Secondly, a wide variety of phosphines can be made synthetically, giving a high degree of control over steric effects; this is especially noticeable when bidentate phosphines are used to completely block off one side of a catalyst. Carbon monoxide is another common ligand; in contrast to phosphines bonded CO frequently takes part in reactions such as carbonylations. The OC–metal bond is complex and consists of a *sigma* bond formed by donation of a lone pair of electrons from the carbon into and empty d-orbital and back-donation (backbonding) from a full d-orbital into an empty antibonding π-orbital on the carbon. This backbonding effectively weakens the C–O bond, making it more reactive, enabling it to take part in reactions such as carbonylations and hydroformylations, which un-coordinated CO could not.

Wilkinson's catalyst (RhCl(PPh)₃, Scheme 4.18) is an excellent early example of the value of homogeneous catalysis. Catalysts related to Wilkinson's are highly valuable for alkene hydrogenation, under ener-getically favourable conditions compared to heterogeneous hydrogen-ation catalysts (often these reactions can be carried out at ambient temperature and pressure). The ligand has a significant influence on reaction rate, with a doubling of the rate being obtained when electron-donating *p*-methoxyphenyl is used in place of phenyl.

The complete catalytic cycle is highly complex with several rhodium-phosphine species being detected, some of which lead to 'dead-end' reactions removing catalyst from the cycle. It has been determined that the mechanism does involve loss of PPh₃ from the catalyst to give a 14-electron co-ordinatively unsaturated species, RhCl(PPh)₂, sub-sequent hydrogenation of which is very rapid.

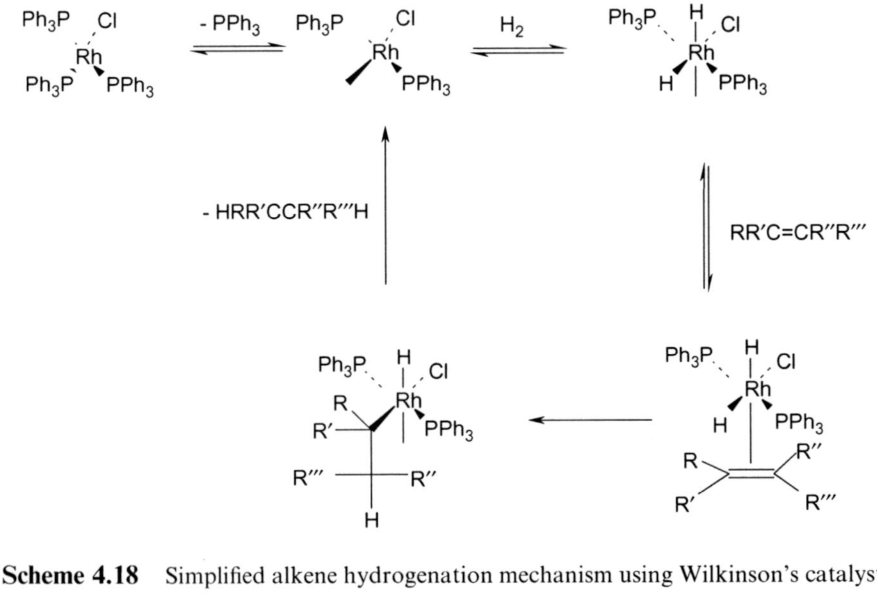

Scheme 4.18 Simplified alkene hydrogenation mechanism using Wilkinson's catalyst.

Hydroformylation is an important industrial process carried out using rhodium triphenylphosphine or cobalt phosphine carbonyl catalysts. It is operated on a large industrial scale for the preparation of detergent intermediates. In general, processes using rhodium operate at lower temperatures and pressure and usually give slightly higher selectivities.

One major industrial process using the rhodium catalyst is hydro-formylation of propene with synthesis gas (potentially obtainable from a renewable resource, see Chapter 6). The product, butyraldehyde, is formed as a mixture of *n-* and *iso-*isomers; the *n-*isomer is the most desired product, being used for conversion into butanol (*via* hydro-genation) and 2-ethylhexanol (*via* aldol condensation and hydrogen-ation). Butanol is a valuable solvent in many surface coating formulations while 2-ethylhexnol is widely used in the production of phthalate plasticizers. Scheme 4.19 shows the main mechanistic elem-ents. Thermodynamics favour formation of the *iso-*product; however, steric bulk of the ligands limits this reaction, with the more valuable *n-*isomer being produced in 70–95% selectivity. The rate of reaction is highly dependent on phosphine concentration, since this determines the amount of active catalyst in the system. Industrially, phosphine

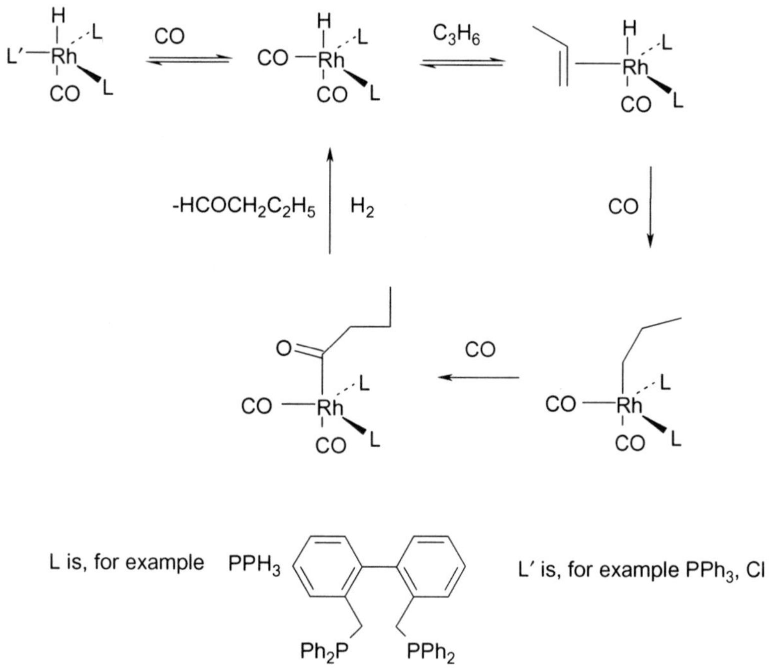

Scheme 4.19   Basic elements of hydroformylation.

concentrations may be up to 75 mol per mol-Rh; at very high concentrations of phosphine rate inhibition takes place through blocking of active catalyst sites. The reaction occurs under the relatively mild conditions of 80 °C and 20–30 atm pressure.

Since butyraldehyde has a low boiling point (75 °C) separation of catalyst from both reactants and product is straightforward. Most of the rhodium remains in the reactor but prior to recovery of propene and distillation of crude product the gaseous effluents from the reactor are passed through a demister to remove trace amounts of catalyst carried over in the vapour. This ensures virtual complete rhodium recovery.

Although rhodium recovery is efficient it is difficult to separate it from 'heavies' that are formed in small amounts, and over time these 'heavies' tend to result in some catalyst deactivation. One answer to this problem has been developed by Ruhrchemie/Rhône-Poulenc. In this process sulfonated triphenylphosphine is used as the ligand, which imparts water solubility to the catalyst. The reaction is two-phase, a lower aqueous phase containing the catalyst and an upper organic phase. Fortunately, the catalyst appears to sit at the interface, enabling reaction to proceed efficiently. At the end of the reaction the catalyst-containing aqueous phase can be separated by decantation, any heavies formed remaining in the organic phase.

Cobalt catalysts such as $HCo(CO)_4$, despite the higher temperatures and pressures required, are widely used for hydroformylation of higher alkenes. The main reason for this is that these catalysts are also efficient alkene isomerization catalysts, allowing a mix of internal and terminal alkenes to be used in the process. Catalyst recovery is more of a problem here and involves production of some waste and significantly adds to the complexity of the process. A common recovery method involves treating the catalyst with aqueous base to make it water-soluble, separation and subsequent treatment with acid to recover active catalyst (4.3):

$$2HCo(CO)_4 + Na_2CO_3 \rightarrow 2NaCo(CO)_4 + H_2O + CO_2$$
$$2NaCo(CO)_4 + H_2SO_4 \rightarrow 2HCo(CO)_4 + Na_2SO_4$$

$$(4.3)$$

Carbonylation of methanol to acetic acid is fully discussed in Chapter 9. Another carbonylation process using a phosphine ligand to control the course of the reaction is a highly atom efficient route to the widely used monomer methyl methacrylate (Scheme 4.20). In this process the catalyst is based on palladium acetate and the phosphine ligand bis(phenyl)(6-methyl-2-pyridyl)phosphine. This catalyst is remarkably (>99.5%) selective in the 2-carbonylation of propyne under relatively mild conditions of <100 °C and 60 bar.

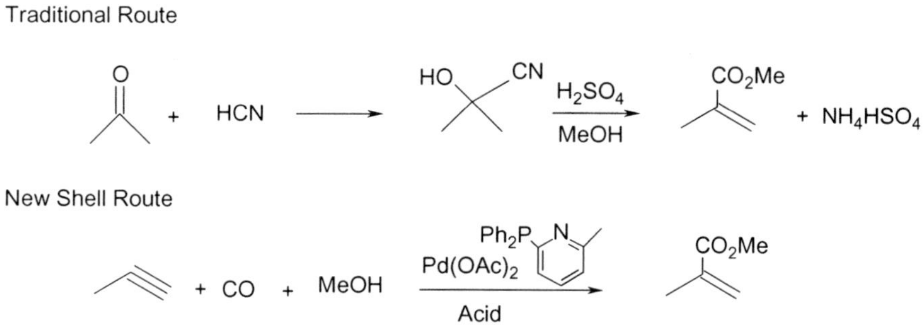

Traditional Route

New Shell Route

Scheme 4.20 Routes to methyl methacrylate.

The environmental benefit of this route is evident when compared to the traditional methyl methacrylate manufacturing process, which uses hazardous hydrogen cyanide and produces stoichiometric amounts of ammonium hydrogen sulfate waste.

4.3.2 Greener Lewis Acids

As mentioned several times, Lewis acids are highly valuable catalysts but the most commonly used ones such as aluminium chloride and boron trifluoride are highly water sensitive and are not usually recovered at the end of a reaction, leading to a significant source of waste. In recent years there has been much research interest in lanthanide triflates (tri-fluoromethanesulfonates) as water stable, recyclable Lewis acid catalysts. This unusual water stability opens up the possibility for either carrying out reactions in water or using water to extract and recover the catalyst from the reaction medium.

Lanthanide triflates have been assessed for a wide range of reactions subject to catalysis by Lewis acids and have generally given positive results. The triflate salts are generally used since their electron-withdrawing properties impart strong Lewis acidity. Activity in individual reactions is dependent on the metal ion; in most cases ytterbium and scandium appear to be good choices. One of the first and most studied reactions is the so-called Mukaiyama aldol; in contrast to conventional base-catalysed aldols this reaction between silyl enol ethers and aldehydes is catalysed by Lewis acids. The usual work-up problems have prevented these potentially valuable carbon–carbon bond forming reactions gaining wider uses. Using as little as 1 mol% ytterbium catalyst in a water–THF solvent mix a yield of well over 90% can be obtained at room temperature. Furthermore, after separation of the organic layer the catalyst can be reused. In contrast to other Lewis acids ytterbium

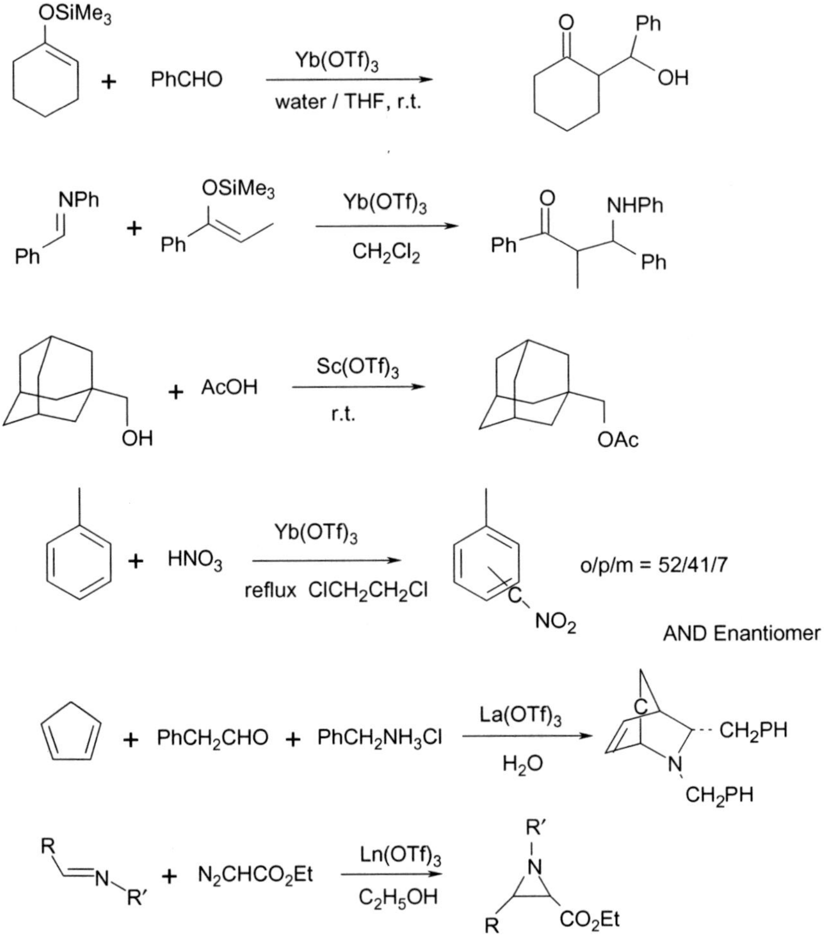

Scheme 4.21   Some reactions catalysed by lanthanide triflates.

triflate has been shown to selectively catalyse reactions between aldi-mines and silyl enol ethers in the presence of aldehydes. Scheme 4.21 shows some examples of reactions catalysed by lanthanum triflates; in all cases the catalyst can, with varying degrees of complexity, be recovered and reused. Many workers are now looking at supported lanthanide triflates to make catalyst recovery easier.

4.3.3   Asymmetric Catalysis

The importance of producing pharmaceuticals in enantiomerically pure forms was brought to the public's attention with the case of thalidomide

(**4.2**) in the early 1960s. In the early 1950s, thalidomide was produced, as a racemic mixture, for use as a sedative and a non-addictive alternative to barbiturates. It was later found that it alleviated many of the unpleasant symptoms of early pregnancy but by 1961 its use had been linked with an increase in the number of severe birth deformities and it was withdrawn. It was later found that it is the (*R*)-enantiomer that is the active drug whilst the (*S*)-enantiomer was responsible for the birth deformities. It is somewhat ironic, and unusual, that even if (*R*)-thalidomide had been the drug administered some birth deformities would have still been likely since racemization of this particular substance occurs within the body.

4.2 (* Chiral centre)

Since then our knowledge of the mode of action of drugs has greatly improved and even more stringent testing of all new drugs takes place before they are put onto the market. Most pharmaceutical products are now marketed as single enantiomers, even though they are often produced as racemic mixtures. This has normally been achieved through extensive resolution and purification of racemic mixtures, resulting in the production of large waste volumes, as well as the loss of at least 50% of the product. Increasingly, pharmaceutical companies are looking for ways to make only the required enantiomer. The two major methods involve use of chiral auxiliaries, which are often not recovered, and asymmetric catalysis. It is not only the pharmaceutical companies who are interested in enantiomerically pure products – they are also being increasingly sought by the agrochemical and flavour and fragrance industries (the enantiomers of limonene smell of oranges and lemons respectively). To be of significant industrial interest the enantiomeric excess should be over 90%, ideally over 95% so that purification can be achieved with a small number of crystallization steps. Unfortunately, although the literature is full of examples of asymmetric catalysis relatively few currently meet this demanding criterion.

L-dopa [(*S*)-enantiomer], an important drug for the treatment of Parkinson's disease, has for many years been made by reducing a pro-chiral alkene with a Wilkinson's-type catalyst based around a chiral phosphine, DIPAMP (Scheme 4.22). The catalytic mechanism is

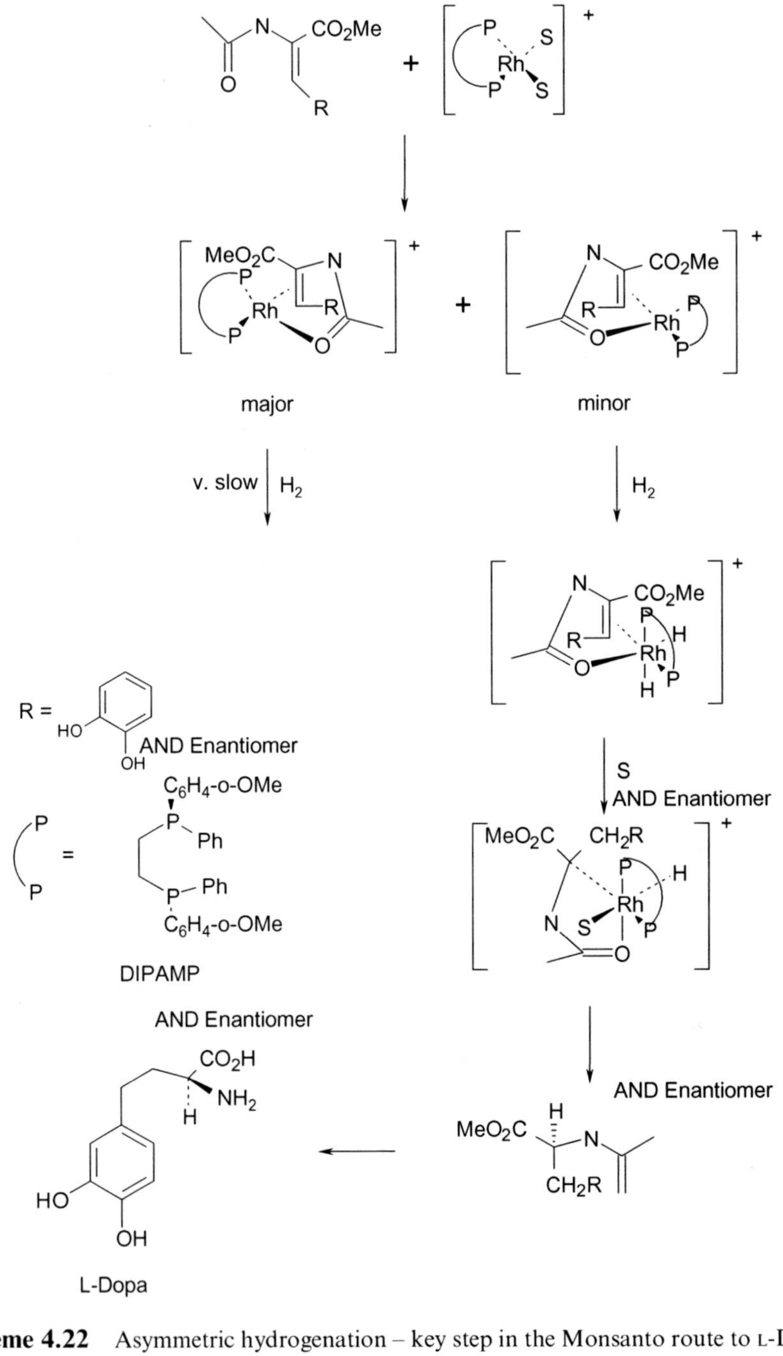

Scheme 4.22   Asymmetric hydrogenation – key step in the Monsanto route to L-Dopa.

interesting in that, under steady state conditions, the diastereomer that would lead to the (*R*)-enantiomer is formed in much higher concentrations. Hydrogenation of this enantiomer (due to differences in ΔG) is, however, some one thousand times slower than for the (*S*)-enantiomer. This results in an enantiomeric excess of $>95\%$. The catalytic activity is very high, resulting in low levels of catalyst being used ($<0.1\%$ of product), and the product can be separated by crystallization from the reaction and the catalyst recycled. A small loss of active catalyst does occur, however; whilst the metal can be readily recovered the loss of even small amounts of chiral phosphine is a significant process cost. Similar processes are used for the production of L-amino acids.

Catalytic hydrogenation of alkenes is widespread but reduction of carbonyl compounds is often carried out with stoichiometric reagents such as sodium borohydride. Ruthenium catalysts are proving useful for carrying out catalytic carbonyl reductions although commercial applications are currently limited. Carbonyl compounds containing other functional groups are more simply hydrogenated due to metal–functional group interactions. Noyori has pioneered this work, which has found industrial use in the synthesis of carbapenem antibiotics (Scheme 4.23) using the catalyst (*R*)-BINAP-Ru. This catalyst can be used for the asymmetric hydrogenation of carbonyls containing a wide

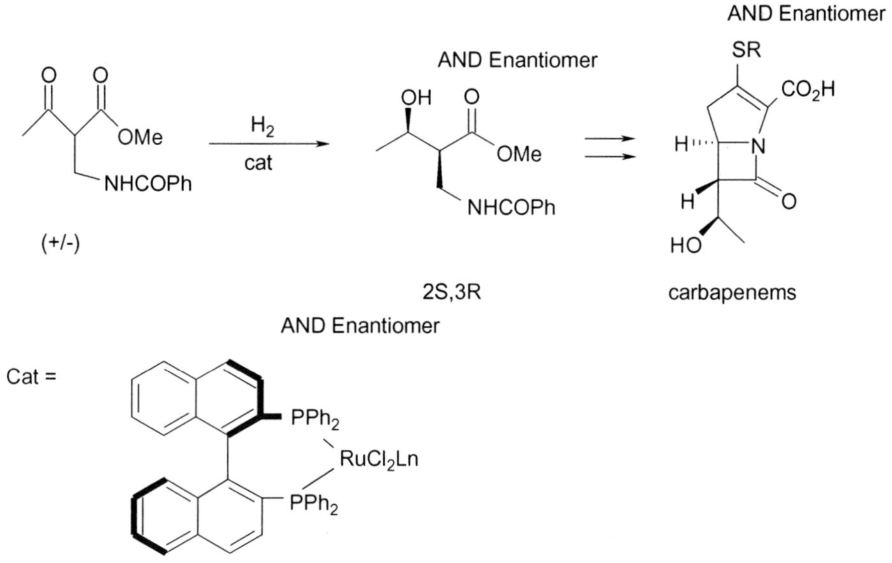

Scheme 4.23 Asymmetric reduction route to carbapenem.

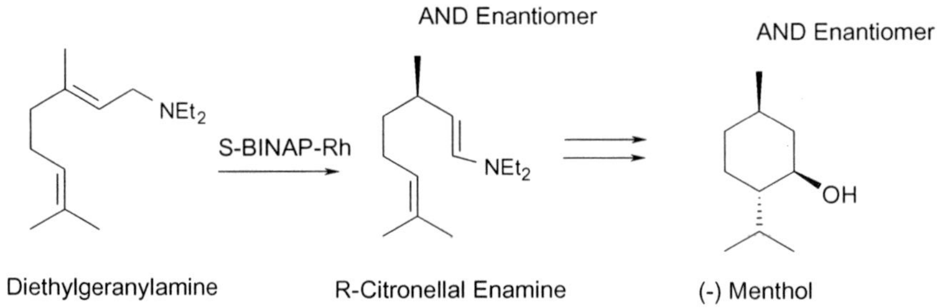

Scheme 4.24   Asymmetric isomerization using (*S*)-BINAP-Rh.

range of alpha functional groups, including amino, hydroxyl, ester, amide, and sulfonate.

The synthetic utility of this reaction has been extended by the finding that simple aromatic ketones can be hydrogenated both selectively (in the presence of C=C bonds) and in high enantiomeric excess using ruthenium catalysts containing chiral phosphines and chiral diamines.

BINAP is a versatile ligand, the (*S*)-enantiomer complexed with rhodium is used in the commercial production of (–)-menthol (Scheme 4.24). In this case the reaction involves isomerization of diethylgeranylamine to (*R*)-citronellal enamine, which proceeds in approximately 99% ee.

Asymmetric epoxidation is another important area of activity, initially pioneered by Sharpless using catalysts based on titanium tetra-isopropoxide and either (+) or (–) dialkyl tartrate. The enantiomer formed depends on the tartrate used. This has been widely used for the synthesis of complex carbohydrates but it is limited to allylic alcohols, with the hydroxyl group bonding the substrate to the catalyst. Jacobson catalysts (**4.3**) based on manganese complexes with chiral Schiff bases have been shown to be efficient in the epoxidation of a wide range of alkenes, normally in >90% ee. The commercial viability of these catalysts has so for been limited due to their relative instability and the need to carry out the reactions at very low temperatures (typically – 78 °C) to get the selectivities required.

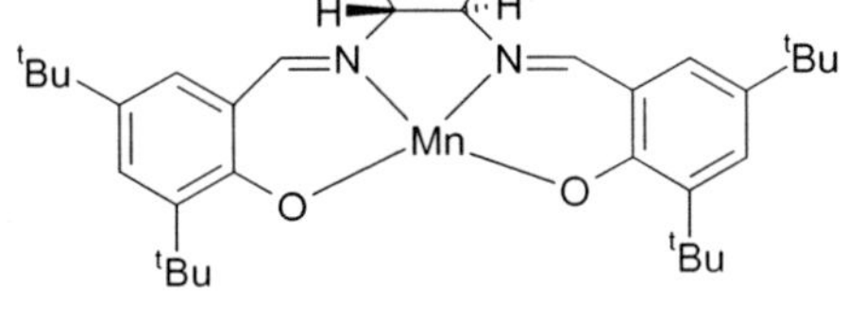

4.3

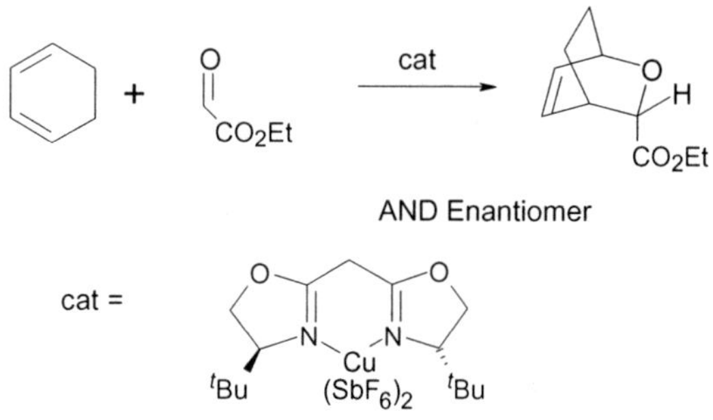

Scheme 4.25 Enantioselective Diels–Alder reaction.

Finally, mention should be made of the extensive research being carried out to synthesize enantioselective *endo*-Diels–Alder products using chiral Lewis acid catalysts. Much work has focussed on using traditional Lewis acid metals (Al, B, Ti, *etc.*) with chiral ligands like 1,1'-binaphthol; however, many of these catalysts suffer significant deactivation. This has been attributed to the high oxophilicity of these 'hard' metals causing catalyst agglomeration. More success has been found using a combination of 'soft' ligands (*e.g.* phosphines) and 'soft' metals (*e.g.* Cu). If such reactions become commercially viable there is the possibility for development of highly green sophisticated reactions having 100% atom economy, high yield, and high selectivity for production of molecules of interest to the pharmaceutical and agrochemical sectors. One example is shown in Scheme 4.25; here a yield of 98% with 95% *endo* selectivity and an ee of 97% has been produced using the chiral BOX-copper catalyst. The reaction is very solvent dependant, only proceeding well in highly polar solvents that stabilize the catalyst. Lactones such as these are interesting for the production of many natural products, including insect pheromones.

Discussion of asymmetric catalysis has been included in this section on homogeneous catalysis to reflect the vast amount of work in this area. Heterogeneous asymmetric catalysis, although less well advanced due to synthetic and selectivity challenges, will certainly have an increasingly important role to play in the development of greener processes and products. Two brief examples of the kind of areas being studied highlight the potential of this area:

1. Hydroformylation of vinyl acetate to give mainly the branched product in >90% ee has been achieved using a rhodium catalyst containing binaphthol and phosphine ligands anchored to polystyrene.

2. Heterogeneous asymmetric hydrogenation is an active area of work. Dimethyl itaconate has been hydrogenated successfully in very high enantiomeric excess using a complex dendrimer-type catalyst. The catalyst is synthesized by building up a dendrimer of ferrocenyl diphosphines, through initial reaction of an amino substituent with benzene-1,3,5-tricarboxylate. Rhodium is then reacted onto this to form the active insoluble catalyst.

4.4 PHASE TRANSFER CATALYSIS

Phase transfer catalysis (PTC) refers to the transfer of ions or organic molecules between two liquid phases (usually water–organic) or a liquid and a solid using a catalyst as a transport shuttle. The most common systems encountered are water–organic, hence the catalyst must have an appropriate hydrophilic–lipophilic balance to enable it to have compatibility with both phases. The most useful catalysts for these systems are quaternary ammonium salts. Commonly used catalysts for solid–liquid systems are crown ethers and poly(glycol ethers). Starks developed the mode of action of PTC 30 years ago (Figure 4.5). In its simplest form it involves transport of the reactive anion X^- from the aqueous phase into the organic phase, reaction with the substrate and return of the catalyst to the aqueous phase often associated with a leaving anion.

There are many possible 'green' advantages in using PTC. These include:

- Higher productivity – PTC reactions are often very rapid; one reason for this is that anions in the organic phase have few associated water molecules, making them highly reactive through a reduction in activation energy. This of course will normally translate into reduced energy usage and greater reactor throughput.

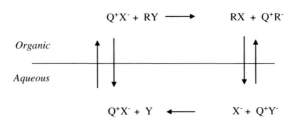

Figure 4.5   Phase transfer catalysis.

- Higher selectivity – due to reduced activation energy these reactions can often be run at lower temperatures, which may reduce by-product formation.
- Ease of product separation – since the organic layer, substantially free from water-soluble contaminants, can simply be decanted off, product separation is often simple, resulting in less waste. Notably, the concentration of X in the organic phase cannot exceed the concentration of catalyst (unless it is soluble in the absence of a catalyst).
- Use of less hazardous solvents – since the reaction is two phase, simple benign solvents can often be used since PTC avoids the need to find a solvent that will dissolve all reactants, *e.g.* dipolar aprotic solvents such as dimethyl formamide. In some cases an organic solvent may not be required at all, with the substrate forming the second phase.

As a cautionary note PTC should not be considered a panacea for all of green chemistry. Two phase reactions involving water are often difficult to deal with industrially, particularly if the water is contaminated with trace amounts of hazardous organic substances. In some cases it may be more practical, cost-effective and environmentally prudent to avoid production of aqueous waste in favour of a recyclable less benign solvent.

As may be expected, the nature of X and Y have a significant impact on the rate of reaction. Anions with very strong hydration energies (high charge densities) have a strong preference for the aqueous phase and hence make a good leaving group Y. Such ions include OH^-, F^-, and Cl^-. Conversely, anions with low hydration energies will readily pass into the organic phase, these include SCN^-, ClO_4^-, and I^-. Consequently, substitution reactions with alkyl iodides, for example, are very difficult.

4.4.1 Hazard Reduction

PTC has been effectively used to reduce the hazardous nature of overall processes; whilst the elimination of hazardous reagents has not been achieved the volume required has been reduced. In Reaction (4.4) sodium cyanide is normally required in excess and so at the end of the process great care must be taken to separate it from the product and either destroy it or recover it. In the PTC system the NaCN is for all practical purposes confined to the aqueous layer, which can be simply separated from the organic layer and reused in the next batch:

$$RCl + NaCN \rightarrow RCN + NaCl \qquad (4.4)$$

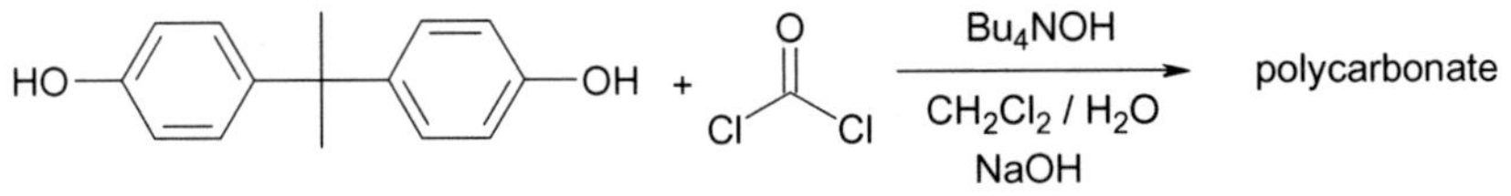

Scheme 4.26   Manufacture of polycarbonates using PTC (phase transfer catalysis).

In the manufacture of polycarbonates (Scheme 4.26) *via* base-catalysed elimination of HCl between Bisphenol A and phosgene, hydrolysis of the phosgene means it is used in considerable excess. By using a PTC (butylammonium hydroxide) the phosgene is prevented from coming into significant contact with aqueous base. This reduces hydrolysis by a factor of 200, enabling phosgene to be used in only 2 mol% excess.

4.4.2   C–C Bond Formation

Owing to the so-called hyperbasic effect sodium hydroxide is able to form carbanions from many organic compounds usually requiring stronger bases. Deprotonation of a carbanion precursor depends on its acidity. In a two phase system the reaction between the organic substrate and NaOH will, to a large extent, take place at the interface. Subsequent reaction with a quaternary ammonium salt will produce a water-soluble sodium salt and a lipophilic quaternary organic complex. Removal of this to the organic phase effectively shifts the equilibrium for the organic/NaOH reaction, driving the reaction forward. The effect of this is that carbanions can be formed from weak acids. This has been used to alkylate phenyl acetonitrile, for example, which normally requires anhydrous conditions and more hazardous bases such as sodium hydride or sodamide. The avoidance of a water quench also reduces the amount of waste compared to the older process. For more acidic substrates such as alkyl cyanoacetates solid potassium carbonate and a crown ether can be used, thereby avoiding the need for an aqueous layer.

PTC has been used extensively for making cyclopropyl derivatives. The most common reaction involves generation of dichlorocarbene from chloroform, using NaOH and a quaternary ammonium hydroxide. The carbene subsequently reacts with an alkene in high yield. Hydrolysis of dichlorocarbene, normally rapid in the presence of water, is minimal. Scheme 4.27 shows an interesting and very efficient example of a Michael addition to produce a cyclopropyl derivative.

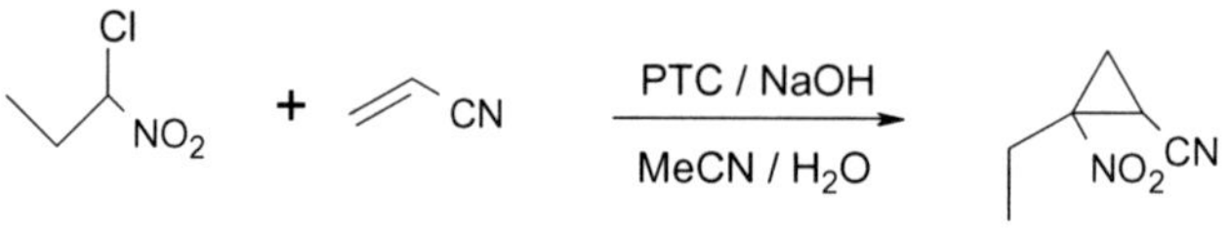

Scheme 4.27 Michael addition employing PTC (phase transfer catalysis).

4.4.3 Oxidation using Hydrogen Peroxide

If hydrogen peroxide can be used in relatively dilute form (30 vol%) it is an excellent environmentally benign oxidant, producing only water as a by-product. One of the major challenges to using dilute aqueous hydrogen peroxide is get the oxidant to the reaction site. The possible advantages of using PTC in these situations are obvious and in many cases may allow hydrogen peroxide to be used in place of higher waste-generating peroxides such as peracetic acid or *t*-butyl hydroperoxide.

There are two possible mechanisms for transferring peroxide to the organic phase. The first, transport of the HO_2^- ion, follows the classical mechanism; however, this anion is strongly hydrophilic and has a high hydration energy and, therefore, does not readily exchange with other anions. There is some evidence that this mechanism operates to a certain extent since addition of base reduces the extraction rate. The major mechanism operating is thought to involve extraction *via* complexes of the type $QX-H_2O_2$. Relatively hydrophobic quaternary salts such as $(C_6H_{13})_4NBr$ are most widely used. In many cases hydrogen peroxide is not involved in the direct oxidation of the organic substrate, the actual oxidant in these cases is a metal complex (commonly W or Mo) that becomes reduced. Here the role of the hydrogen peroxide is to re-oxidize the metal complex *in situ*, enabling the process to become catalytic in metal. In these cases the mechanism becomes complex, with several alternatives being possible, *e.g.* the low-valent metal species may be returned to the aqueous layer for oxidation or it may take place in the organic medium.

The huge potential environmental benefits of this technology are best exemplified through the pioneering work of Noyori and alkene cleavage. Cyclohexene can be reacted with 30% hydrogen peroxide using catalytic amounts of sodium tungstate and methyl(tricetyl)ammonium hydrogen sulfate at moderate temperatures ($\sim 90\,^\circ$C) to give adipic acid in 93% yield (Scheme 4.28). This process is much cleaner than the current method for producing adipic acid (Chapter 6), the commercial viability of this route rests on the long-term price and availability of hydrogen peroxide (conversion of all adipic acid plants would consume several times the current production), but the potential is obvious.

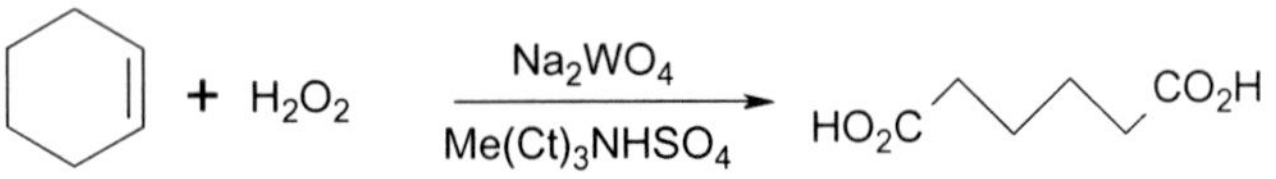

Scheme 4.28   Noyori synthesis of adipic acid.

Table 4.2   Examples of phase transfer catalysed hydrogen peroxide reactions.

Reaction
Epoxidation of alkenes
Oxidative cleavage of α,β-unsaturated ketones
Oxidation of alcohols to carbonyls (secondary alcohols can be selectively oxidized)
Synthesis of aromatic nitro compound from aromatic amines
Oxidation of sulfides to sulfones
Oxidation of benzene to phenol
Conversion of alkynes into carboxylic acids
Bayer–Villiger oxidations to lactones
Conversion of benzylic chlorides into benzyl aldehydes

The scope of reactions involving hydrogen peroxide and PTC is large – some idea of the versatility can be found from Table 4.2. A relatively new combined oxidation/phase transfer catalyst for alkene epoxidation is based on $MeReO_3$ in conjunction with 4-substituted pyridines (*e.g.* 4-methoxypyridine), the resulting complex accomplishing both catalytic roles.

4.5   BIOCATALYSIS

Biocatalysis refers to catalysis by enzymes; the enzyme may be introduced into the reaction in a purified isolated form or as a whole cell microorganism. Enzymes are highly complex proteins, typically made up from 100 to 400 amino acid units. The catalytic properties of an enzyme depend on the actual sequence of amino acids, which also determines its three-dimensional structure. In this respect the location of cysteine groups is particularly important since these form stable disulfide linkages that hold the structure in place. This three-dimensional structure, though not directly involved in the catalysis, plays an important role by holding the active site or sites on the enzyme in the correct orientation to act as a catalyst. Table 4.3 summarizes some important aspects of enzyme catalysis relevant to green chemistry.

Since enzymes are composed of amino acids they may be assumed to act as either acid or base catalysts through groups such as $-COOH$, $-NH_2$, and $-CONH_2$. The scope of activity, however, is enhanced

Table 4.3 Aspects of enzyme catalysis relevant to green chemistry.

Property	*Green chemistry relevance*
Fast reactions due to correct catalyst orientation	Faster throughput
Orientation of site gives high stereospecificity	Possibility for asymmetric synthesis
High degree of substrate specificity due to limited flexibility of active site	High degree of selectivity
Water soluble	Opportunity for aqueous phase reactions
Naturally occurring	Non-toxic, low hazard catalysts
Natural operation under conditions found in body	Energy efficient reactions under moderate conditions of pH, temperature, *etc.*
Possibility for tandem reactions when using whole organisms	Possibility for carrying out sequential one-pot syntheses

considerably through co-ordination with metallic ions found in the body such as Mg^{2+}, Fe^{2+}, Fe^{3+}, Ca^{2+}, and Zn^{2+}. Enzymes have been classified into six functional types according to the reactions they catalyse:

1. Oxoreductases include enzymes such as dehydrogenases, oxidases, and peroxidases, which catalyse transformations such as oxidation of alcohols to carbonyls and dehydrogenation of functionalized alkanes to alkenes.
2. Hydrolases such as the digestive enzymes amylase and lactase catalyse hydrolysis of glycosides, esters, anhydrides, and amides.
3. Transferases include transmethylases and transaminases and transfer a group (*e.g.* acyl) from one molecule to another.
4. Isomerases catalyse reactions such as cis–trans isomerization or more complex transformations such as D-glucose into D-fructose.
5. Lyases catalyse group removal such as decarboxylation.
6. Ligases catalyse bond-forming reactions, typified by condensation reactions.

The diversity and complexity of enzymatic reactions are too great to do justice within the space available; the utility of such reactions will merely be illustrated through three examples. In Chapter 2 the recent, 'greener', commercial synthesis of ibuprofen was described. This synthesis produces ibuprofen for sale in the racemic form and, as pointed out above, it is frequently only one optical isomer of a drug that has the desired therapeutic effect. In this particular case it is the (*S*)-enantiomer that has the desired effect (Scheme 4.29). Hence the commercial route is not as 'green' as may have been thought from the discussion presented in Chapter 2 since 50% of the product is effectively waste. Whilst the (*R*)-enantiomer has no

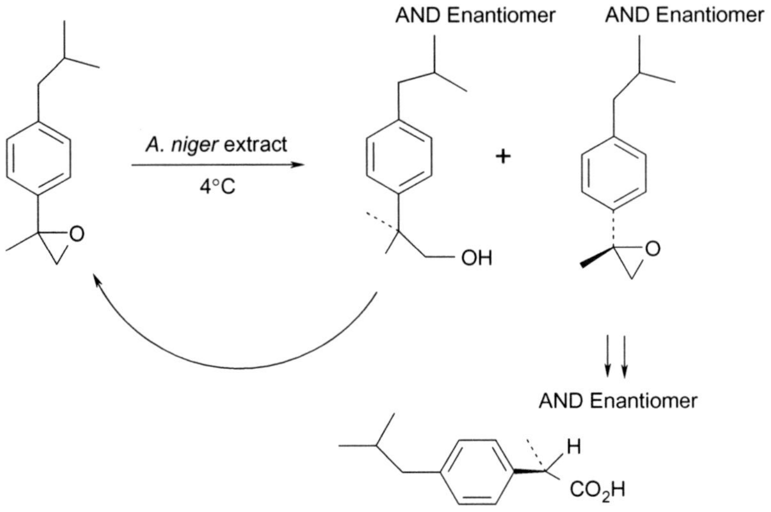

AND Enantiomer　　　AND Enantiomer

+

AND Enantiomer

Scheme 4.29　Synthesis of (*S*)-ibuprofen.

$$HO_2C(CH_2)4CO_2H + HO(CH_2)_6OH \xrightarrow[60°C]{CAL-B} -OC(CH_2)_6OCO(CH_2)_4CO_2-$$

Scheme 4.30　Enzyme-catalysed polyester production.

known adverse effects it does accumulate in fatty tissues. Enzyme extracts from *Aspergillus niger* contain an epoxide hydrolase that has been used to selectively catalyse hydrolysis of the (*R*)-enantiomer of 4-isobutyl-α-methylstyrene oxide, leaving the (*S*)-enantiomer intact for conversion into (*S*)-ibuprofen. Overall this route may not be commercially attractive but it is a useful illustration of the utility and selectivity of enzymes.

Condensation reactions involving lipases are unusual in that they do not work effectively in aqueous environments. The use of such enzymes in condensation polymerizations has been well studied since they offer the potential for efficient low temperature processes; however, the use of expensive solvents often limited the commercial viability. Recently, Baxenden chemicals have produced commercial quantities of poly (hexamethylene) adipate in a solvent-free process using the supported, and hence recoverable and reusable, lipase derived from *Candida antarctica*, CAL-B (Scheme 4.30).

Other than fermentation-type processes the largest commercial use of enzyme catalysis is in the production of acrylamide (Scheme 4.31). In this process the lyase, nitrile hydratase, is used to convert acrylonitrile into acrylamide. The original process developed by Nitto Chemicals

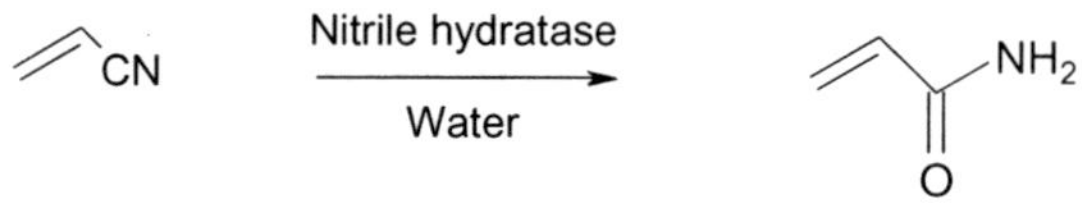

Scheme 4.31 Acrylamide production.

in 1985 used *Rhodococcus* sp. *N-774* but this suffered from thermal instability as well as not being stable at high acrylamide concentrations. Later developments have involved use of immobilized *Rhodococcus rhodococcus J1*, with acrylamide concentrations over 50% being achievable. Many enzymes can carry out this transformation; commercial selection has been limited by the need to avoid acrylic acid by-product formation through amidase activity, which such enzymes also often have. Compared to the conventional copper-catalysed process, the biotechnology route operates at lower temperature, avoids use of pressure reactors, and produces a product with lower acrylic acid residues (this means that more valuable high molecular weight polymers can be made).

Although enzymes are valuable components of the green chemistry toolkit their commercial utility is currently limited by several factors. Most enzymes are only stable under mild conditions, hence moving outside a relatively narrow temperature and pH range often destroys them. Using whole microorganisms (*e.g.* fermentation) is relatively inexpensive but the overall process is often not as selective as ones using single enzymes (or synthetic catalysts). Single enzymes are expensive to isolate and purify and are often not economically recoverable at the end of a reaction. Many enzymes are also easily poisoned and are only effective in highly dilute solutions; both of these factors add to the cost of using biocatalysts. In an attempt to overcome this last limitation work has been carried out using enzymes in organic aqueous tuneable solvents, *e.g.* kinetic resolution of *rac*-1-phenylethyl acetate to (*R*)-1-phenylethanol with lipase B. The product can be recovered with an ee of over 99% using a 30% solution of 1,4-dioxane. Following the reaction CO_2 is bubbled into the reaction mix, causing two layers to form, the hydrophobic product partitions into the organic phase and can be separated and the hydrophilic enzyme partitions into the water phase for reuse.

4.6 PHOTOCATALYSIS

Photocatalysis usually refers to the activation of semiconductors by light; light is not acting as a catalyst but as a source of energy to activate

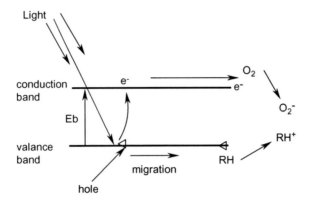

Figure 4.6   Hydrocarbon oxidation using photocatalysis.

the catalyst. When light of a suitable wavelength falls on a semi-conductor it may promote an electron from the valence band to the conduction band (Figure 4.6). For this to happen, the photon must have an energy greater than the band gap energy (E_b). The result of this is that positive holes are generated in the valence band that may migrate to the surface and act as strong oxidizing agents. Titanium dioxide is the most popular semiconductor used due the band gap energy (3.2 eV) being highly accessible by UV light.

Owing to the strong oxidizing potential of photocatalysts in the presence of oxygen and water they are generally not good catalysts for synthesis of chemicals, but they can catalyse the complete mineralization of a range of organic materials. Well over 200 organic compounds have been shown to undergo complete mineralization according to Equation (4.5). Heteroatoms such as chlorine or sulfur are converted into the corresponding mineral acid during the process. This process has huge potential for both *in situ* waste treatment and pollution clean up:

$$C_xH_yCl + O_2 \xrightarrow{\text{photocatalyst}} CO_2 + H_2O + HCl \qquad (4.5)$$

Recently, Japanese companies have pioneered several interesting and valuable commercial applications. Cleaning windows is time consuming, wasteful of water, and can be relatively dangerous when dealing with high tower blocks. By adding a very fine transparent (*i.e.* the particles are smaller than the wavelength of visible light) coating of TiO_2 the to the glass surface the 'ever-clean' window has been developed. In the presence of air and sunlight dirt that normally adheres to windows is completely oxidized to carbon dioxide and water (and trace amounts of inorganic acids). In a similar way self-cleaning paving stones are also being used in some cities. Similar technology is being developed to remove odours rather than dirt; hence bathroom tiles are now available with a thin TiO_2 coating, and air fresheners for cars are being developed. One of the unresolved issues regarding these developments concerns catalyst poisoning and lifetime, since the 'ever-clean' window, for example, may be expected to last 20 years or more to be commercially viable.

In terms of more conventional pollution clean-up a significant amount of research is being carried out into areas such as destruction of explosive waste, oil spill clean up, and removal of NO_x by conversion into nitric acids. Waste water clean up using photochemical reactors has been studied extensively. One of the main issues concerns the nature of the photocatalyst. Reactors with fine suspended micro-particles have generally proved more efficient than ones with supported catalysts; however, the former are difficult to remove by normal filtration techniques. Although many types of photocatalytic reactor have been developed, simple fixed bed reactors using supported catalyst over which the contaminated water flows as a thin film are popular due to the low cost. This type of reactor has proved relatively efficient in treating waste water from a phenol production plant.

Rutile TiO_2 has been shown to be effective for the oxidation 4-methoxybenzyl alcohol to the aldehyde. Conversions up to 74% have been achieved in aqueous suspension, but conversion is very sensitive to the mode of preparation of the photocatalyst.

4.7 CONCLUSIONS

Catalysts are extensively used and have played a huge role in making bulk chemical manufacturing technology more competitive and environmentally friendly. Undoubtedly, catalysis will continue to provide the answer to many economic and environmental challenges currently faced by industry. As indicated above, catalysts now required by the fine chemical and pharmaceutical industries need to be robust, selective, recoverable, and reusable.

REVIEW QUESTIONS

1. With reference to the mechanism of cracking dodecane assess the relative environmental merits of the thermal and catalytic cracking processes to give gasoline grade products.

2. Oxidation reactions are frequently used in the production of both bulk and fine chemicals. Review the main differences in the processes usually used in each sector, discussing these differences in terms of the 12 principles of green chemistry.
3. Devise a synthesis of (S)-ibuprofen that, in terms of the principles of green chemistry, compares favourably with the commercial route outlined in Chapter 2.
4. With reference to two products of your choice discuss the role catalysts have played in the development of a more economic and environmentally friendly bulk chemical industry. Why has catalysis had less impact on the development of fine and pharmaceutical products?

FURTHER READING

S. Bomarius and B. R. Riebel, *Biocatalysis*, Wiley-VCH Verlag, Weinheim, 2004.

G. P. Chiusoli and P. M. Maitlis (ed.), *Metal-Catalysis in Industrial Organic Processes*, Royal Society of Chemistry, Cambridge, 2008.

J. H. Clark and C. N. Rhodes, *Clean Synthesis Using Porous Inorganic Solid Catalysts and Supported Reagents*, Royal Society of Chemistry, Cambridge, 2000.

J. Hagen, *Industrial Catalysis: A Practical Approach*, Wiley-VCH Verlag, Weinheim, 2006.

R. Noyori and T. Ohkuma, Asymmetric catalysis by architectural and functional molecular engineering: practical chemo- and stereoselective hydrogenation of ketones, *Angew. Chem. Int. Ed.*, 2001, **40**, 40–73.

CHAPTER 5

Organic Solvents: Environmentally Benign Solutions

5.1 ORGANIC SOLVENTS AND VOLATILE ORGANIC COMPOUNDS

Organic solvents have played a key role in the development of many useful products. They are used, for example, to produce pharmaceuticals of the required purity, to ensure the easy flow and good finish of gloss paints, to formulate inks that dry successfully, and in aerosol applications. In the latter three consumer applications all the solvent is lost to the atmosphere whereas in industrial applications 'end of pipe' solutions can be installed to recover much of the solvent for reuse, safe disposal, or for energy recovery. In chemical manufacture, organic solvents are widely used in various unit operations, including extraction, recrystallization, and the dissolution of solids for ease of handling. One of the key roles organic solvents play in the chemical industry, however, is that of reactant solvent, allowing the homogenization of a reactant mixture, speeding up reactions through improved mixing, and heat transfer, thereby reducing energy consumption. Solvents also contribute to safety by acting as a heat sink for exothermic reactions. Many of the applications discussed above use 'volatile organic compounds' (VOCs) as solvents due to their ease of removal or evaporation. VOCs have a significant vapour pressure at room temperature and are released from many sources, including process industries and most forms of transport, the latter being responsible for most VOC emissions.

Green Chemistry: An Introductory Text, 2nd Edition
By Mike Lancaster
© Mike Lancaster 2010
Published by the Royal Society of Chemistry, www.rsc.org

The main environmental issue concerned with VOCs is their ability to form low-level ozone and smog through free radical air oxidation processes. The EPA (US Environmental Protection Agency) has published a list detailing several adverse health effects which are now thought to originate from the presence of VOCs in the environment. These include:

- conjunctival irritation,
- nose and throat discomfort,
- headache,
- allergic skin reaction,
- dyspnea,
- declines in serum cholinesterase levels,
- nausea,
- fatigue,
- dizziness.

Owing to these adverse health effects, stringent legislation and, in addition, voluntary control measures are in place to reduce VOC emissions. Nevertheless the worldwide value of the volatile organic solvents market is some £4000 million p.a. Since this figure excludes in-house recycled material it is evident that the vast amount of this solvent is released to atmosphere, or into effluent water. Closer analysis of VOC emission data from solvents reveals that almost half of the emissions come from the surface coating industry, including automotive finishing. Non-industrial processes account for around 40% of emissions; these include pesticide applications as well as consumer solvent and surface coating applications, such as painting. VOC emissions from chemical manufacturing processes are a relatively small fraction of overall emissions, but where these materials are used they often make up a sizeable percentage of the total process waste coming from chemical factories. In addition, flammability and worker exposure concerns entail significant capital expenditure on control measures. Many technical solutions to the VOC problem are being developed, each potentially likely to find niche applications. Some of the more common alternatives to using VOCs are discussed in this chapter, including the use of:

- solvent-free processes,
- benign non-volatile organic solvents,
- water based processes,
- ionic liquids,
- supercritical fluids,
- fluorous biphasic solvents.

Table 5.1   Properties of volatile organic solvents.

Solvent	Boiling point (°C)	Flash point (°C)	TLV-TWA (mL m⁻³)ᵃ	Hazard indicator
Isopropanol	96	15	400	–
Ethyl acetate	76	− 2	400	–
2-Butanone	80	− 3	200	Irritant
1-Butanol	117	12	100	Harmful
Toluene	110	4	100	Harmful
Tetrahydrofuran	65	− 17	200	Irritant
Methanol	64	11	200	Toxic
Dichloromethane	40	None	100	Harmful; suspected carcinogen
Hexane	68	− 22	50	Harmful
Chloroform	61	None	10	Possible carcinogen

ᵃTLV-TWA = threshold limit values – time weighted average in vapour.

Although total elimination of volatile organic solvents from all chemical manufacturing processes is a worthy goal, the pursuit of this goal must be subject to some caution. Alternative organic solvent-free processes may have poor heat and/or mass transfer and/or viscosity limitations, which could result in excessive energy use or the production of less pure products, needing large amounts of organic solvents in subsequent purification steps. Not all volatile organic solvents are equally bad. Table 5.1 is an attempt to broadly rank some of the more common ones in terms of boiling point (ease of containment), flash point, and hazardous nature. Although the choice of solvent will depend on many other factors, not least ease of recycle, ability to handle low flash point materials, price, and, of course, performance, if organic solvents are to be used then ones towards the top of the table are generally preferred on health and environmental grounds.

5.2   SOLVENT-FREE SYSTEMS

It is a misconception that most chemicals are manufactured in organic solvents. Most high volume bulk chemicals are actually produced in solvent-free processes, or at least ones in which one of the reactants also acts as a solvent. Typical examples of such large-scale processes include the manufacture of benzene, methanol, methyl *t*-butyl ether (MTBE), phenol, and polypropylene. In addition, some heterogeneous gas-phase catalytic reactions, a class of solvent-free processes, are discussed in Chapter 4.

Although now not widely used (Chapter 3), MTBE has made a significant contribution over the last 20 years to the lowering of VOC

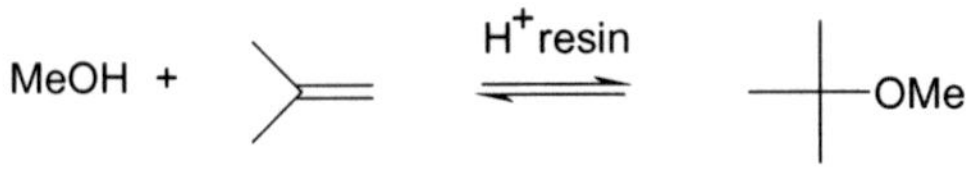

Scheme 5.1   Synthesis of MTBE (methyl *t*-butyl ether).

emissions from car exhausts. This is due to its clean burn properties (producing fewer hydrocarbon by-products). MTBE is commonly produced in a fixed-bed reactor by passing a mixture of 2-methylpropene and excess methanol over an acidic ion exchange resin (Scheme 5.1). The reaction takes place in the liquid phase at temperatures between 70 and 90 °C and pressures upwards of 8 atm. Depending on the feed, conversions of over 90% are achieved. The 2-methylpropene feed comes from two sources. The majority comes from the cracking of crude oil, from which the resulting mixed C_4 stream produced is used directly. This contains around 30% 2-methylpropene, the remainder being unreactive hydrocarbons such as butane. A significant amount of 2-methylpropene now comes from the co-production of propylene oxide *via* epoxidation of propene with *t*-butyl hydroperoxide. In this case pure 2-methylpropene is used in the production of MTBE and lower reaction pressures are required.

In general, many reactions involving miscible or partially miscible reagents can proceed under solvent-free, often mild conditions. Solvents are sometimes unnecessarily used due to the process being directly scaled up from laboratory studies with inadequate process development. This is an excellent example of how multi-disciplinary teams could be critical in helping to devise alternative options. The challenge now being addressed by many researchers is the solvent-free synthesis of more complex fine chemicals that are solids, and that involve the use of some solid reagents with high melting points. If such solvent-free reactions also produce yield or selectivity improvements then there are obviously additional triple bottom line advantages to be gained. One quite remarkable example of this is Raston's synthesis of complex pyridines (Scheme 5.2). The route involves sequential solvent-free aldol and Michael addition reactions, which both proceed in high yield. The aldol reaction is carried out by grinding together solid sodium hydroxide with a benzaldehyde and acetyl pyridine (both usually liquids). The product of the reaction is, however, a solid that forms after a few minutes. On further grinding of the un-purified product, with either the same or a different acetyl pyridine, a Michael reaction takes place. Both reactions proceed quantitatively, compared to the less than 50% yields normally achieved in solvent based (ethanol) synthesis. One of the advantages of this

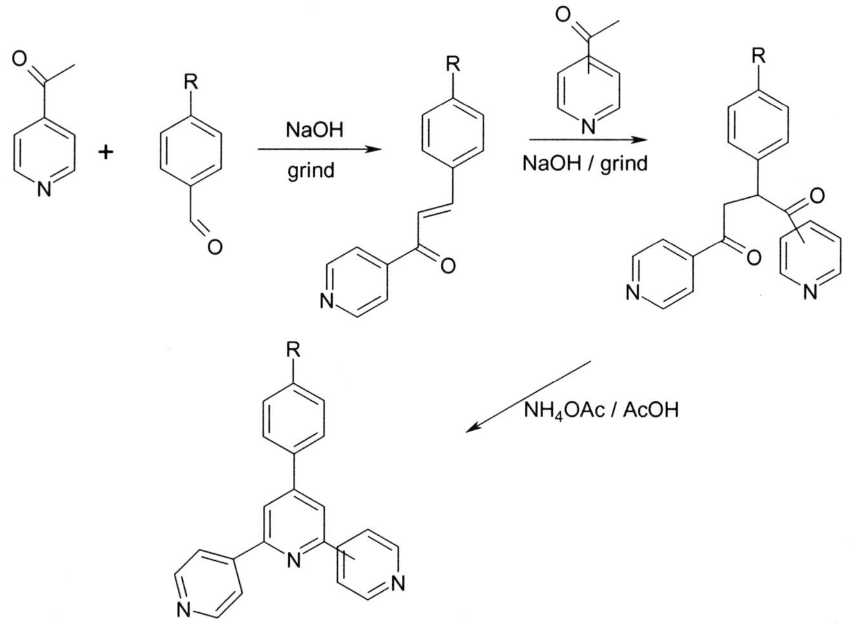

Scheme 5.2 Solvent-free pyridine synthesis.

approach is that purification by recrystallization in organic solvents is not required before proceeding with the final synthesis step, which involves treatment with ammonium acetate in acetic acid. In some instances such methods have yielded products that do not form when solvents are used.

Grinding with a pestle and mortar has become the established laboratory technique for many solvent-free syntheses. Friedel–Crafts reactions are mentioned many times in this book as examples of environmentally unfriendly reactions usually employing non-recoverable Lewis acid catalysts and chlorinated solvents. The laboratory grinding technique has been successfully used to carry out the reaction between benzene and 2-bromopropane using solid aluminium chloride as a catalyst – to give the tri-alkylated product. Whilst this avoids the use of chlorinated solvents, organic solvents are still required to extract the product following the water quench. Few, if any, of these 'grinding' reactions have reached commercial scale; however, one possible commercial reactor type is the ball mill, which is widely used for grinding ores. Some work has been carried out in this type of reactor, *e.g.* methyl methacrylate has been polymerized by grinding with a talc catalyst.

The potential for solvent-free synthesis is relatively large with examples of many well-known reaction types proceeding quite well

under this type of regime; these include transesterification, condensation, and rearrangement reactions. Many workers have moved away
from conventional thermal sources for providing the energy needed for
these reactions – since heat transport is often poor. Microwave and
photochemical sources of energy have proved valuable, often further
enhancing yield and selectivity through better energy targeting. Several
examples of these emerging technologies will be discussed later. Whilst
more commercial applications for solvent-free processes will be
developed for fine and pharmaceutical chemicals the use of some kind of
solvent will continue to be required for other chemical and practical
reasons; here the emphasis will be on finding suitable more environmentally benign solvents.

5.3 SUPERCRITICAL FLUIDS

A supercritical fluid (SCF) can be defined as a compound that is above
its critical pressure (P_c) and above its critical temperature (T_c). Above T_c
and P_c the material is in a single condensed state with properties between
those of a gas and a liquid. Simplistically, the process can be viewed as
the coming together of the densities of the liquid and gaseous phases
along the co-existence line shown in Figure 5.1. As the temperature of a
liquid rises it becomes less dense and as the pressure of a gas rises it
becomes more dense, at the critical point the densities become equivalent. In general SCFs have densities nearer to liquids and viscosities
similar to gases, leading to high diffusion rates. The properties of the

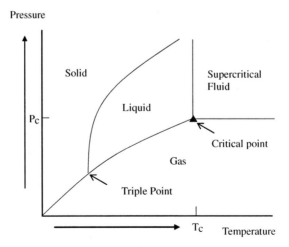

Figure 5.1   Phase diagram showing the supercritical fluid region.

Table 5.2 Critical points of some common supercritical fluids.

Material	T_c (°C)	P_c (bar)
Ammonia	132.4	113.2
Carbon dioxide	31.1	73.8
Ethane	32.2	48.7
Ethene	9.2	50.4
Fluoroform	25.9	48.2
Propane	96.7	42.5
Water	374.2	220.5

fluid can be adjusted by altering the temperature and pressure, as long as they remain above their critical points. In principle, by increasing the pressure at the critical temperature solids can be formed, for most materials the pressures required to do this are very high, *e.g.* 5700 bar for CO_2. The critical temperature and pressure vary very widely from material to material (Table 5.2) and for many applications this has historically limited the practical utility of materials, notably water, which has a particularly high critical temperature and pressure.

Supercritical fluids have been known for well over a 100 years but it is only in the last 20 years or so that their huge potential has been recognized. That said, a small number of high temperature and pressure processes have been operated in the supercritical region for many years. Two of the most widely known ones are:

- High temperature and pressure free radical polymerization of ethene, to produce low density polyethylene (LDPE).
- The Haber process for ammonia manufacture, which operates above the critical point of ammonia.

Until the mid-1980s these were two of the few processes operating under supercritical conditions. Although these processes were not specifically developed to operate under supercritical conditions, nevertheless the advantages have since become clear. Typically the key advantages of carrying out a process under supercritical conditions include:

- improved heat & mass transfer due to high diffusion rates and low viscosities;
- the possibility of fine tuning solvent properties by varying temperature & pressure;
- a potentially large operating window in supercritical region;
- easy solvent removal and recycle.

As noted previously, different materials have very different critical points, which are more or less accessible: So what is the real value of supercritical fluids in Green Chemistry? It can, somewhat simplistically, be argued that if an advantage such as increased reaction rate or increased solubility is brought about by the use of a solvent in the supercritical state then this can be considered a 'green' improvement. However, it may be the case that the, often significant, energy requirements of operating supercritical processes outweigh any in-process environmental benefits. In most cases the reference to supercritical fluids being 'green' refers to the replacement of an organic solvent with a more benign supercritical fluid, notably carbon dioxide or water; here real environmental as well as technical advantages can be obtained. The remainder of this section is restricted to a discussion of these two fluids.

5.3.1 Supercritical Carbon Dioxide (scCO$_2$)

Recent interest in SCFs and scCO$_2$ in particular arose out of the 1970s energy crisis. Separation processes involving distillation are amongst the most energy intensive processes operated by the chemical industry. If the separation can be carried out by extraction into a solvent that does not need to be removed by distillation there is the potential for saving energy. Carbon dioxide has many ideal characteristics for this type of application, not least its relatively accessible triple point and the fact that it can be removed simply by reducing the pressure. Notably, unlike CO$_2$ produced from burning fossil fuels, CO$_2$ released from its use as a solvent does not give a net contribution to global warming. This is because the CO$_2$ used is an industrial by-product (often from ammonia manufacture, or even breweries) that would normally have been released to atmosphere.

The two main uses for scCO$_2$ are its use as an extraction solvent and as an in-process solvent. Another, as yet small scale but environmentally significant, use is as a solvent/dispersion medium for spray coating. Table 5.3 shows some of the many advantages and a few disadvantages of using scCO$_2$ for these applications. In terms of reaction chemistry one of the greatest advantages of scCO$_2$ is its miscibility with gasses arising from the gas-like nature of the fluid. This can lead to significant rate enhancements in reactions such as hydrogenation compared to the use of conventional organic solvents that are relatively poor solvents for hydrogen. Whilst this is a definite advantage it can pose heat transfer challenges for particularly exothermic reactions. Carbon dioxide has interesting and unusual solvent properties. Being non-polar it may be expected to be a good solvent for hydrocarbons, but this is generally not

Table 5.3 Advantages and disadvantages of using scCO$_2$ as a solvent.

Advantages	Disadvantages
Non-toxic	Relatively high pressure equipment
Easily removed	Equipment can be capital intensive
Potentially recyclable	Relatively poor solvent
Non-flammable	Reactive with powerful nucleophiles
High gas solubility	Possible heat transfer problems
Weak solvation	
High diffusion rates	
Ease of control over properties	
Good mass transfer	
Readily available	

the case due to its high quadrupole moment. Several ways have been identified to overcome this limitation, including the use of small amounts of co-solvent. However, the high quadrupole moment means that scCO$_2$ is a relatively good solvent for small polar molecules like methanol and caffeine. The lack of polarity does have a positive influence on many reaction rates since CO$_2$ does not readily co-ordinate to many catalysts or solvate complexes. Although CO$_2$ is relatively inert it does react with good nucleophiles such as amines, which means it can not be used as a solvent for certain reactions.

5.3.1.1 Extraction Processes. One of the most widely established processes using scCO$_2$ is in the decaffeination of coffee. Prior to widespread use of this process in the 1980s the preferred extraction solvent was dichloromethane. The potential adverse health effects of chlorinated materials were realised at this time and although there was no direct evidence of any adverse health effects being caused by any chlorinated residues in decaffeinated coffee there was always the risk, highlighted in some press 'scare' stories. Hence the current processes offer health, environmental, and economic advantages.

Figure 5.2 depicts the basic process outline; moist un-roasted coffee beans and CO$_2$ are fed counter currently into the extractor under supercritical conditions. Caffeine is selectively extracted into the CO$_2$ and this stream is lead to a water-wash column to remove caffeine, at a reduced pressure, with the CO$_2$ being recycled back to the extraction column. Extraction of the caffeine into water is necessary to avoid dropping the CO$_2$ pressure too low, since compression is energy intensive. There is now the problem of separating the caffeine (which is used in soft drinks and pharmaceuticals) from the water. Distillation is also an energy intensive process; hence many modern plants use reverse

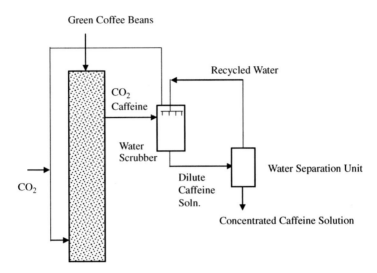

Figure 5.2   Essentials of coffee decaffeination process.

osmosis or membrane technology to concentrate the aqueous solution.
There are two other interesting observations to make about the process.
First, whilst scCO$_2$ selectively extracts caffeine from green coffee beans it
co-extracts many of the aroma oils produced on roasting if carried out
after the roasting process. Second, some moisture is required in the
extraction process, this is thought necessary to free the caffeine from
complexes within the bean in some way; moisture was also a requirement in the dichloromethane process.

Owing to the benign nature and efficiency, this type of small- to me-
dium-scale extraction process is becoming increasingly common in the
food and flavour and fragrance industries. Examples include extraction
of flavours from hops, extraction of essential oils, and defatting nuts and
fried goods. As an example of the latter, the current demand for healthy
tasty snacks like potato crisps is growing rapidly. Conventional potato
crisps are highly calorific, with some brands containing over 45% fat.
Within the normal cooking and processing of crisps there is limited
scope for significantly reducing the fat content. By using scCO$_2$ the fat
content can be reduced by up to 50% without, it is claimed, any loss of
flavour. This relatively high solubility of fatty acid triglycerides in scCO$_2$
is also central to several studies looking at separating vegetable oils from
soya protein. This can be efficiently achieved provided the cell walls are
broken down prior to the extraction process.

One other successful commercial application for high pressure CO$_2$
technology that may be considered an extraction process is dry cleaning.
Traditionally most dry cleaning processes have used chlorinated

solvents, initially highly hazardous carbon tetrachloride and now per-chloroethylene (perc). Even though solvent recovery and recycling is efficient there are many environmental and health concerns surrounding the process. Contaminated land from previous dumping of used perc (resulting in contaminated drinking water) and the fact that it is a suspected carcinogen are two big concerns. With some 180 000 dry cleaning units worldwide using perc, adoption of CO_2-based cleaning processes could make a significant overall impact. Carbon dioxide has some technical advantages over perc: items that cannot be dry cleaned with perc, such as leather, fur, and some synthetics, can be safely cleaned with CO_2. A second claimed advantage is the improved colourfastness of some garments. Repeated traditional dry cleaning does remove small amounts of certain dyes, gradually altering the colour of the garment over time; this is claimed not to be the case with the CO_2 system.

Strictly speaking most high pressure CO_2 cleaning process operate at sub-critical temperatures and pressures, in liquid CO_2, for economic reasons, but as the liquid is relatively close to the critical point many of its properties, such as low viscosity and low surface tension, are similar to $scCO_2$. It is these two properties that are primarily responsible for the efficiency, since the CO_2 is readily able to penetrate the fibrous structure of the garments being cleaned. Because of the solvent properties of $scCO_2$ it may be expected to be a good dry cleaning solvent for fatty and greasy stains but less good for stains originating from red wine, egg yoke, *etc.* To overcome this commercial systems also use a surfactant. These surfactants differ from those used in wet cleaning processes since they must be compatible with the stain and CO_2. The actual surfactants used are proprietary but are thought to contain fluorinated functions to make them CO_2 compatible; such surfactants are expensive and are largely recycled. Related cleaning processes are now being developed for other applications, some of these are far reaching and could provide significant environmental advantage, *e.g.* degreasing of electronic components.

5.3.1.2 Supercritical CO_2 as a Reaction Solvent.

The use of $scCO_2$ as a reaction solvent is an area of significant research activity. The previous lack of attention is at least in part due to the difficulties of carrying out such high pressure experiments in university laboratories. One of the most studied areas is polymerization; here supercritical fluids afford the possibility of obtaining polymers of different molecular weights by altering the density of the medium through simple variation of pressure. Polymerization of fluorine- and silicon-containing monomers has been particularly well studied; this has largely been driven by the lack of solubility of these materials in organic solvents.

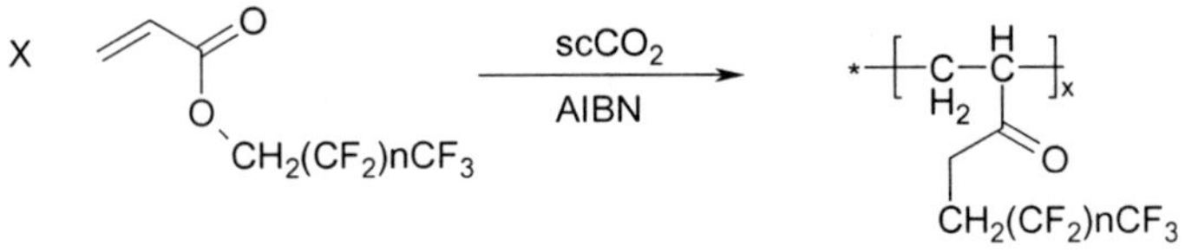

Scheme 5.3   Fluorinated polyacrylate synthesis in scCO$_2$.

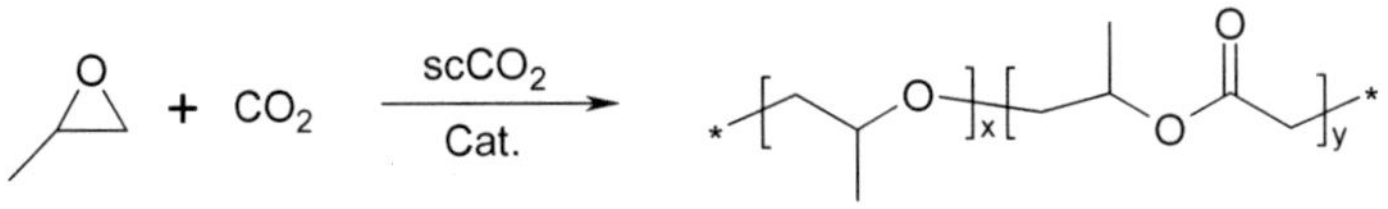

Scheme 5.4   Polyether–polycarbonate synthesis in scCO$_2$.

Free radical polymerization (using a free radical initiator such as AIBN) of acrylate monomers containing perfluoro-ponytails proceeds well at a temperature of around 60 °C and pressures over 200 bar (Scheme 5.3). Interestingly, the initiator decomposes more slowly in scCO$_2$ than in more conventional solvents; however, overall it is more efficient due to the lack of cage effects in the low viscosity medium. One of the challenges of polymerizing polar materials such as acrylates in scCO$_2$ is to build up the molecular weight to the stage where they become commercially attractive before they fall out of solution. By use of small amounts of dispersing agents, essentially surfactants, having CO$_2$ compatible groups (*e.g.* siloxanes) the polymer can be kept in solution until useful molecular weights are obtained. ScCO$_2$ is also an ideal inert solvent since there is virtually no chain transfer, even from highly electrophilic radicals. This effect has been put to use in the polymerization of tetrafluoroethene. Dupont have recently commercialized a process for producing PTFE in scCO$_2$, thereby replacing the use of ozone-depleting chlorofluorocarbon solvents. Although more soluble in scCO$_2$ than the acrylate system discussed above, the build up of sufficient molecular weight is still a concern. In this case small amounts of co-monomer can be added to disrupt the crystallinity, the amorphous polymer being more soluble, particularly at high temperature.

Several examples of polymerization of non-fluorinated materials have been developed. Through a thorough analysis of polymer–scCO$_2$ interactions, involving both the entropy and enthalpy, it has been possible to prepare 'designer' polymers that may lead to the development of more commercially useful systems. One example of this is the synthesis of polyether-polycarbonates from propylene oxide, using CO$_2$ both as a solvent and reagent, in the presence of a heterogeneous zinc or aluminium catalyst (Scheme 5.4). This polymer is highly flexible (favourable entropy of mixing), due to the presence of carbonate groups, and only has weak inter-chain polymer–polymer interactions, both features aiding solubility in the solvent.

PET, from polymerization of ethylene terephthalate, has also been synthesized in $scCO_2$. In this case the CO_2 swells the polymer, acting as a plasticizer, aiding the removal of co-produced ethylene glycol. Several other polymerization processes have also been found to proceed well, often advantageously in $scCO_2$; these include ring-opening metathesis polymerization of norbornenes and the cationic polymerization of iso-butene, which is conventionally carried out in chlorinated solvents. Owing to the high pressures involved, supercritical conditions may offer significant rate advantages for addition polymerization processes, since these have negative intrinsic activation volumes. Possibly the greatest benefit of carrying out polymerization reactions under supercritical conditions is the ability to fractionate polymers of a narrow molecular weight range. This arises from the ability to closely control the density of the medium and hence polymer solubility through variation of pressure.

Palladium-catalysed carbon–carbon bond forming reactions such as the Heck and Suzuki reactions are versatile and efficient methods for synthesis of fine and pharmaceutical intermediates, but such reactions often suffer from catalyst separation problems. As well as avoiding the use of organic solvents, by use of carefully designed fluorine-containing phosphine ligands, $scCO_2$ offers a potential solution to this problem. In principle, the product can be separated from the reaction mix whilst active catalyst remains in solution. The main draw back with this approach is that the fluorinated phosphine ligands are very expensive and difficult to prepare. Recent work has, however, shown that similar results can be obtained using more conventional phosphines with fluorinated Pd sources [*e.g.* $Pd(OCOCF_3)_2$]. Scheme 5.5 shows some examples of reactions carried out. The first example, involving amino-iodobenzene, is interesting in that the CO_2 affords protection to the amino group *via* formation of carbamic acid, thereby avoiding the need for an additional reaction step involving an ancillary reagent.

Hydrogenation is one of the most well studied synthetic reactions in $scCO_2$, one of the main technical reasons for which is the high miscibility of H_2 and $scCO_2$ compared with its solubility in organic solvents. This high miscibility overcomes the mass transfer limitations that often control the overall rates of many hydrogenation processes. A consequence of this, however, is the often reduced solubility of reagents that now may become rate determining. In the presence of a suitable catalyst CO_2 itself may be hydrogenated to formic acid; in most cases this will be an unwanted side-reaction and catalysts should be chosen to avoid this

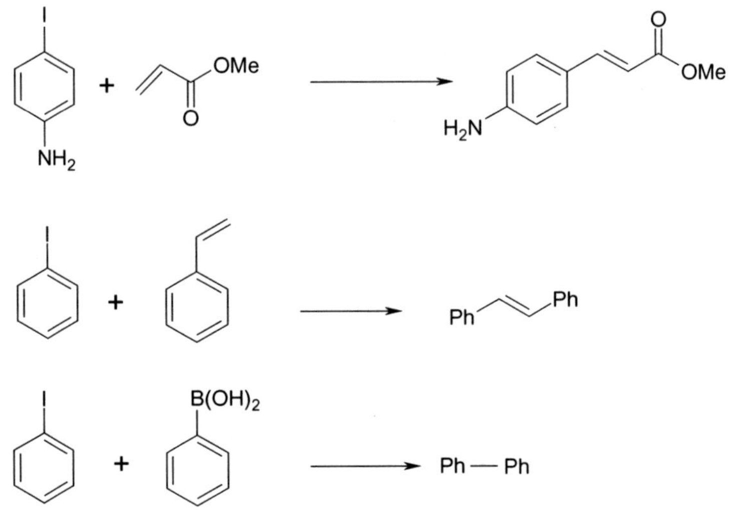

Scheme 5.5   Examples of Pd carbon–carbon bond formation in scCO$_2$.

complication. On the other hand, if the reaction can be made to go efficiently it could become a very 'green' source of formic acid, using up unwanted CO$_2$. Ruthenium/phosphine catalysts are efficient for carrying out this reaction but the insolubility of triarylphosphine ligands in scCO$_2$ has proved problematic. More success has been obtained using the more soluble P(Me)$_3$ ligand.

Many homogeneous and heterogeneous-catalysed hydrogenation reactions have been studied in scCO$_2$ with high yields generally being obtained. The scope of homogeneous-catalysed reactions can be limited by catalyst solubility problems highlighted above; however, heterogeneous catalysts have been found to be very effective. Rate enhancement in heterogeneous systems is thought to be largely due to improved mass transport within catalyst pores brought about through the low viscosity of the medium. Several enantioselective hydrogenations have also been carried out using chiral catalysts. Although the enantioselectivity can be optimized to a certain degree by control of reactions conditions there are few cases in which the use of supercritical conditions has significantly enhanced enantioselectivity. Scheme 5.6 highlights some of the reactions that have been successfully carried out.

One of the main commercial blocks to more widespread use of supercritical fluids has been the cost of large high pressure reactors. Owing to the rapid reaction rates found, high throughputs can be obtained from bench top size reactors, significantly reducing overall plant cost. By having several of these small reactors operating in series

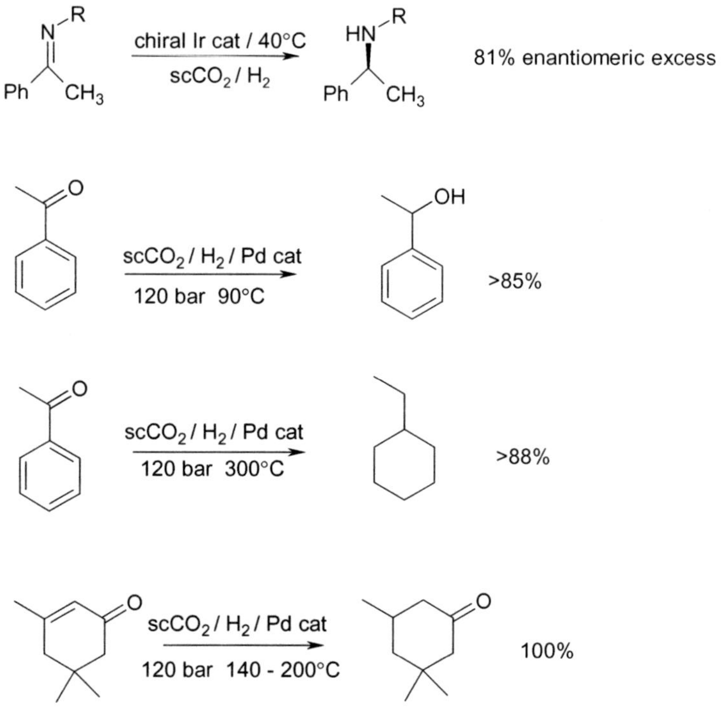

Scheme 5.6 Some high yielding hydrogenation reactions carried out in scCO$_2$.

throughputs of tens of thousands of tonnes per year can be achieved in a continuous process over a fixed-bed catalyst. Figure 5.3 shows a schematic of this type of reactor. Such reactors are, of course, much simpler to operate if both product and starting material are liquids and the starting material is soluble in the reaction medium. In situations were this is not the case a co-solvent such as methanol may be employed; this of course does raise some environmental concerns since total containment of a volatile organic solvent when the pressure is dropped from something over 100 bar is not simple. As will be evident from the hydrogenation of acetophenone shown in Scheme 5.6; a great deal of reaction control may be exerted to obtain the product of choice. This 'tunability', which may be achieved through control of pressure, temperature, and CO$_2$/H$_2$ ratio as well as catalyst and co solvent, is one of the overriding aspects to the commercial viability of these flexible, multi-purpose plants. In the case of acetophenone, total hydrogenation to ethylcyclohexane is achieved through variation of temperature and CO$_2$/H$_2$ ratio.

The scope of reactions that can be potentially advantageously carried out in a scCO$_2$ medium is too large to fully cover in the space

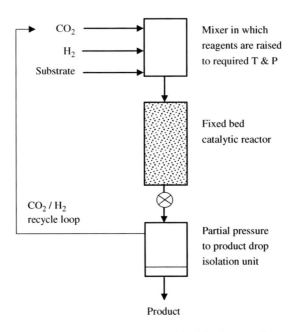

Figure 5.3 Schematic of a continuous supercritical hydrogenation reactor.

available – these reactions include rearrangements, hydroformylation, Friedel–Crafts, esterification, and chlorination. The reader is referred to more thorough texts in the Further Reading section. Discussion will be limited to two other important reaction types, namely, the Diels–Alder and oxidation reactions. The green benefits of Diels–Alder reactions are one of the underlying themes of this book; in some cases solvent-free conditions can be used, in others organic solvents that have to be re-moved at the end of the reaction are employed. Other than avoiding the use of organic solvents it has been found that employing $scCO_2$ affords the possibility for tuning the *endo:exo* ratio for many reactions, by varying the pressure and hence the density of the reaction medium. Although it is not always clear why this ratio should vary it has been postulated that it is related to differences in partial molar volumes of *endo* and *exo* isomers. One of several examples of this is the reaction of cyclopentadiene with *t*-butyl acrylate (Scheme 5.7). In organic solvents such as toluene or chloroform the *endo:exo* ratio is around 10:1 whilst in $scCO_2$ ratios as high as 24:1 can be achieved. Conversely, by carrying out the reaction at low densities (around $0.3\,\mathrm{g\,mL^{-1}}$) lower ratios than those obtained in organic solvents can be obtained. This tunability has obvious economic and environmental advantages.

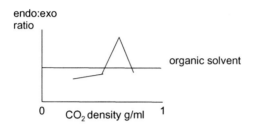

Scheme 5.7 Tuning *endo:exo* ratios using scCO$_2$ density.

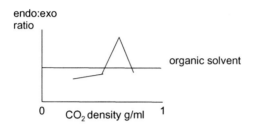

up to 95% diastereomeric excess

Scheme 5.8 Selective oxidation using scCO$_2$.

The non-flammability of scCO$_2$ makes it an attractive medium for carrying out oxidation reactions outside the explosive limit. Selective propane oxidation to acrylic acid or acrolein is of commercial interest but currently selectivity to acrolein is relatively low even at low conversions. The air oxidation of propene has been carried out in scCO$_2$ at temperatures up to 250 °C using cobalt catalysts. At low conversions total oxidation to CO$_2$ and water can be controlled but a whole range of oxidation products is often formed, with acetic acid being a major product. Other than providing an inert reaction medium it is unclear what role the scCO$_2$ is playing. Oxidation reactions using *t*-butyl hydroperoxide have proved more successful in scCO$_2$; for example, several olefins have been converted into epoxides almost quantitatively. Oxidation of cysteine derivatives to the sulfoxide (Scheme 5.8) proceeds with high diastereoselectivity in scCO$_2$, the actual diastereomeric excess being tunable with solvent density. Under conventional conditions (dichloromethane and Amberlyst 15 resin at 25 °C) no diastereomeric excess was observed.

5.3.1.3   Other applications.   Before leaving this topic it is worth briefly mentioning two other applications for scCO$_2$. Several techniques use the properties of supercritical fluids for control of particle size during crystallization procedures. One of these has been termed RESS (rapid expansion of supercritical solution). As the name implies the process consists of rapidly dropping the pressure of a saturated solution of scCO$_2$. This is normally done by passing the solution *via* a pressure relief valve and nozzle into a collection chamber. As the solid comes out of solution very rapidly there is no time for nucleation or crystal growth, hence the particle size is small and uniform. Such processes are particularly useful to the pharmaceutical industry since no contaminating solvent is involved and have been used to produce fine drug particles for injection.

Union Carbide has a successful patented process for spray coating from scCO$_2$. The environmental advantages of this are particularly good for situations in which it is impractical to recover solvent. The process, which is quite complex from an engineering viewpoint, uses decompression atomization technology to produce fine droplets. The technique is currently used by the US Navy for spray coating ships and is being evaluated by several car manufacturers. The coating equipment can be made portable, opening up possibilities for 'painting' bridges and other large outdoor structures.

5.3.2   Supercritical Water

In contrast to scCO$_2$, the conditions required to obtain scH$_2$O are harsh. In particular the temperature requirement of 374 °C precludes it synthetic utility for most organic compounds. That said syntheses in subcritical but high temperature water are well studied and will be covered in the following section. Since many natural minerals and precious stones formed in water at high temperature and pressure in the Earth's crust the synthesis of inorganic solids in scH$_2$O has been relatively well studied. The most notable commercial success in the area is the synthesis of quartz crystals for mobile phones. Typically such reactions between silica and sodium hydroxide are carried out around 400 °C and 700 bar. Under these conditions water is highly corrosive to most steel types; fortunately, in the quartz process an inert compound, NaFeSiO$_4$, rapidly forms a coating on the reactor wall, affording protection.

Many attempts have been made to use similar technology to produce valuable gems. In fact graphite has been converted into diamond at 800 °C and 1700 bar. The reaction is, however, very slow and the overall cost is many times that of mined diamonds. Emerald is thought to be commercially produced using scH$_2$O by reacting alumina, silica, and beryllium hydroxide with hydrochloric acid. Other valuable minerals may also be made economically under what has become called 'hydrothermal' conditions. For example, a phosphate called KTP ($KTiOPO_4$) is being increasingly used in electronic applications, especially solid state lasers.

Since most organic compounds are not stable in scH_2O under oxidizing conditions it has potential use in remediation and waste treatment applications, the technique being referred to as SCWO. Although waste treatment using high temperature water oxidation processes is relatively well established, the use of scH_2O increases the scope of products that can be mineralized and speeds up the process to the extent that most organic materials can be completely oxidized within 2 min. Decontamination of soil impregnated with 'difficult to treat' industrial waste such polyaromatic hydrocarbons and polychlorinated biphenols has been carried out efficiently. In most cases the organic species can be removed to an extent greater than 99.95%. Many organic pollutants also contain heteroatoms such as P and S as well as metals; any waste treatment process must also convert these into benign materials. Pyridine is one of the more resistant heterocyclic materials to deal with; however, it is readily destroyed in SCWO reactors at around 600 °C. An additional benefit of using SCWO technology to destroy nitrogenous pollutants is that any nitrogen oxides formed are reduced to nitrogen at temperatures close to 600 °C.

Reactor cost, due to the requirement for specialist materials such as Inconel 625 or titanium, to prevent corrosion is significant. This is, however, partially offset by the high throughput that can be obtained. The cost of equipment is also related to the operating temperature and pressure. To reduce the temperature required for complete mineralization to below 400 °C whilst still maintaining a high throughput several heterogeneous oxidation catalysts have been assessed. Most of these have been transition metal oxides, *e.g.* TiO_2, V_2O_5, and Al_2O_3. The commercial catalyst Carulite (MnO_2/CuO on alumina) has shown exceptional performance for the complete rapid oxidation of phenol and other 'difficult' substrates at temperatures just above T_c. The first full-scale SCWO plant has been commercialized by Huntsman and it is expected that the technology will now become more mainstream as the value of different kinds of supercritical fluid technology becomes generally more widely appreciated and cost-effective.

5.4 WATER AS A REACTION SOLVENT

Although the corrosive properties of supercritical water combined with the high temperatures and pressures required for its production have

Table 5.4 Advantages and disadvantages of using water as a solvent.

Advantages	*Disadvantages*
Non-toxic	Distillation is energy intensive
Opportunity for replacing VOCs	Contaminated waste streams may be
Naturally occurring	difficult to treat
Inexpensive	High specific heat capacity – difficult to
Non-flammable	heat or cool rapidly
High specific heat capacity – exothermic	
reactions can be more safely controlled	

limited its use as a reaction solvent, the study of subcritical water as a solvent is of growing interest. From a 'green chemistry' viewpoint the use of water as a solvent has many advantages but also some disadvantages (Table 5.4). Notably, it is important to study the whole manufacturing process not just the reaction stage. Production of a contaminated aqueous waste stream can have significant environmental and economic impacts, *e.g.* concentration of contaminated water streams by distillation is very energy intensive compared to, say, concentration of a propanol waste stream. The merits of replacing organic solvents by water should be viewed on a case by case basis. In many processes aqueous effluents are created through vessel cleaning; in these cases an aqueous reaction effluent may not pose any additional problems.

Even though synthesis of organic chemicals in nature occurs very efficiently in water, traditionally chemists have been taught that water is not generally a good solvent for carrying out synthetic reactions, either due to its poor solvent properties or the hydrolytic instability of reagents or products. Water has many interesting properties though that are now being exploited in synthetic chemistry. As the temperature of water is raised the ionic product increases whilst its density and polarity decrease. Above 200 °C (in the liquid state) water starts to take on many of the properties of organic solvents and at the same time becomes a stronger acid and base, *e.g.* at 300 °C water has solvent properties similar to acetone. These effects are related to the reduction in hydrogen bonding of water at higher temperatures.

Replacement of organic solvents by water may be made for environmental, cost (*e.g.* reduction in raw materials and VOC containment costs), or technical reasons. In the flavour and fragrance industry the presence of even trace amounts of volatile impurities can be detected by the expert 'nose'. Here significant process costs are entailed in ensuring

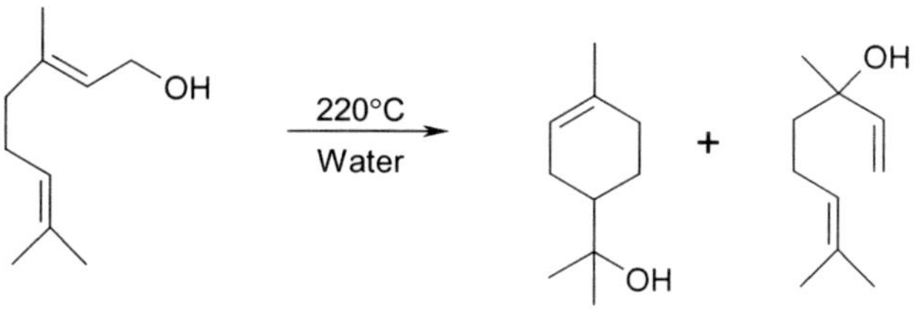

Scheme 5.9 Isomerization of geraniol in water to terpinol and linalol (right-hand product).

complete removal of solvent. If reactions can be carried out in water then these additional costs can be saved. As an example geraniol can be isomerized to the important fragrance intermediates α-terpinol and linalol in water at 220 °C (Scheme 5.9).

For some reactions selectivity improvements and/or significant rate enhancements can be obtained by conducting the reaction in water. An important example of the latter, which sparked much interest in the use of water as a solvent for Diels–Alder reactions, was the finding by Breslow that reaction between cyclopentadiene and butenone was over 700 times faster in water than in many organic solvents. This increased rate has been attributed to the hydrophobic effect. Owing to the difference in polarity between water and the reactants, water molecules tend to associate amongst themselves, excluding the organic reagents and forcing them to associate together, forming small drops, surrounded by water. A further method of increasing the rate of Diels–Alder reactions in water is the so-called salting-out effect. Here a salt such as lithium chloride is added to the aqueous solution. In this case water molecules are attracted to the polar ions, increasing the internal pressure and reducing the volume. This has the effect of further excluding the organic reagents. For reactions such as the Diels–Alder, which have negative activation volumes, the rates are enhanced by this increase in internal pressure, in much the same way as expected for an increase in external pressure (Scheme 5.10).

The increase in acidity and basicity of water at high temperatures often means that lower amounts of acid or base can be used in the process, which in turn results in lower salt waste streams. Scheme 5.11 illustrates such a hydrolysis process and highlights the synthetic versatility and tunability that can be obtained. At 200 °C indole carboxylic acid esters are rapidly hydrolysed in high yield in the presence of small amounts of base but at 255 °C the resulting carboxylic acid is decarboxylated in over 90% yield in under 20 min. This decarboxylation step also has environmental advantages when compared to more usual

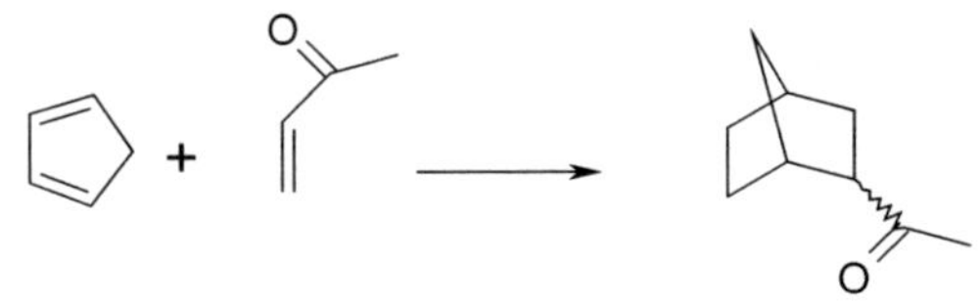

Solvent	Approx. relative rate
isooctane	1
methanol	12.5
water	740
water / LiCl	1800

Scheme 5.10   Enhancement of Diels–Alder rates in water.

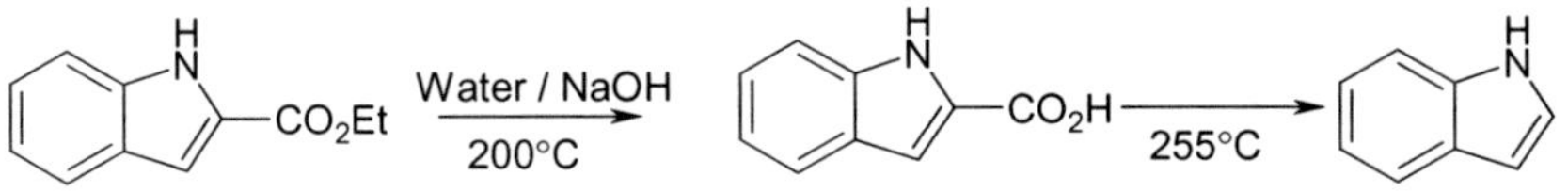

Scheme 5.11   Indole synthesis in high-temperature water.

methods involving use of copper catalysts in non-volatile organic bases at high temperatures.

Some reduction reactions, notably involving iron and hydrochloric acid or sodium dithionite, have been carried out in water for many years, but the significant by-product streams from these processes negate many of the environmental advantages of using water. Recently, examples of heterogeneously catalysed reduction processes using water as a solvent have been developed. In some of these processes sodium formate has been used as the reducing agent, which, in some circumstances, may be a safer and more convenient source of hydrogen than using the gas directly. Scheme 5.12 shows two examples. The example of geraniol is somewhat unusual since many alcohols, perhaps surprisingly, are dehydrated by high temperature water; for example, good yields of cyclohexene can be obtained by treating cyclohexanol with water above 275 °C.

Addition reactions to carbonyl compounds, typified by the Grignard reaction, frequently require an anhydrous VOC solvent and are relatively hazardous to carry out on an industrial scale. Alternative procedures using water stable reagents based on tin, zinc, and especially indium have now started to be developed for many allylic substrates to replace processes using magnesium or lithium. Apart from the obvious safety, environmental, and cost advantages of not having to use anhydrous organic solvents there is potential to broaden the scope of the

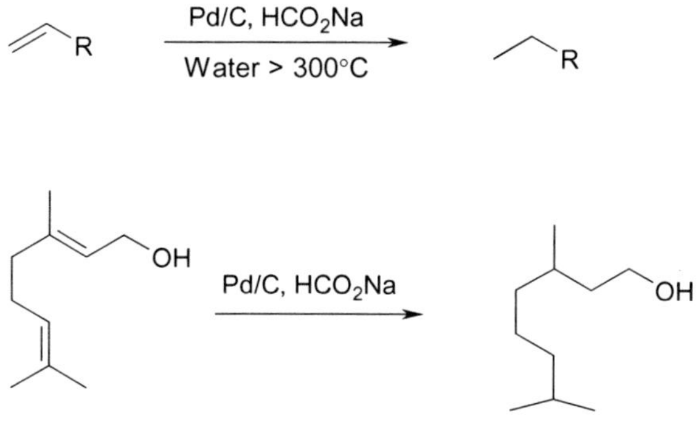

Scheme 5.12 Reduction in high-temperature water.

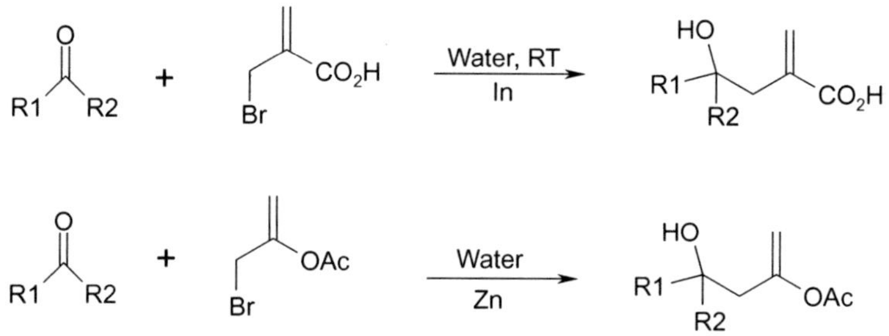

Scheme 5.13 Metal-mediated addition to carbonyls.

reaction since reagents with acidic hydrogens are often stable under these conditions. Frequently, such reactions take place at relatively low temperatures (25–40 °C) and, as such, reagent solubility can be difficult; to overcome this water–tetrahydrofuran mixtures have often been used. Such mixtures may pose significant problems if used industrially; hence the two examples shown in Scheme 5.13 all take place in water as the sole solvent. Whilst most such reactions currently require stoichiometric amounts of metal, which limits the commercial viability for indium based processes, catalytic processes are starting to be developed.

The scope of possible reactions using water as a solvent is quite remarkable and water is much under utilized as a solvent in many academic and industrial research institutions, largely through lack of knowledge and a culture of using organic solvents. Several other examples will be found elsewhere in this book but space is not available to do sufficient justice to the scope of possible reactions; however, a study of Scheme 5.14 will give the reader an indication of the versatility of this undervalued solvent.

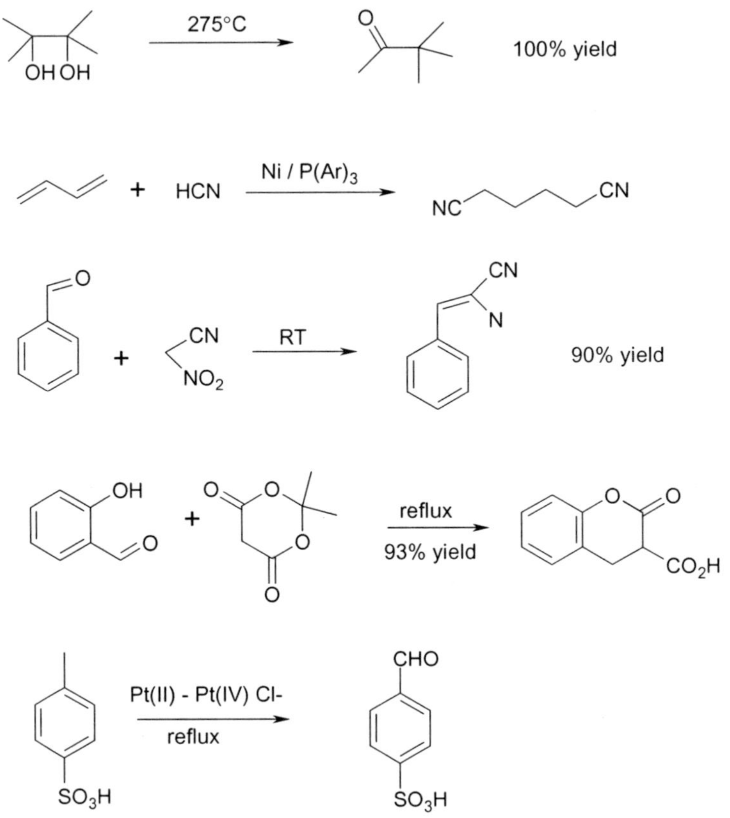

Scheme 5.14   Some examples of water as a reaction solvent.

5.4.1   Water Based Coatings

Although water is likely to be used more frequently as a reaction solvent, it is its use as a solvent for coatings that will bring most environmental benefits. Water based coatings have been around for many years but new formulations are continually being developed to meet more demanding applications. Replacing an organic solvent by water is not simple and often requires development of new additives and dispersants as well as reformulation of the polymeric coating materials themselves. Table 5.5 shows some of the main advantages and challenges that need to be met in development of new water-based coatings. Although much progress has been made in recent years to meet all these challenges for some applications, there is still a consumer choice to be made between environmental and technical performance.

Most water based paints use copolymers of poly(vinyl acetate) (often with butyl acrylate) and they do contain small amounts of higher boiling

Table 5.5 Advantages and challenges faced in development of water based coatings.

Advantages	Challenges
Reduction in VOC emissions	Stability of formulation at low temperatures
Reduced user exposure to harmful materials	Acceptability of drying rate
Reduced hazardous production waste	Energy costs for drying
Possibly less expensive	Adequacy of corrosion resistance
	Wear properties
	High gloss properties
	Storage stability
	Water resistance

organic solvents but VOC emission and health risks are minimal; these products are now widely accepted by the public for home use. A typical household emulsion paint will contain around 30% polymer, 25% pigments, 15% dispersant, and 20% water; the remainder is made up of a range of additives such as antifoams and in particular around 6% high boiling organic solvents, mainly ethylene and propylene glycols.

5.5 IONIC LIQUIDS

From overall environmental emissions and toxicity viewpoints the systems discussed so far (solvent free, $scCO_2$, and water) offer obvious advantages to use of organic solvents. The case is not quite so clear for ionic liquids. The major advantage of using ionic liquids as solvents is their very low vapour pressure; this coupled with the fact that they can often act both as catalyst and solvent has sparked considerable interest. The main issues still to be resolved centre on their toxicity (most ionic liquids have not been assessed), and, despite the potential for recycling, the fate and protocol for their ultimate disposal has not been established.

Ionic liquids (or more correctly non-aqueous ionic liquids) have been known for many years. In broad terms they can be viewed like common ionic materials such as sodium chloride, the difference being that they are liquid at low temperatures, this being due to poor packing of the respective ions. To achieve this poor packing requirement, room-temperature ionic liquids are generally made from relatively large, non-coordinating, asymmetric ions. Invariably at least one of these ions is organic in nature. Based on this criterion there are many thousands of possible ionic liquids, of which only a few have been studied in any detail, Figure 5.4 shows some of the more common ones. Ionic liquids are gradually becoming commercially available, but they are more expensive

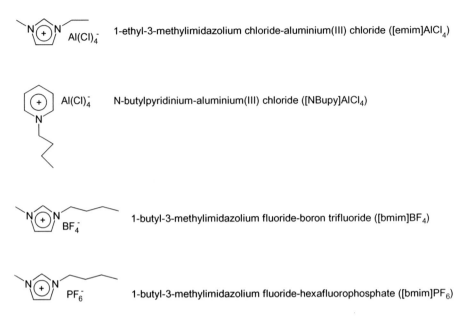

Figure 5.4 Some common room-temperature ionic liquids.

than organic solvents and not available in such large volumes. Preparation is relatively straightforward, although synthesis of very pure
materials is more difficult. For example, [NBupy]AlCl₄ can simply be
prepared by mixing the imidazolium chloride with aluminium chloride,
the resulting exothermic reaction producing the liquid product. Metathesis reactions are also widely used for preparing ionic liquids, *e.g.*
[emim]BF₄ can be prepared from reaction of [emim]I with ammonium
fluoroborate in acetone, with the ammonium acetate remaining in the
solvent. When assessing the overall greenness of a process using an ionic
liquid its synthesis should also be taken into account since many literature routes employ volatile and chlorinated solvents in synthesis or
purification routes. The problem of using a volatile organic solvent may
just be being moved back down the overall process chain. For industrial
use it is often vital to remove contamination by chloride ions, since these
may form trace chlorinated by-products (particularly a problem for the
pharmaceutical industry) and give excessive corrosion in stainless steel
reactors. Because purification is quite difficult it is often vital to start with
highly pure reagents. For example, for the preparation of imidazolium
halides, 1-methylimidazoline is distilled over NaOH and the haloalkane
washed with sulfuric acid then with sodium bicarbonate and finally with
water before being dried and distilled, all of which have an impact when
assessing the overall environmental impact of the ionic liquid.

The types of ionic liquids shown in Figure 5.4 have been most extensively studied, especially ones based on chloroaluminate. These chloroaluminate materials display useful Lewis acid properties but they are highly air and moisture sensitive, which renders them relatively commercially unattractive. Newer ionic liquids containing ClO_4^- and NO_3^- anions, for example, which are less air and moisture sensitive are now being more widely studied, but these are often less catalytically active. Other than lack of vapour pressure and catalytic properties there are several other features common to most ionic liquids that make them attractive reaction solvents, these include:

- Tunability – by varying the cation to anion ratio, type, and alkyl chain length properties such as acidity/basicity, melting temperature, and viscosity can be varied to meet particular demands.
- Many ionic liquids are stable at temperatures over 300 °C, providing the opportunity to carry out high temperature reactions at low pressure.
- Ionic liquids that are not miscible with organic solvents or water may be used to aid product separation or used in liquid–liquid extraction processes.
- For a given cation the density and viscosity of an ionic liquid depend on the anion – in general density increases in the order $BF_4^- <$ $PF_6^- < (CF_3SO_2)_2N$ and viscosity increases in the order $(CF_3SO_2)_2$-$N < BF_4^- < PF_6^- < NO_3^-$.

5.5.1 Ionic Liquids as Catalysts

The highly acidic properties of [emim]AlCl$_4$ have been used in several cationic alkene oligomerization reactions. For example, refinery raffinate streams containing high levels of butenes can be oligomerized/polymerized to give materials ranging in molecular weight from 600 to 100 000; such materials are used as lubricating oils. Such reactions are rapid, with conversions well over 90% being achieved within 30 min. The major benefit of such reactions lies in product work-up. Whilst no solvent is required for butene oligomerization a homogeneous ethyl aluminium chloride catalyst is often used, and is largely lost in the heavy residue by-products and aqueous wash. In contrast the organic product is not miscible with the ionic liquid and can be readily decanted from the catalyst, which can be reused. No aqueous wash is thought to be required. A commercial-scale process (Difasol) for the nickel-catalysed dimerization of butane using a chloroaluminate ionic liquid as solvent has been developed. The main advantage of this process is that the octenes can be decanted from the reaction vessel since they are not soluble in the ionic liquid.

Chloroaluminate ionic liquids are also highly active Friedel–Crafts catalysts. In fact with active alkyl halides mono-alkylation is often difficult to achieve due to the very high activity of the catalyst. More success has been achieved using alkenes as the alkylating agents. One potential application of this that has been demonstrated at laboratory scale is the alkylation of benzene with dodecene, with the ionic liquid being readily recoverable. This route again offers advantages compared to traditional processes using $AlCl_3$ (which is lost as waste) or HF (hazardous) although there are no clear environmental advantages compared to the latest zeolite-catalysed processes. Unfortunately, ionic liquid catalysts have not yet provided a solution to the need for truly catalytic acylation catalysts for un-reactive substrates. As with traditional Lewis acid catalysts the product complexes to the ionic liquid and can only be separated by quenching with water, destroying the catalyst.

The acidic and catalytic properties of chloroaluminate(III) ionic liquids are frequently said to arise from the species $Al_2Cl_7^-$, which is generated according to Equation (5.1):

$$2AlCl_4^- \leftrightarrow Cl^- + Al_2Cl_7^- \tag{5.1}$$

It has been shown, however, that such catalysts may contain protons, either by design or because of the difficulty in removing all traces of moisture; these protons have been shown to be superacidic with Hammett acidities up to −18. These protons will also play some role in the catalytic activity of these ionic liquids in practical situations. Ionic liquids in which superacidic protons have deliberately been generated by addition of small amounts of water, HCl, or H_2SO_4 have been used to catalytically crack polyethylene under relatively mild conditions. The main products are mixed C_3–C_5 alkenes, which would be a useful feedstock from waste polyethylene recycling. In contrast to other cracking procedures no aromatics or alkenes are produced; however, small amounts of polycyclic compound are obtained.

5.5.2 Ionic Liquids as Solvents

If ionic liquids become widely used it is likely to be the less water and air sensitive ones that prove to be more commercially attractive. Here the main emphasis will lie with their use as a solvent rather than as a catalyst. An interesting contrast between catalytic and non-catalytic

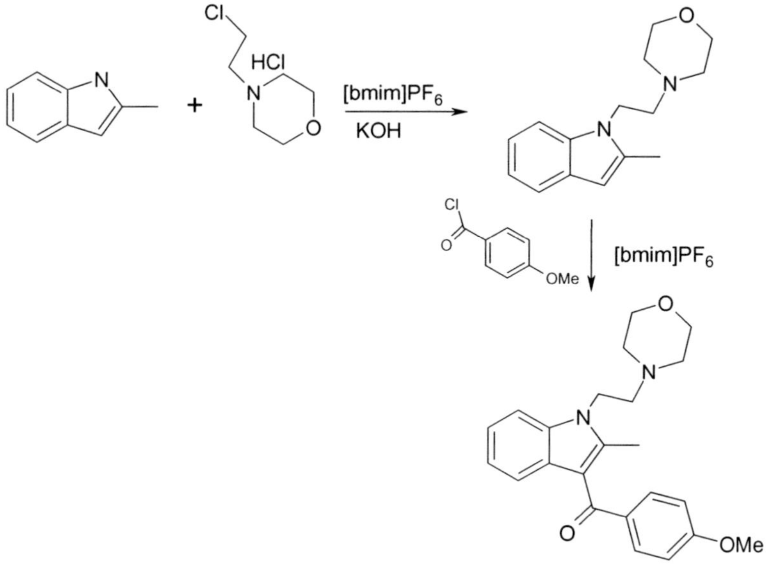

Scheme 5.15 Synthesis of pravadoline in ionic liquid.

ionic liquids is provided in Seddon's synthesis of pravadoline, a potential non-steroidal anti-inflammatory drug (Scheme 5.15). When the second stage acylation reaction was carried out using [emim]AlCl$_4$ the reaction product co-ordinated to the Lewis acid catalyst, requiring a water quench to isolate the product. However, it was found that due to the high activity of the indole group towards acylation no Lewis acid catalyst was required. The reaction could be carried out in high yield in the ionic liquid [bmim]PF$_6$, although a temperature of 150 °C was required compared to 0 °C when using a catalyst. The non-catalysed route afforded simple product separation and a recoverable solvent.

The Diels–Alder reaction is receiving much attention due to findings of rate enhancements similar to those that occur when using water as a solvent. The *endo* selectivities are also found to be generally good. Of course, ionic liquids could be used to enhance the rate of Diels–Alder reactions involving water sensitive reagents. Scheme 5.16 shows some of the many examples of the types of reactions carried out. In some instances, *e.g.* the aza-Diels–Alder reaction illustrated, Lewis acid catalysts are additionally required but the use of ionic liquids greatly enhances the ease of recovery and recycle.

Ionic liquids are proving to be valuable solvents for carrying out what usually would be considered to be homogeneously catalysed reactions since in many cases the catalyst stays in the ionic liquid and can be

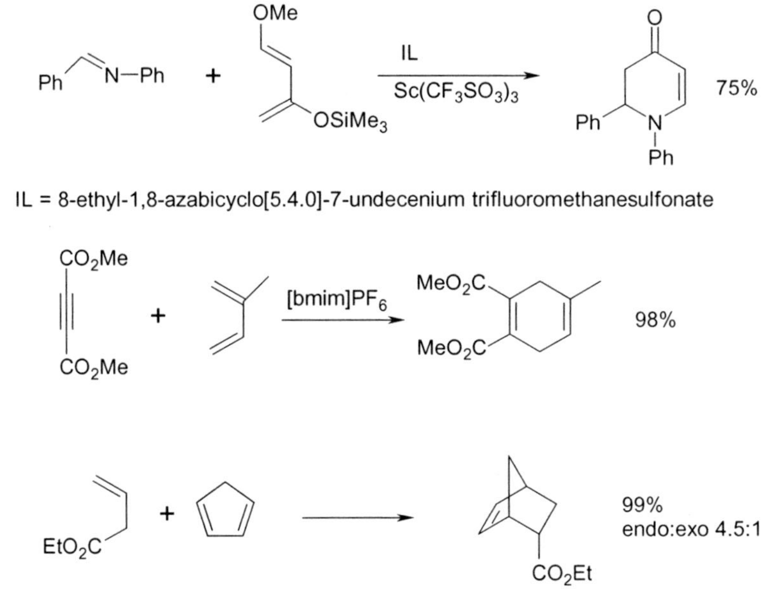

IL = 8-ethyl-1,8-azabicyclo[5.4.0]-7-undecenium trifluoromethanesulfonate

Scheme 5.16 Examples of Diels–Alder reactions in ionic liquids.

readily be reused, bringing in many of the advantages of heterogeneous reactions. Where reported, catalyst leaching is normally very low. Hydrogenation reactions have been well studied, with rates several times those found in organic solvents often being obtained. This has been attributed to stabilization of intermediate catalyst species by the ionic liquid. The scope of the reaction is wide and by tuning reaction conditions, catalyst, and ionic liquid high levels of product selectivity can be obtained. Pentene, for example, can be hydrogenated to pentane using $Rh(nbd)(PPh_3)_2$ [nbd = norbornadiene) as catalyst in [emim]PF_6. Dienes are more soluble in ionic liquids than mono-alkenes, which has led to a highly selective conversion of cyclohexadiene into cyclohexene, whilst butadiene can be selectively reduced to bute-2-ene using [bmim]$_3$-$CO(CN)_5$. There are several reported examples of enantioselective hydrogenations being carried out, *e.g.* hydrogenation of 2-phenylacrylic acid (Scheme 5.17) and the related synthesis of (*S*)-naproxen.

Owing to the potential for easy recycle and low catalyst losses Pd-catalysed reactions have received much attention. In principle, use of ionic liquids could overcome many of the catalyst loss issues that have prevented wider commercial use of valuable Pd-catalysed coupling reactions such as those developed by Heck or Suzuki. Possibly the simplest (and least expensive) example of an ionic liquid being used in a Heck

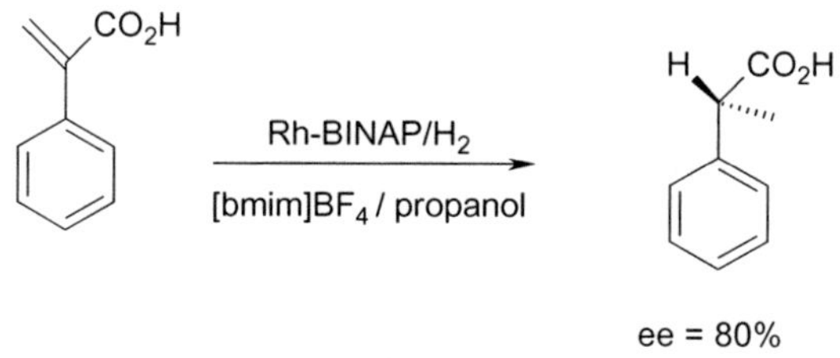

Scheme 5.17   Enantioselective hydrogenation of 2-phenylacrylic acid.

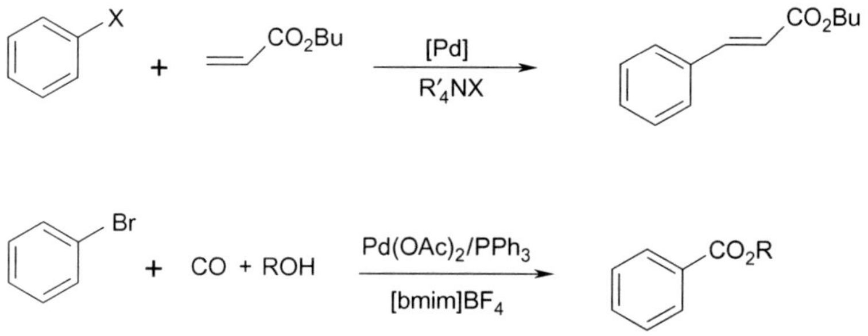

Scheme 5.18   Examples of Pd-catalysed reactions in ionic liquids.

reaction is the synthesis of *trans*-cinnamic acid derivatives (Scheme 5.18) using simple molten tetraalkylammonium bromides. Although these are not liquids at room temperature many melt below typical reaction temperatures of 100 °C. Use of these salts has led to excellent yield improvements, by a factor of 5 in some cases, compared to traditionally employed solvents like DMF. The product can usually be isolated by vacuum distillation, and the molten salt/catalyst mix reused without loss of activity. Palladium-catalysed carbonylation of aryl bromides has also been studied (Scheme 5.18); whilst in limited studies the reactions proved to be significantly more efficient than those run without solvent catalyst activity did decline on subsequent reuse.

Similarly, hydrosilation reactions involving addition of a Si–H group to an alkene are catalysed by H_2PtCl_6. Usually the catalyst is decomposed during work-up at the end of the reaction and the platinum is lost during filtration and work-up. By using an ionic liquid, such as trimethylimidazolium methylsulfate, as solvent the catalyst was immobilized in the ionic liquid. The reaction product could be decanted from the solvent at the end of the reaction, leaving the catalyst ready for reuse.

From reviewing much of the literature it is easy to conclude that ionic liquids are excellent solvents for catalysts and reagents but not for products. This is obviously not the case. While some products can be decanted from the liquid and others can be recovered by distillation there are many useful reactions in which removal of the product (or residual reactants) from the ionic liquid is challenging. Extraction with an organic solvent, or even water, would reduce the overall eco-efficiency. Initial studies in which $scCO_2$ has been used to extract material from various ionic liquids show promise, with, for example, naphthalene being readily extracted from [bmim]PF_6.

Before leaving ionic liquids it is worth mentioning their potential value in separation processes. Organic solvents are frequently used in multi-phase extraction processes and pose the same problems in terms of VOC containment and recovery as they do in syntheses, hence ionic liquids could offer a more benign alternative. Interesting applications along this line that have been studied include separation of spent nuclear fuel from other nuclear waste and extraction of the antibiotic erythromycin A.

5.6 FLUOROUS BIPHASE SOLVENTS

The term fluorous biphase has been proposed to cover fully fluorinated hydrocarbon solvents (or other fluorinated inert materials, *e.g.* ethers) that are immiscible with organic solvents at ambient conditions. Like ionic liquids the ideal concept is that reactants and catalysts would be soluble in the (relatively high boiling) fluorous phase under reaction conditions but that products would readily separate into a distinct phase at ambient conditions (Figure 5.5).

The C–F bond is very stable and is responsible for imparting the benign properties possessed by perfluoro-compounds, which include non-toxicity, high levels of inertness, high thermal stability, non-

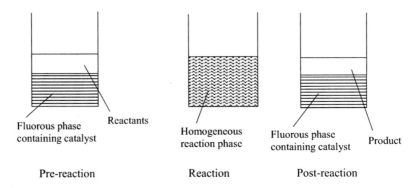

Pre-reaction   Reaction   Post-reaction

Figure 5.5   Concept of fluorous biphasic reactions.

flammability, and hydrophobicity. The main issues regarding their use are cost of the material (and that of compatible catalysts), 'greenness' of their manufacture, and, to a lesser extent, their ultimate effect on the environment after disposal. Regarding the latter point, volatile C_1 and C_2 fluorocarbons are potent greenhouse gasses but fluorous phase solvents typically containing 6–10 carbons are much less volatile and therefore pose less threat. The longer term consequences of significant amounts of these products entering the environment are still unknown; they are very persistent and there is some evidence of accumulation in flora and fauna. Development of biodegradable fluorinated solvents is being undertaken.

To gain maximum compatibility between the non-polar fluorous phase and reactants as well as ease of separation of products the use of fluorous biphasic systems may be expected to work best where the reactants are non-polar but in which the polarity of the product is relatively high. This philosophy has been demonstrated in the hydroformylation of long-chain alkenes (Scheme 5.19). The main issue in such reactions is ensuring catalyst solubility in the fluorous phase. This is usually achieved by incorporating ligands containing highly fluorinated 'ponytails'. C–F bonds, however, have very strong electron-withdrawing effects and frequently reduce the catalytic effect of transition metal centres; incorporating 'spacers' consisting of 1 or 2 –CH_2– groups next to the metal centre minimizes this. In the hydroformylation reaction shown the product could be separated from the fluorous phase at room temperature but at the reaction temperature of 100 °C the substrate was completely soluble. Catalyst losses after nine runs were only just over 4%. Whilst the linear to branched ratio was relatively poor it could be improved by using high ligand to metal ratios.

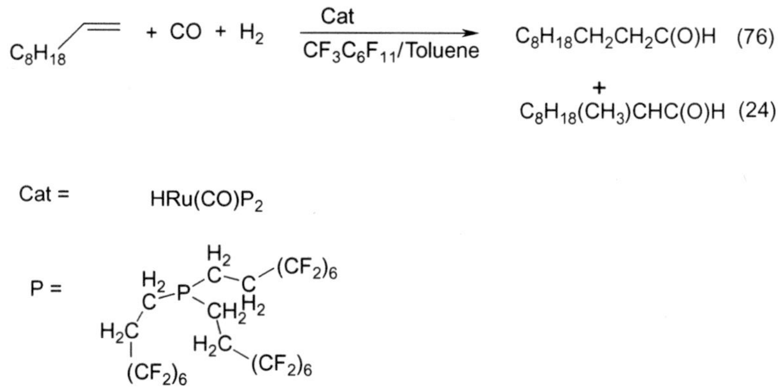

Scheme 5.19 Fluorous phase hydroformylation.

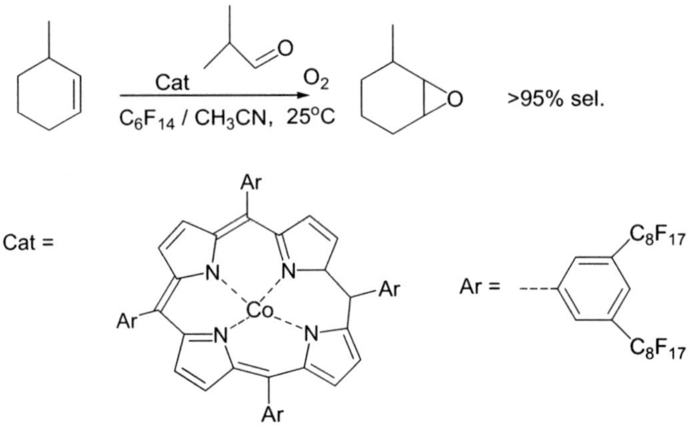

Scheme 5.20   Alkene epoxidation using fluorinated co-porphyrin catalyst.

Removing the toluene and using a catalyst based on [Rh(acac)(CO)]$_2$ and P(C$_6$H$_4$-pC$_6$F$_{13}$)$_3$ resulted in faster rates and a higher linear to branched ratio.

Such biphasic systems are now being applied to many reaction types, including many oxidation reactions since, importantly, oxygen is much more soluble in fluorinated solvents than in most organic ones; this could have significant advantages for mass transport limited reactions. Cobalt-fluorinated porphyrin complexes have been used successfully for the selective epoxidation of several internal alkenes (Scheme 5.20) using oxygen in the presence of 2-methylpropanal as reducing agent. High catalyst turnover numbers were found compared to similar reactions carried out in organic solvents.

5.7   COMPARING GREENNESS OF SOLVENTS

It is now widely accepted that no solvent is *a priori* green. Its greenness is a combination of its inherent properties, like acute toxicity, ozone-depleting potential and persistency in the environment, its method of production, and the application it is being used in, including the method of recovery and reuse. A number of solvent selection tools, essentially taking a life cycle assessment (LCA) approach and taking into account the above are available, but are often complex to use and compare. Currently, there is no simple guide to enable a green solvent to be selected for a particular reaction type. For a major new process a full LCA should be carried out but this may not be warranted for smaller scale processes. In these cases the R&D chemist is advised to try various potential solvents, including those discussed in this chapter, paying particular attention to the ease of recovery and reuse as well as the inherent health, safety, and environmental aspects of the solvent.

5.8 CONCLUSIONS

Using the principles of green chemistry, all other things being equal, it is preferable to avoid use of any kind of solvent altogether, both in the reaction and subsequent isolation and purification stages. This goal has been achieved for many bulk chemicals. Practically, this is not currently achievable for most fine, speciality, and pharmaceutical products, with, until recently, there being little choice other than to use organic solvents. As described above, a whole range of solvent options are now becoming available but caution is required before a particular solvent type is described as being 'green' or otherwise.

When choosing a particular solvent for a specific application a wide range of factors should be considered, including some not directly related to the specific application. Obviously cost, efficacy, and safety need to be considered first; this will generally rule out some options. Following this a more detailed assessment of additional factors should be carried out, ideally including:

- Full life cycle assessment of solvent. Does its manufacture use more noxious materials than it replaces? What is the ultimate environmental fate of the solvent?
- Effect on overall energy requirement for the application, including recompression energy for $scCO_2$ and any recycle/re-purification stages.
- Overall quantitative and qualitative assessment of waste generated from application.

Few solvents are inherently green or otherwise, despite some misleading literature assertions. Unfortunately, comparative data on solvent performance (technical and environmental) is lacking, making an informed choice very difficult. Whilst certain organic solvents are not desirable, on both health and environmental grounds, most organic solvents, if they are handled safely in well engineered plants with good recovery and recycle facilities, in some instances, may be the preferred option. That said the practising technologist should no longer be restricted to using noxious or volatile organic solvents – there are now many alternatives, organic and non-organic, that deserve assessment. With current knowledge the advice should be 'try it and see' then critically assess.

REVIEW QUESTIONS

1. Discuss the advantages and disadvantages of using supercritical carbon dioxide and water as solvents in place of organic solvents.
2. Surfactants are often used in polymerization of fluorinated materials using $scCO_2$. What is the role of the surfactant? Describe the mechanism of one such reaction.
3. Carry out a literature survey on the use of solvents, including organic, water, ionic liquids, and CO_2, for Diels–Alder reactions (either generic or a reaction of your choice). Compare each solvent in terms of yield, selectivity, ease of isolation of product, approximate energy use, and overall cost. From the advantages and disadvantages of each solvent type rationalize your choice of preferred solvent for carrying out a Diels–Alder reactions on a 10 000 tpa scale.
4. In most solvent-containing consumer products the solvent is lost to the environment on use. List three consumer products that contain non-aqueous solvents and assess the likely longer term environmental impacts of continued use of these products. Critically assess more benign alternatives.

FURTHER READING

J. M. DeSimone and W. Tumas, *Green Chemistry Using Liquid and Supercritical Carbon Dioxide*, Oxford University Press, Oxford, 2003.
F. M. Kirton, *Alternative Solvents for Green Chemistry*, Royal Society of Chemistry, Cambridge, 2009.
K. Mikami (ed), *Green Reaction Media in Organic Synthesis*, Blackwell Publishing, Oxford, 2005.

CHAPTER 6

Renewable Resources

6.1 BIOMASS AS A RENEWABLE RESOURCE

All living material may be considered as biomass but, commonly, only non-animal renewable resources such as trees and crops that may be harvested for energy or as a chemical feedstock tend to be referred to as biomass. Although the total amount of biomass available at any one time is relatively small, some estimates putting the figure at under 2000 billion tonnes (including water contained in the cells), unlike fossil resources it is readily renewable on a time scale useful to mankind. The energy for this renewal process comes from the sun, with around eight times our annual energy consumption (5×10^{20} J) being fixed each year through photosynthesis. This is equivalent to the generation of some 70×10^9 tpa of organic matter. For most of the history of mankind biomass was the only source of energy, and even with our current heavy reliance on fossil fuels biomass still accounts for around 15% of the world's energy use (Figure 6.1). This use is heavily concentrated in developing countries like India and southern Africa but many well-forested developed countries still use significant amounts of wood for burning.

6.2 ENERGY

6.2.1 Fossil Fuels

The era of society's almost total dependence on fossil resources is likely to come to an end during the twenty-first century. This is the startling

Green Chemistry: An Introductory Text, 2nd Edition
By Mike Lancaster
© Mike Lancaster 2010
Published by the Royal Society of Chemistry, www.rsc.org

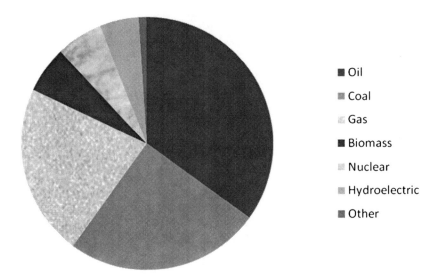

Figure 6.1   Sources of world energy.

conclusion now being reached by a growing number of experts. Indeed, as far as oil is concerned, many now believe that a rapid decline in supply will be evident before the middle of this century. This, coupled with the fact that consumption of fossil fuels is a significant cause of climate change and global warming, has led to the growing need for alternative feedstocks to replace fossil resources. For energy use there are several options, including wind, water, solar, and nuclear as well as biomass. For organic chemical production the only non-fossil option is biomass (with possibly a small contribution from atmospheric CO_2). In the medium term, as oil prices continue to rise, increased use of gas and possibly coal for both energy and chemicals production is likely.

Other than longer term supply issues the main driver for moving away from fossil resources is pollution and climate change. Since pre-industrial times the level of atmospheric CO_2 has risen from 280 to 380 ppm, whilst some observers believe this may be a natural cycle in the Earth's history the vast majority believe it is a direct consequence of burning fossil fuels. This additional CO_2 is now thought to be the main cause of global warming *via* the greenhouse effect (Box 6.1).

Many policymakers and scientists now believe that climate change, largely brought about by fossil fuel burning, is the biggest problem currently facing mankind. Although CO_2, from combustion of fossil fuels, is thought to be the major contributor to global warming there are many more anthropogenic (man-made) greenhouse gases that have far more potent global warming potentials (GWPs) than CO_2, (Table 6.1).

Box 6.1 The greenhouse effect

Contrary to popular belief the greenhouse effect is a natural phenomenon that enables life to survive on planet Earth. The main components of the atmosphere, nitrogen and oxygen, do not absorb or emit thermal radiation. If the atmosphere was completely made up of these gases radiation from the sun, which peaks at 400–750 nm, would hit the surface of the earth and be re-radiated at lower energy, in the infrared region (4000–50 000 nm) back into space. A suitable analogy for this is a greenhouse with all the roof and side windows open.

It is fortunate, however, that there are small amounts of other gasses in the atmosphere, notably carbon dioxide and water vapour. These gases both absorb strongly in the infrared region of the reflected radiation, keeping the heat energy close to the Earth's surface. The analogy for this is the greenhouse with the windows closed.

The natural consequence of the green house effect is to keep the Earth's temperature over 20 °C above that which it would be if the atmosphere were completely composed of N_2 and O_2, thereby enabling life to survive.

From this natural effect it is evident that small amounts of trace gases present in the atmosphere can have a significant effect on the Earth's temperature. As an extreme example Venus, with an atmosphere containing 96% CO_2, has a temperature of almost 500 °C compared to around 35 °C if no greenhouse gasses were present. We now believe that other, very potent, man-made greenhouse gases such as nitrous oxide and CFCs are contributing to global warming in addition to the build up of CO_2. Although the exact degree of warming is difficult to quantify a doubling of CO_2 concentration in the atmosphere may cause a global temperature rise of 2 °C.

Whilst the effects of global warming are still somewhat controversial it is likely to cause more storms and flooding, melting of the polar ice caps, and cause changes in animal distributions.

Table 6.1 Some anthropogenic greenhouse gases.[a]

	CO_2	CH_4	N_2O	CFC-11	CF_4	HCFC-22
Pre-industrial concentration	280 ppm	700 ppb	275 ppb	0	0	0
Current concentration	380 ppm	1720 ppb	312 ppb	268 ppb	72 ppt	110 ppt
GWP	1	56	280	N/a	4400	9100

[a]ppm = parts per million, ppb = parts per billion, ppt = parts per trillion.

International agreements and legislation are now in place to limit global production and release of many of these materials. In addition to these restrictions, global emissions of CO_2 need to be reduced by 60% to avoid the worst consequences of climate change.

Oil is the most widely used of the fossil resources, supplying almost 40% of the world's energy and virtually all chemical feedstocks. By far the greatest user of crude oil, some 90%, is the energy sector, with the chemicals sector accounting for a further 8%. Hence pursuit of a more sustainable agenda, by cutting CO_2 emissions and conserving valuable fossil resources for future generation, will require increased focus on renewable energy. There is still a significant difference of opinion with regard to future oil supply both in terms of reserves and demand. What is clear is that there has been a gradual growth in demand over recent years (Figure 6.2) that is likely to continue to rise in the longer term, as more nations join the developed world.

According to UN figures the world's population is likely to reach 10 billion people before 2050, a rise of some 3.5 billion on the current level. Assuming that energy demand per capita remains roughly constant (which is highly unlikely) oil consumption can be expected to rise to rise from around 30 billion barrels to 46 billion barrels per year by 2050. Estimating proven oil reserves is an inexact science, with published figures varying widely; most estimates, however, fall mainly in the range 900–1300 billion barrels. This equates to between 30 and 43 years consumption at present levels or between 20 and 28 years at 2050 levels. Of course new reserves continue to be found but in recent years oil has been consumed at roughly twice the rate that new reserves have been discovered. If this trend continues prices will steadily rise. From an

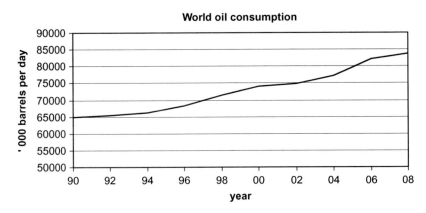

Figure 6.2   Rising world oil consumption.

economic viewpoint the important date will be when production starts to fall off and demand starts to exceed production. At this point, unless energy and feedstock demands can be met from elsewhere, prices will start to rise rapidly. The consensus of evidence suggests that this will happen before 2050 unless viable alternatives are found in the meantime.

6.2.2 Energy from Biomass

Most of the biomass used for energy is burned either directly to provide heat or in a power station to provide electricity. Although biomass is a complex mixture of starch, cellulose, *etc.*, in simple terms the burning process can be viewed as being represented by Equation (6.1). The CO_2 output can be considered as being essentially neutral since a similar amount of CO_2 is consumed in growing the biomass:

$$[CH_2O] + O_2 \rightarrow CO_2 + H_2O \qquad (6.1)$$

The actual energy content of even dried biomass is typically in the range 14–17 GJ per tonne, this being roughly one-third that of oil and significantly less than natural gas ($55 \, GJ \, te^{-1}$), the most calorific fuel commonly used.

To compete with fossil fuels, biofuels must become available in a variety of guises to meet the needs of our energy hungry society. To meet these needs several primary conversion processes have been developed as summarized in Figure 6.3.

6.2.2.1 Thermolysis and Pyrolysis. Both thermolysis and pyrolysis involve heating biomass (mainly wood), largely in the absence of oxygen at temperatures from a few hundred degrees centigrade (thermolysis) up to 1500 °C (pyrolysis). At the lower temperature char or charcoal is a major product and has become a valuable export commodity for many wood rich countries. The other main product is a fuel oil but this usually is quite acidic, requiring treatment before it can be used. Overall, thermolysis is relatively inefficient, with well over 50% of the energy content of the original biomass being lost. At higher temperatures the char content is considerably reduced, the main product being a gas mixture rich in H_2, CO and acetylene, the composition of which varies with temperature.

6.2.2.2 Gasification. Gasification differs from the purely thermal processes in that air and steam are used to give a product richer in oxygen. Other than nitrogen, the main products are CO (around 17%) and H_2,

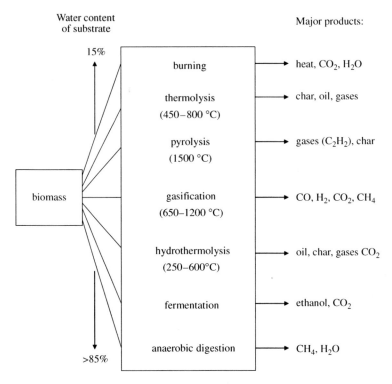

Figure 6.3   Biomass conversion processes.(Reproduced with permission of the Royal
Society of Chemistry from C. Okkerse and H. van Bekkum, *Green
Chemistry*, 1999, **1**, 107–114.)

with varying amounts of methane and CO_2 being formed. The product
can be used for electricity generation or to generate synthesis gas as a
chemical feedstock.

6.2.2.3   Hydrothermolysis.   This is a specific process developed by Shell
for producing an oil-like material, called bio-crude, with a low oxygen
content. The process is so called due to treatment of the biomass with
water at temperatures of 200–330 °C and pressures over 30 atm. Recent
studies suggest that agricultural waste such as wheat straw and corncob
can be treated to give useable concentrations of sugar solutions that may
be fermented to bioethanol. At temperatures above 260 °C large amounts
of furfural are also produced, which limit the fermentation process.

6.2.2.4   Anaerobic Digestion.   Anaerobic digestion is an extension of
anaerobic treatment of waste discussed in Chapter 2; it is also similar to
the natural process operating in landfill sites that evolves methane.

By treatment of biomass with bacteria in the absence of air a gas rich in methane can be produced – a typical digester may produce over $300\,m^3$ of gas containing over 50% methane per tonne of dry biomass. In developed countries, plants have been built because of the need to treat waste such as sewerage sludge and provide little energy compared to the national needs. For developing countries the United Nations see this as a sustainable technology to generate electricity on a local basis using farm waste.

6.2.2.5 Bioethanol. Production of ethanol by fermentation of glu-cose-based crops such as sugar cane and corn starch using *Saccharomyces* yeasts has been known since pre-industrial times. As a result of the oil crisis in the mid-1970s some countries, notably Brazil, began commercial production of bioethanol to use as a fuel in place of gasoline. Bioethanol may be used either alone or as a blend with gasoline. Production volumes in Brazil have reached some 16 billion litres per annum; almost half of this is hydrated ethanol which, due to the relatively high water content, cannot be mixed with gasoline and must be used as a single fuel in specially adapted engines. In the US the use of anhydrous bioethanol is growing, in part due the replacement of MTBE previously used as an octane enhancing supplement. This bioethanol can be mixed with gasoline at levels up to 22% (typically 10% is used) and used in conventional engines. Brazil and the US are responsible for about 80% of the world's bioethanol production.

The so-called first-generation processes for the production of ethanol from corn uses only the corn kernels; these are taken from the corn plant and only the starch, which represents about 50% of the dry kernel mass, is transformed into ethanol. Two types of second-generation processes are under development. The first uses enzymes and yeast to convert the plant cellulose into ethanol while the second type uses pyrolysis to convert the whole plant into either a liquid bio-oil or a syngas. Second-generation processes can also be used with plants such as grasses and wood or agricultural waste material such as straw, helping ensure bioethanol production does not impact adversely on the food supply chain.

Although based on a renewable feedstock there are significant environmental issues raised by large-scale production of bioethanol using first-generation processes:

- In the production process an acidic by-product, called vianesse, largely consisting of organic and amino acids, is produced at a level of over ten times the amount of ethanol produced. This material has a high COD and BOD and poses significant problems if it finds its way into the watercourse.

- Traditional fermentation processes can only make use of glucose, leaving other sugars present, notably xylose, untouched. This both increases the 'waste' element and lowers the competitiveness of the process compared to petrochemical based ethanol. Recent advances in technology are making it commercially viable to ferment starch and other sugars, but the real goal is to be able to convert cellulose into ethanol by fermentation.
- Fermentation processes typically give a product with an ethanol concentration of between 7% and 15%; above this level the yeast find it difficult to perform. The usual method of concentration is by distillation, which is a very expensive and energy intensive process. This high energy use reduces the overall 'CO_2 neutral' benefit often claimed for renewable feedstocks.

Several possible solutions to the first problem have been commercially developed. These include restricted use as a fertilizer since it is a valuable source of both nitrogen and potassium, and aerobic degradation to single cell protein for use as an animal feed. The second issue has proved more problematical, and has largely prevented cellulosic waste being used for ethanol production. The problem is that *Saccharomyces* yeasts do not have an enzyme for converting xylose into xyulose, a key step in the metabolic pathway. In the last few years developments in genetic engineering have enabled yeast strains to be produced that overcome this problem, at least in the laboratory. In Brazil recent developments have moved away from using yeast to using the bacterium *Escherichia coli* (K011). In this process cellulosic waste from sugar cane is hydrolysed to pentose and hexose, which are converted into ethanol by the bacterium. Innovations in membrane separation processes have provided an energy efficient route to partial concentration of dilute ethanol streams.

In the US the first small production unit for converting cellulose into ethanol has been built.

There are several stages to produce ethanol using a biological approach:

1. a 'pretreatment' phase, to make the lignocellulosic material such as wood or straw amenable to hydrolysis;
2. cellulose hydrolysis (cellulolysis), using enzymes from the fungus *Trichoderma reesei* to break down the molecules into sugars;
3. separation of the sugar solution from the residual materials, notably lignin;
4. microbial fermentation of the sugar solution;

5. distillation to produce roughly 95% pure alcohol;
6. dehydration by molecular sieves to bring the ethanol concentration to over 99.5%.

6.2.2.6 Biobutanol. Formation of acetone, butanol, and ethanol from fermentation of starch using *Clostridium* bacteria has been known for many years and was once an industrial process before the advent of the modern chemical industry based on oil. Recent renewed interest has led to modern technology using a fibrous bed bioreactor that produces predominantly butanol. Very little acetone and ethanol are produced, the main by-products being butyric acid and hydrogen. These can be converted back into butanol in another bioreactor. This process is competitive in terms of price with bioethanol production and is being commercialized by BP and Dupont. Because of its lower solubility in water it is more compatible with current gasoline transport and storage systems than ethanol. On the negative side it does have a lower octane rating than ethanol and, as has been found with *t*-butyl methyl ether in the USA, if storage systems leak there is the possibility of small amounts of material entering into the water supply.

6.2.2.7 Biodiesel. The first engines invented by Rudolf Diesel ran on groundnut oil; however, due to the advent of relatively cheap oil this type of biodiesel never became commercially viable. In the last 70 years or so the diesel engine has been refined and fine tuned to run on the diesel fraction of crude oil which consists mainly of saturated hydrocarbons. For this reason the modern diesel engine can not run satisfactorily on a pure vegetable oil feedstock due to problems of high viscosity, deposit formation in the injection system, and poor cold-start properties. Today, environmental issues have caused renewed interest in biodiesel, with commercial sources being available since the late-1980s. The main problems with petroleum based diesel include:

- depletion of natural resources
- CO_2 emission
- contribution to smog
- NO_x and SO_x emission.

The latter two issues have been overcome to a large extent by post combustion designed to remove NO_x in particular. Most recent research into biodiesel has focussed on vegetable oils such as soybean, sunflower, palm, and rapeseed. Although animal fats have been considered, their availability in the quantities required have precluded serious utilization.

Table 6.2   Main fatty acid components of vegetable oils (% of total fatty acids).

Fatty acid	Soybean	Sunflower	Palm	Rapeseed (HEAR)[a]
Palmitic	11	7	42.8	3
Stearic	4	5	4.5	1
Oleic	23	18	40.5	11
Linoleic	54	69	10.1	12
Linolenic	8	0	0.2	9
Erucic	0	0	0	52

[a]HEAR = high erucic acid rapeseed.

$$
\begin{bmatrix} -O_2CR_1 \\ -O_2CR_2 \\ -O_2CR_3 \end{bmatrix} + 3\,ROH \underset{}{\overset{catalyst}{\rightleftharpoons}} \begin{bmatrix} -OH \\ -OH \\ -OH \end{bmatrix} + \begin{matrix} R_1CO_2R \\ R_2CO_2R \\ R_3CO_2R \end{matrix}
$$

Scheme 6.1   Transesterification of triglycerides.

The main constituents of vegetable oils are triglycerides, which are esters of glycerol with long-chain saturated and unsaturated fatty acids. Table 6.2 shows the fatty acid components of the triglycerides found in the oils. In part it is the high levels of unsaturated acids, particularly polyunsaturated acids, that prevent these oils being used directly. Oxidation and polymerization of these molecules can lead to a viscosity increase on storage and gum formation both in storage and use.

To convert the raw oils into useful material transesterification technology is used. The oil is reacted with a low molecular weight alcohol, commonly methanol, in the presence of a catalyst to form the fatty acid ester and glycerol (Scheme 6.1). The ester is subsequently separated from the glycerol and used as biodiesel, the glycerol being used as a raw material for fine chemicals production. Although the chemistry is simple, to make biodiesel commercially viable the process must be operated as economically as possible; this entails fast reactions, driving the reaction to completion and efficient separation of products.

Both alkaline and acid catalysts can be used but industrially alkali catalysts such as sodium or potassium hydroxide or methoxide (at around 0.5 wt%) are preferred due to the speed of the reaction.

A blend of up to 20% biodiesel can usually be incorporated into conventional engines. In Europe the Renewable Transport Fuel Obligation requires a minimum of 5% biodiesel to be incorporated into all commercial diesel from 2010. Late-stage tests are also underway to use biodiesel in rail and air transport. Amongst the benefits of biodiesel are its enhanced lubrication and higher cetane rating, on the negative side its calorific value is about 9% lower than oil based diesel. From a commercial perspective one of the original benefits of biodiesel was the co-production of glycerol. However, at the volume now produced the market for glycerol has been swamped and new uses for this inexpensive raw material are now being sought.

6.2.2.8 Issues with Biofuels. The growing use of biofuels has been criticised by some as contributing to food shortages and the rise in prices of food crops, which has become a particular problem in some third world countries. This is due in part due to land previously growing edible crops being diverted to biofuel crops, encouraged by government subsidies, and direct competition between the two markets for crops such as rape, corn, and sunflower. This growth is also held responsible for the increase in plantations growing palm, many of which are replacing forests and endangering rare species. One potential way of preventing these issues becoming more severe is the production of biofuel from algae, which grow in water, *e.g.* at waste water treatment plants. Feasibility studies show that algae can be produced that contain around 50% oil that is suitable for biodiesel use.

6.2.2.9 Generations of Biofuels. As technology advances people are starting to differentiate between the various kinds of biofuels and use the terminology first, second and third generation.

First-generation biofuels are fuels based on sugar, starch, and vegetable oil using conventional technology. The feedstocks are seeds, grain, and corn and the products are bioethanol and biodiesel. The main concern about first-generation technology is that they use food crops and are only able to utilize a small amount of the crop, such C_6 sugars or vegetable oil.

Second-generation fuels are being developed to overcome the limitations of first-generation fuels and use non-food crops, including waste biomass like wheat and corn stalks as well as specially grown crops like miscanthus. Because of the differing feedstocks, the technologies and fuel produced are also different from the first generation. Technologies include biomass to liquid, *i.e.* conversion into a hydrocarbon feedstock *via* syngas and Fischer–Tropsch technology as discussed below, as well as materials under development such as biomethanol and cellulosic ethanol production. The latter uses lignocelluloses, the inedible woody part of the plant. Technically this is quite difficult and research is being aimed at mimicking the process that goes on in a cow's stomach, which converts grass into sugar using enzymes. More recently a fungus has been discovered with the potential to convert cellulose into a diesel-like substance.

Third-generation biofuels use algae that can produce up to 30 times more energy per unit area than other crops. Although algae are easy to grow extraction of the fuel component is often difficult. Some algae have now been cultivated that excrete ethanol that can be harvested without killing the crop.

6.2.3   Solar Power

There are two main methods for harnessing the sun's energy directly. The first, most direct and simplest method is to use sunlight as a source of heat to warm up water, for example, this is usually termed solar heating. The second more complex method involves converting the sun's energy into electricity and is known as photovoltaics. Efficient solar heating relies on the ability of glass and certain plastics such as the polyester Mylar to be transparent to light and short-wave infrared but opaque to the reflected longer wave infrared. Solar panels or collectors often consist of a series of fine copper, water-carrying, tubes encased in glass. As the sun heats the water in the tubes a siphon system is set up (Figure 6.4) that causes the hot water to rise into a heat exchanger being replaced by colder water. This system is used successfully to provide domestic hot water in many countries. Overall there are well over 20 million square meters of panels like this in use globally. This simple method can be extended, by use of a sophisticated mirror or lens system, to 'concentrate' the power to provide sufficient energy to drive steam turbines.

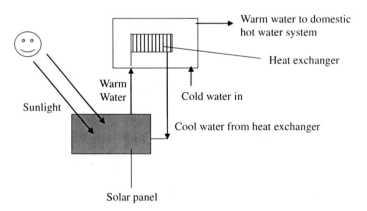

Figure 6.4   Solar heating.

Photovoltaic (PV) cells capable of directly converting sunlight into electricity are made of semiconducting materials, of which silicon is the most common. Although these cells have been available for almost 50 years it is only in the last 15 years, since efficiencies have gone up and costs have come down, that they have become commercially attractive. Growth in PV cell now runs at around 40% per year, making it the world's fastest growing energy technology. PV cells use very high purity silicon (99.9999%). Silicon has four valence electrons and in a crystal of pure silicon each Si atom is joined to four other Si atoms by sharing two electrons. By doping the silicon crystal with a small amount of phosphorous, which has five valence electrons, an n-type semiconductor is produced. This is used to form the negative side of the solar cell and is so called because of the excess of negatively charged electrons. Conversely, positive or p-type semiconductors can be made by doping the silicon with boron, which only has three valence electrons, leaving holes in the structure.

By joining n- and p-type semiconductors together a junction is set that in simple terms enables electrons to move to fill the holes. When light above a certain wavelength hits the p–n junction it gives electrons enough energy to escape the valence band (band gap energy, see Chapter 4) and leaves them free to move around the material, conducting electricity. Contacts on the front of the cell (Figure 6.5) collect the current generated and deliver it to an external circuit, and electrons are returned to the p-layer *via* a back contact. A top anti-reflective coating helps reduce reflectance and improve efficiency (defined as the percentage of sunlight converted into usable electricity).

Although the costs of silicon solar cells are falling due to improved manufacturing techniques several other cell types are either commercially available or are being studied. Cells based on gallium-arsenide are highly efficient, but very expensive and are generally only used in

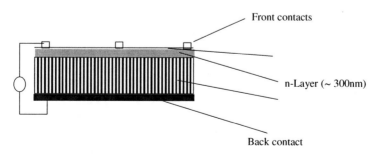

Figure 6.5 Simple schematic of a solar cell.

applications such as space, were cost is not a major concern. At the other end of the cost spectrum are cadmium telluride cells; whilst these are relatively inexpensive they have lower efficiencies than silicon cells and the presence of toxic cadmium has implications for manufacturing and eventual disposal. Although not yet commercialized, a whole range of conducting polymers is being assessed in PV cells, these include poly-anilines, poly(*p*-phenylene vinylenes) and polythiophenes. Although PV cells will continue to provide a higher proportion of electricity demand, at current prices pay back is around 12–20 years, which is often con-sidered too long for business or domestic customers. To provide sig-nificant amounts of power PV cells also require a lot of land space; one of the largest plants to be built, 550 MW in California, will take up approximately 10 square miles. Ideally very large PV plants should be built in deserts to maximize solar energy and minimize competition for land use.

6.2.4 Other Forms of Renewable Energy

The market for wind turbines is increasing at almost the same rate as solar cells. Like PV cells, this growth is a direct consequence of improved efficiency and reduced cost. Nevertheless the cost of energy generated is still high compared to conventional sources. The engineering of wind turbines is highly complex, with a whole variety of different designs being available to meet different conditions. By most standards the turbines are a very environmentally friendly but as their use grows public opposition to land based turbines is mounting. This opposition is centred around both the noise aspects, and their presence on highly visual high ground in the countryside. In many cases they are claimed to be a potential threat to migrating birds. Increasingly turbine systems are being placed offshore to avoid public opposition and harness the high winds found there.

Hydroelectricity is a major competitive source of energy for many countries. Essentially, the power generated (P) is a function of the height (H) through which water falls and the flow rate (R), as shown in Equation (6.2):

$$P(\text{kW}) \cong 10 \times P(\text{m}) \times R(\text{m}^3\,\text{s}^{-1}) \qquad (6.2)$$

Hence to generate large amounts of energy significant volumes of water must flow from great heights; this has lead to some of the largest hydroelectric power plants being amongst the largest man-made struc-tures in the world. This in turn has raised many local environmental issues connected with water diversion, flooding, *etc.* and resulted in public opposition to many new projects.

6.2.5 Fuel Cells

A fuel cell is an electrochemical system for converting hydrogen and oxygen directly into electricity, the only by-product being water and some excess heat. Figure 6.6 shows the basic design of a fuel cell. For practical purposes in many types of cell the anode and catalyst may be assumed to be a continuous layer, usually consisting of platinum on an inert support such as carbon. Hydrogen entering the anode is converted into hydrogen ions, which subsequently pass into the electrolyte layer. The released electrons, which can not enter the electrolyte, enter an external circuit generating electricity. At the cathode, protons, returning electrons, and oxygen combine to form water.

Fuel cells are the subject of vast amounts of research and most experts now predict that within the next 20 years or so they will be widely used for mass transportation. The main hurdles preventing commercial introduction are too high cost, lack of durability, system complexity, and a lack of fuel infrastructure. There are four major potential benefits to using fuel cell technology compared to more conventional sources of energy:

1. The overall efficiency of fuel cells is higher than the conventional heat engines. Running on pure hydrogen, fuel cells have a theoretical efficiency of up to 80%, and in some kinds of fuel cell practical efficiencies of over 70% have been reported. For most practical purposes modern fuel cells generally have efficiencies of over 40%.
2. Pollution, especially at the local level is minimized since, from a hydrogen-powered cell, the only by-product is water.

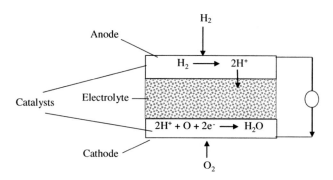

Figure 6.6 Schematic of a fuel cell.

3. There is the possibility of obtaining the fuel from renewable resources.
4. They are more reliable since there are no moving parts.

The theoretical efficiency of a fuel cell is given by the ratio between the Gibbs free energy (ΔG), which is the maximum electrical work that can be obtained, and the enthalpy (ΔH) of the fuel (6.3):

$$\text{Efficiency of a fuel cell} = \Delta G / \Delta H \qquad (6.3)$$

The Gibbs free energy for the reaction is related to the equilibrium cell potential (E_O) [Equation (6.4)]. For the reaction between hydrogen and oxygen to produce water, the number of electrons (n) per molecule participating in the electrochemical reaction is 2 and ΔG has a value of $-37.2 \, \text{kJ mol}^{-1}$, giving E_O a value of 1.23 V:

$$\Delta G = -nFE_O \qquad (6.4)$$

where $F =$ Faraday's constant.

In comparison the theoretical efficiency of a conventional combustion engine is limited by the Carnot-cycle efficiency. This efficiency [Equation (6.5)] is a function of the operating temperature (T_2) and the temperature of the surroundings (T_1):

$$\text{Carnot efficiency} = (T_2 - T_1) / T_1 \qquad (6.5)$$

Since for an engine with moving parts the operating temperature is subject to practical limitations the efficiency of the engine is usually around 20%, *i.e.* less than half that of a fuel cell.

Although hydrogen is the ultimate fuel for fuel cells it is not ideal. In the gaseous state it has a very low fuel value on a volume basis compared to other fuels and liquefied hydrogen is difficult to store and transport. Hence, particularly for use in transport applications, its use poses many practical problems, not least the weight of the storage container required. One commercial source of 'solid' hydrogen is being marketed under the name Powerball™; powerballs are sodium hydride pellets coated with polyethylene for safe transport. The hydrogen generation unit consists of a tank containing powerballs, water, and a pressure regulated cutting device. Below a certain pressure a single powerball is cut to allow the sodium hydride to react with water to produce hydrogen and sodium hydroxide. If the hydrogen is not immediately required once the pressure reaches a certain limit no more balls are cut. Each liter of powerballs contains the equivalent of around 4 kWh of electricity. The

$$CH_3OH \rightarrow CO + 2H_2$$
$$CH_4 + H_2O \rightarrow CO + 3H_2$$
$$CO + H_2O \rightarrow CO_2 + H_2$$

Scheme 6.2 Methanol and methane reforming.

sodium hydroxide and cut polyethylene balls produced can be returned to the filling station for recycling. Recent advances in nanotechnology offer potential longer term storage options, *e.g.* graphite based nano-fibers can take up 65% of their own weight as hydrogen.

Not being a natural resource hydrogen needs to be produced; the most common method of production being steam reforming of methanol, natural gas, or even petroleum. The reforming process consists of treating the methanol or methane with steam at high temperature over a catalyst, usually nickel, the initial products being hydrogen and carbon monoxide (Scheme 6.2). Since CO is both toxic and a poison for most fuel cell catalysts it must be removed. This is achieved by reacting the CO with water, usually over an iron based catalyst at $400\,^{\circ}C$, to produce more hydrogen and carbon dioxide. Even traces of CO, and other materials such as sulfur commonly found in the feedstocks, cause detrimental effects to the efficiency and lifetime of the fuel cell if left in the hydrogen.

In principle, biomass is a useful fuel for fuel cells; many of the technologies discussed above for using biomass as a fuel produce either methane or hydrogen directly and as highlighted below synthesis gas production from biomass for conversion into methanol is an attractive option. Cellulose based material may be converted into a mixture of hydrogen (70% hydrogen content recovered), CO_2, and methane by high temperature treatment with a nickel catalyst.

The ultimate renewable long-term fuel source could be water. In an ideal system water could be electrolyzed into hydrogen and oxygen using photochemically generated electricity. The hydrogen and oxygen could then be used as fuel for the fuel cell, which produces water for recycle. Although efficiencies need to be improved this option offers an extremely safe transport fuel. An alternative approach being extensively studied is biophotolysis in which water is enzymatically split into hydrogen and oxygen; although many years away, this may be a lower energy alternative. Hydrogenase enzymes, which can carry out this process, are found in certain types of green and blue-green algae. The process occurs under anaerobic conditions but, unfortunately, oxygen produced in the process tends to limit productivity. Genetic engineering techniques are being sought to overcome this problem.

Table 6.3 Fuel cell types.

Fuel cell type	Electrolyte	Operating temperature (°C)
Phosphoric acid	Conc. H_3PO_4	~200
Solid oxide	Ceramic/ZrO_2	~1000
Proton-exchange membrane	Conducting polymer	80–90
Molten carbonate	Ca or Li/K carbonates	>600
Alkaline	25–85% KOH	80

There are a whole variety of types of fuel cell, named after the electrolyte used, each operating at a preferred temperature range with its own feedstock purity criteria (Table 6.3).

6.2.5.1 Phosphoric Acid Fuel Cell (PAFC). The PAFC is the most established type of cell, and is used in some power generating projects, *e.g.* in hospitals and offices with outputs of a few megawatts of electricity. The latest cells now use 100% phosphoric acid as the electrolyte supported between polytetrafluoroethylene (PTFE) plates; this high concentration of electrolyte allows the cell operate at just over 200 °C and above atmospheric pressure, giving relative efficiencies of over 40%. The cell operating temperature is low enough to enable the steam produced to be efficiently used in cogeneration (*i.e.* to drive a steam turbine) but it can also be used for hot water generation and space heating, raising the overall efficiency of the cell. Platinum supported on carbon black and PTFE is commonly used as the electrode and whilst this cell may also be used in conjunction with an external methane or alcohol reformer the anode is readily poisoned by low levels of CO. Current research is targeted at cost reduction and efficiency improvements, especially through reduction of the amount of Pt used in the cell, and trying to overcome corrosion problems (of the carbon black) at high cell voltages.

6.2.5.2 Solid Oxide Fuel Cell (SOFC). This type of cell operates at the highest temperature (1000 °C) of all current fuel cell types. This high operating temperature has many advantages, including producing steam suitable for cogeneration, the ability to reform many fuels efficiently, and the overall fast kinetics of the system; however, operating at high temperatures severely restricts the type of material that can be employed. This cell uses a ceramic electrolyte composed of zirconia stabilized with yttria (~8 mol.%), this system having oxygen ion vacancies within the lattice. This solid electrolyte has many practical cell design advantages compared to the PAFC. The anode is composed mainly of zirconia with some nickel. At high temperatures the Ni tends to agglomerate, although the zirconia does help prevent this it is still a cause of concern for long-term efficient operation of these cells. The cathode is composed of doped lanthanum manganite ceramic. Owing to high temperature steam production overall efficiencies of SOFC can reach 70% or more and hence they are mainly being targeted at large volume electricity generation projects, although use in transport applications is also being studied.

6.2.5.3 Proton Exchange Membrane Fuel Cell (PEMFC). This type of cell is sometimes referred to as the 'polymer electrolyte fuel cell' or the 'solid polymer fuel cell'. Perfluorinated sulfonic acid–tetrafluoroethylene copolymers (Nafion) are used as membrane materials due to their relatively high thermal and chemical stability as well as their ability to conduct protons efficiently. Dehydration is a concern with these membranes and water management must be carefully controlled. Because of the low operating temperature (80 °C) highly efficient electrodes made of platinum impregnated onto carbon or PTFE are required. These are sensitive to impurities in the fuel; hence pure hydrogen and oxygen (or air) are the preferred choice. Because of the low operating temperature the steam produced is of little value, hence this cell is largely being studied for transport applications were this is not an issue. The light weight of PEMFCs enhances their suitability for transport applications.

6.2.5.4 Molten Carbonate Fuel Cells. As the name suggests these cells use an electrolyte of molten carbonates (generally of lithium and potassium). To keep the carbonates molten and provide good conductivity the cells need to operate at around 650 °C. This type of cell is becoming increasingly favored for commercial power production. The moderate operating temperature means that construction materials do not need to be as exotic as for SOFC, although corrosion may be a problem. The steam generated is at more useful temperature than that generated in PEMFCs, giving this type of cell a significant economic advantage. The basic cell is around 55% efficient in electricity production, with total efficiencies of over 80% being achievable if all forms of steam generated heat are used efficiently. The charge transfer species in this cell is the carbonate ion rather than the proton. Early expensive electrodes based on precious metals have more recently been replaced with nickel alloys for the anode and lithium oxides for the cathode. Internal natural gas reforming is a real possibility for this type of cell.

6.2.5.5 Alkaline Fuel Cells (AFCs). AFCs were initially used in space and military applications because of their high efficiencies ($\sim 70\%$) and even today the space shuttle is powered by an AFC, which also provides a pure source of drinking water. They are quite versatile and can operate within a wide range of electrolyte concentrations (aqueous KOH), temperatures, and electrode types. The main problem for general commercial applications is that they are sensitive to CO_2, which reacts with the electrolyte to form potassium carbonate, which reduces cell efficiency. Hence highly pure oxygen must be used, which considerably adds to the cost of running the cell. Although a prototype taxi is powered by an AFC most experts agree that other types of fuel cell offer greater commercial possibilities.

Fuel cells for transport applications are the subject of intensive research by vehicle manufacturers, oil companies, and cell manufacturers; they are generally considered to be the best option for future pollution-free transport. To be competitive with conventional petrol/diesel/electric power units mass production will be needed to bring down the price of a fuel cell. Many in the industry believe it will be at least 2020 before this can be achieved. There are, however, numerous demonstration projects being conducted throughout the world, including major long-term trials with buses in Iceland and Chicago. The NECAR demonstration projects run by Daimler Benz have been going on since 1994, using PEMFC cells. Initially, a stack of 12 cells, occupying most of a Mercedes Benz 190 van, was required to deliver 50 kW of power. By 1997 technology had been improved to the extent that the cells would comfortably fit inside a Mercedes Benz A-class car. The fuel had also been changed from hydrogen to the more convenient methanol, a tank of 40 liters providing enough fuel for 400 km.

6.3   CHEMICALS FROM RENEWABLE FEEDSTOCKS

Other than for energy use the main current and future applications for renewable feedstocks are likely to remain in the following areas:

- lubricants,
- fibres & composites,
- polymers,
- solvents,
- speciality chemicals & surfactants, including dyes & paints,
- agrochemicals,
- pharmaceuticals.

The overall market for crop derived materials is estimated to rise by over 50% in the next 10 years, with products based on vegetable oils becoming increasingly important. A large proportion of the growth will be in the lubricants area – a combination of stringent legislation on the use of mineral oil lubricants coupled with performance advantages are driving the change in areas as diverse as chainsaw lubricants and hydraulic fluids in farm vehicles.

Nevertheless, despite Government and society's requirement for sustainable development, interest and commercialization of products (other than biofuel) from renewable resources is not well advanced. This is largely due to several significant barriers preventing effective competition between renewable and non-renewable resources. These barriers occur throughout the manufacturing and product supply chain. The largest barrier is concerned with cost of product, especially the difference between the cost to manufacture and the perceived value of the product. The basic chemical difference between oil based and crop based feedstocks is that the former are devoid of oxygen whilst the latter are oxygen rich. This presents a technical barrier since many large-scale oxidation processes are now well developed and competitive whilst reduction (of carbon–oxygen bonds) is much less well advanced on an industrial scale. To minimize this effect, different, oxygen rich, products have been developed from renewable resources but the cost of developing markets for new products is a significant entry barrier. This market entry barrier is compounded by the general lack of consistency in the basic feedstock ingredients; hence customer confidence that they will be able to purchase consistent quality product as and when needed is low. Improvements in understanding and controlling plant metabolic pathways should help overcome this issue. Currently market pull for these new products is not a significant driver in most instances.

Transport and distribution are often significant barriers; the chemicals infrastructure has been built to handle crude oil, and hence many processing plants are built near the sea. Generally most crops are grown away from coastal and industrial regions; either expensive transport is required to deliver the products to current processing plants or else new processing plants need to be built in the regions where the crops are grown. Developments in intensification and cost reduction of production plants should help this situation. Because of the complex nature of the various mixtures that can be derived from renewable feedstocks, separation processes can be costly and inefficient, especially when coupled with the highly dilute solutions often encountered. Advances in processing systems such as reactive catalytic distillation and efficient membrane separation processes are needed to reduce this technical barrier. One possible solution is to convert all the biomass into a universal consistent feedstock such as synthesis gas or methanol, which can then be used with current technologies.

Box 6.2   Fatty acid nomenclature

Fatty acids are often described by numbers such as 16:0, 18:1, or 18:3. The first number (16 or 18) is used to describe the number of carbon atoms in the chain whilst the second (0, 1, 3) gives the number of C=C double bonds in the molecule. Palmitic acid can, therefore, be shortened to 16:0 whilst oleic acid is 18:1.

6.3.1   Chemicals from Fatty Acids

Whilst the fatty acid components of vegetable oils may be used to provide a renewable energy source in the form of biodiesel they may also be used as valuable chemical feedstocks (see Box 6.2 for common nomenclature). Currently over 100 million tonnes of oils and fats are produced annually, nearly 25% of this coming from soya. Most of the oils and fats produced go into human or animal food production but some 14% are used in chemicals manufacture. The mix and concentration of acids present in most crop derived oils does raise some economic issues, since the desired product is often only present in relatively small amounts, making extraction complex and expensive. Natural breeding methods have been used for many years to try and increase the desired components in industrial crops; in recent times this method has been supplemented by genetic engineering techniques. By using these techniques to produce variant strains of rapeseed, for example, the erucic acid (*cis*-13-docosenoic acid) component in the fatty acids can be varied from 0% to over 50%, for example, whilst the lauric acid component can be varied from 0% to some 37%. With sunflower seeds the oleic acid content has been increased to over 92%, providing a highly valuable source of this material.

   As with biodiesel, the fatty acid component needs to be isolated from the naturally occurring triglyceride. Early methods based on saponification – heating with sodium hydroxide – are no longer used commercially due to difficulties in obtaining the useful by-product glycerol. Direct hydrolysis methods sometimes using acid catalysts are preferred. The triglycerides are quite stable in water, several days at 100 °C being required to obtain significant hydrolysis. In the absence of catalysts commercial processes run at temperatures over 210 °C at high pressure. If small amounts of sulfuric acid or, more usually, zinc oxide are added the temperature may be reduced to around 150 °C. Following hydrolysis, water and low-boiling components are removed by distillation, usually in the absence of air to prevent oxidation of the unsaturated acids. A series of vacuum distillations is then carried out to obtain

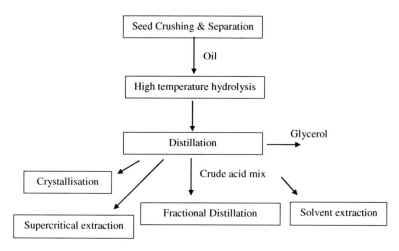

Figure 6.7 Typical fatty acid isolation procedure.

various fatty acid fractions (Figure 6.7). If high purity products are required fractional distillation, crystallization, or solvent extraction techniques are additionally employed. Separation of oleic and stearic acids is a common problem; the use of solvents such as methanol or acetone can preferentially extract the oleic component. Supercritical fluid (CO_2 or ethene) extraction processes are also available and also preferentially extract the unsaturated component.

Some more recent processes have been developed that involve direct hydrogenation of the oil to the fatty acid and 1,2-propanediol. These high temperature ($>230\,^{\circ}$C) and high pressure processes generally use a copper chromium oxide catalyst.

6.3.1.1 Some Chemical Transformations. Fatty acids, especially unsaturated ones, offer many possibilities for conversion into a range of chemical feedstocks, as exemplified in Schemes 6.3 and 6.4. Most of the fatty acids produced are either converted into the alcohol for subsequent conversion into surfactants or else transformed into metal salts for uses as 'soaps'.

Whilst the acids and many of there derivatives currently find niche applications in the market sectors identified above, factors related to price, volume of supply, and consistency have all limited commercial viability. Longer term reduced cost and improved consistency through improved growing and harvesting techniques coupled with increased requirement for biodegradability will increase demand for fatty acids.

By way of a specific example let us consider erucic acid. The main commercial source of erucic acid is a specially bred form of rape seed

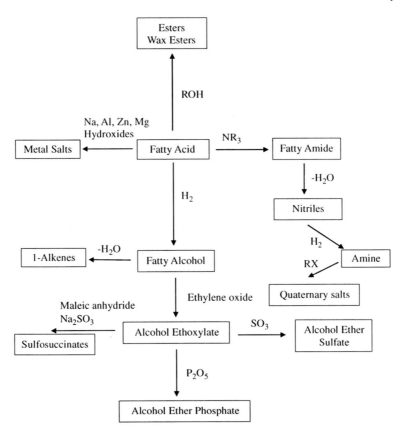

Scheme 6.3   Some transformations of the acidic function of fatty acids.

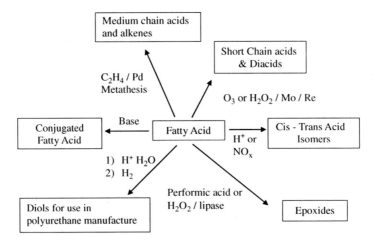

Scheme 6.4   Some transformations of the alkene function of fatty acids.

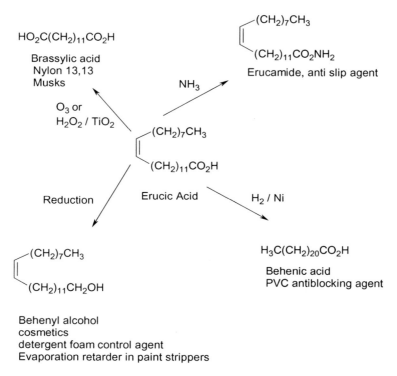

Scheme 6.5 Some uses of erucic acid.

(HEAR), as pointed out above. With European consumption being around 70 000 tpa, almost 45 000 ha of land is used to grow rapeseed for erucic acid production in Europe. The high level of erucic found in this type of rape seed oil make it unsuitable for human consumption due to the indigestibility of such large amounts of this acid. Erucic acid is also the major fatty acid to be found in nasturtium and crambe seeds (up to 75% and 56%, respectively), and it is also found in the salad herb rocket.

There are several applications for the acid (Scheme 6.5). The current major use is in the production of erucamide, a 'slip agent' used in the manufacture of polythene bags to make them open more easily.

The important, but unusual fatty acid, ricinoleic acid or 12-hydroxyoleic acid is a major component of castor oil (>87%) and is also found in useful quantities in ergot. The metal salts of the acid find use in dry cleaning soaps but the majority is converted into aminoundecanoic acid (Scheme 6.6) which is used to make nylon 11. Nylon 11 has very good chemical and shock resistance properties, which have led to it being used in the automotive industry. Ricinoleic triglyceride is initially transesterified to the methyl ester, which is heated to 300 °C, at which temperature it is cleaved into methyl undecanoate and an aldehyde. The

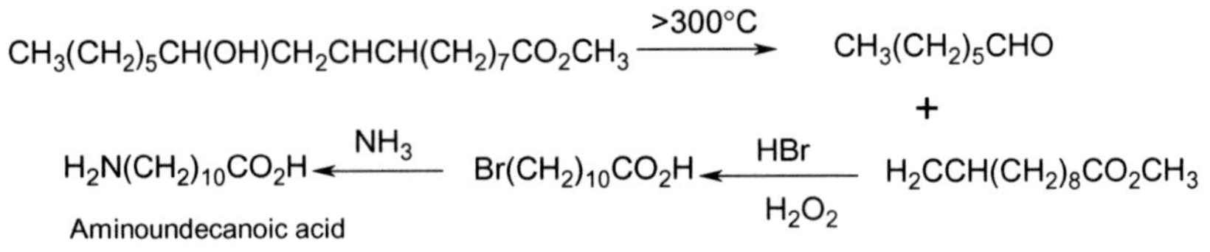

$$CH_3(CH_2)_5CH(OH)CH_2CHCH(CH_2)_7CO_2CH_3 \xrightarrow{>300°C} CH_3(CH_2)_5CHO$$

$$+$$

$$H_2N(CH_2)_{10}CO_2H \xleftarrow{NH_3} Br(CH_2)_{10}CO_2H \xleftarrow[H_2O_2]{HBr} H_2CCH(CH_2)_8CO_2CH_3$$

Aminoundecanoic acid

Scheme 6.6 Conversion of ricinoleic acid into aminoundecanoic acid.

ester is then converted into 11-bromoundecanoic acid using HBr and hydrogen peroxide. The bromide is subsequently substituted with an amine group to give the desired product.

Linolenic acid is also important industrially; it is the major constituent of linseed oil (*ca* 47%), which is obtained from flax. The high degree of unsaturation present in this acid makes the oil an excellent drying agent for use in paints, varnishes, and inks.

Fatty acids are also valuable starting materials for the production of fatty nitrogen compounds, which are useful surface agents and find use in everything from shampoos to treating textiles. One example is distearyl(dimethyl)ammonium chloride, which is used as a fabric conditioner.

6.3.1.2 Lubricants. Lubricants are one of the fastest growing markets for natural fatty acid based products although at present this market only consumes about 2% of the non-food, non-fossil oil used in Europe. By lowering the coefficient of friction between moving parts, thereby allowing surfaces to slide over one another more easily and hence preventing wear and increasing component lifetime, lubricants have an important role to play in the development of sustainable products. Mineral oil lubricants are increasingly becoming environmentally unacceptable due to their low levels of biodegradability and potential for causing long-term damage, particularly in 'total loss' situations such as chain saw use, railway rails, and cables on cranes. Legislation is now being introduced in some countries, *e.g.* Germany, that effectively bans use of non-biodegradable lubricants in these 'total loss' applications. As well as having good lubricant properties the oil also needs to have good oxidative, temperature, and ageing properties. In the most stringent applications such as in jet engines, hydrocarbon oils leave something to be desired and aliphatic esters of dicarboxylic acids such as bis(2-ethylhexyl) sebacate and the bis-C_8 oxo alcohol ester of adipic acid have become the lubricants of choice. Although some sebacic acid, $HO_2C(CH_2)_8CO_2H$, is manufactured commercially from castor oil it is currently more economically obtained synthetically from 1,7-octadiene and a competing electrochemical route

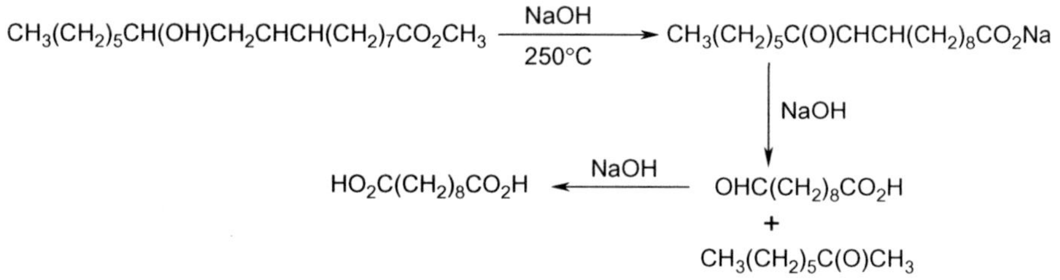

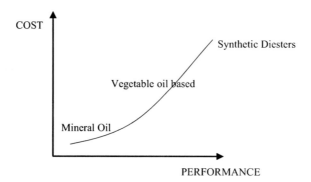

Scheme 6.7   Synthesis of sebacic acid.

Figure 6.8   Cost–performance for lubricants.

(Chapter 7). The commercial process from castor oil involves alkaline fusion of ricinoleic acid and proceeds along the lines of Scheme 6.7.

Natural fatty acid based lubricants fall between mineral oil and the synthetic acid diester lubricants in both performance and price (Figure 6.8); as such they find niche applications. Compared to mineral oils vegetable oil base lubricants offer several advantages, notably:

- higher lubricity,
- higher viscosity index, *i.e.* the viscosity changes less with temperature,
- lower evaporation losses,
- lower toxicity,
- higher biodegradability.

The physical properties of a lubricant, such as pour point, are as important as chemical properties such as oxidative stability; both of these are determined by the structure of the fatty acid. The most common vegetable oils used in lubricant applications are rapeseed, canola, and soybean; these contain mixtures of 16:0, 18:0, 18:1, 18:2, 18:3, and 22:1 acids. Increasing levels of di- and tri-unsaturation reduce the oxidative and thermal stability of the oils whilst high levels of unsaturated acids raise the pour point, limiting low temperature performance. Some of these challenges can be met by using chemical additives such as anti-oxidants or by chemically modifying the oil by partial hydrogenation, for example. There is also a significant amount of biotechnology research taking place to tailor the perfect lubricating oil, by having a high oleic acid content with just the right amount of saturated components to impart stability.

6.3.1.3 Surfactants. The surfactants market is very large – including soaps it is approximately 3 million tpa in Europe. This offers significant scope for use of renewable feedstocks. Most commonly surfactants are encountered in the form of household detergents. As discussed in Chapter 2 early detergent formulations contained non-biodegradable alkylbenzene sulfonates made from propene tetramer. Most of the detergent actives produced end up in the sewerage system or are expelled directly into rivers and the sea. It is, therefore, important that these detergents should be benign and degrade within a reasonable timescale. There are over 60 different classes of surfactant and many of these are generally considered to be benign and adequately degradable. There is continuing pressure to produce more readily degradable surfactants. In some Scandinavian countries, for example, residual amounts of linear alkylbenzene sulfonates, the highest volume surfactant after soap, have been detected in cold, slow flowing rivers in amounts that may cause harm to fish, resulting in calls to replace these materials by even more degradable ones.

There are many types of surfactants based on renewable feedstock such as starches and sugars; this section discusses only those based on fatty acids. In total the European surfactants sector consumes over half a million tpa of oils from renewable sources. Fatty alcohol ethoxylates are the largest class of surfactants based on renewable oils. Although the market for fatty alcohol ethoxylates is growing, less expensive synthetic alcohols are being increasingly used. The procedure for producing fatty alcohols from acids is outlined in Scheme 6.3 and involves hydrogenation of the acid or methyl ester at 200 bar and over 250 °C using a catalyst based on copper and chromium oxides. Ethoxylation with ethylene oxide (EO) is often carried out at 130–180 °C, usually using sodium or potassium hydroxide as a catalyst. The ethoxylation reaction has some similarities with a step polymerization reaction and gives a product with a molecular weigh distribution due to different molecules having different degrees of ethoxylation. The degree of ethoxylation quoted, *e.g.* 7 or 9, is the average degree of ethoxylation of that product.

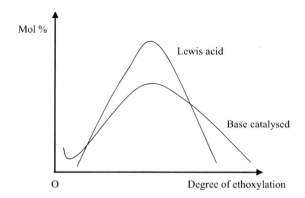

Mol %

Lewis acid

Base catalysed

O   Degree of ethoxylation

Figure 6.9 Ethoxylation molecular weight distribution with base and Lewis acid catalysts.

For some applications products with a narrow molecular weight distribution are required. To some extent this is controlled by the reaction conditions and more importantly by the catalyst. Lewis acid catalysts give narrower molecular weight distributions and lower levels of residual alcohol than bases (Figure 6.9), but difficulties in removal from the final product have largely prevented their commercial use. Several heterogeneous catalysts have been developed in attempts to overcome this problem; those based on calcined hydrotalcite [$Mg_6Al_2O_5(OH)_2$] show particular promise.

The surfactant properties of simple materials like fatty alcohol ethoxylates depend, to a considerable extent, on the HLB (hydrophilic–lipophilic balance), with ethylene oxide being the hydrophile and the carbon backbone the lipophile. Most detergent applications require materials with a HLB in the range 11–15. The approximate HLB value of an alcohol ethoxylate can be determined by dividing the percentage weight of EO in the molecule by five. Common alcohol ethoxylates manufactured include those from C_{12}–C_{14} alcohols with 7 moles of EO. For performance reasons these materials are gradually replacing C_{16}–C_{18} alcohols with 11 moles EO. The lack of natural materials, except coconut and palm oils, containing high levels of C_{12}–C_{14} acids is one reason why use of renewable feedstocks in this area is declining. As well as being valuable surfactants in their own right fatty alcohols and their ethoxylates may be reacted with sulfur trioxide to give the corresponding sulfates. One common such material is sodium dodecyl sulfate, which is made by hydrogenation of the C_{12} acid from coconut oil followed by sulfonation. Quaternary ammonium salts derived from fatty acids (Scheme 6.3) by a process of amination and reduction are also valuable surfactants.

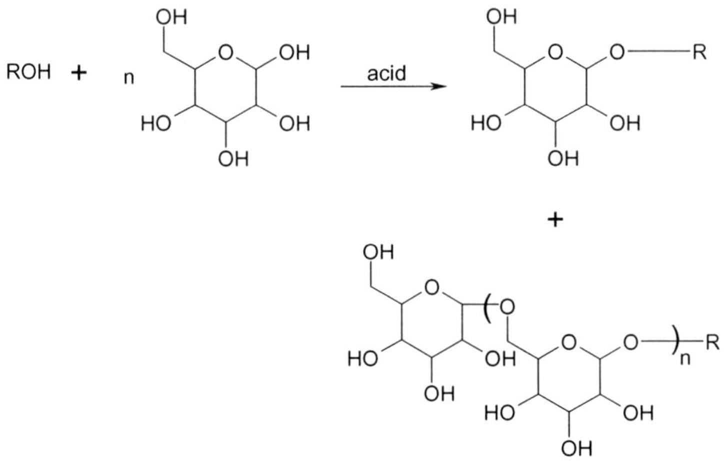

Scheme 6.8   Synthesis of alkyl polyglucosides.

Although partially naturally derived surfactants like fatty alcohol ethoxylates have environmental advantages over their petroleum based competitors the use of carcinogenic materials such as ethylene oxide detracts from their greenness. There has been considerable recent interest in a class of surfactants called alkyl polyglucosides. These materials are derived from fatty alcohols (hydrophobic part) and glucose (hydrophilic part). The compounds are formed by an acid-catalysed etherification known as the Fisher glycosidation (Scheme 6.8). The reaction is complex, partially due to the insolubility of glucose in fatty alcohols. The product of these reactions usually has, on average, between 1.2 and 1.5 glucose molecules reacted to each alcohol molecule, these being referred to as the degree of polymerization.

Alkyl polyglucosides (APGs) have very good biodegradability, very low toxicity, and are extremely mild, making them suitable for use in cosmetics. By altering the alkyl group and controlling the degree of polymerization APGs with a range of properties can be obtained, *e.g.* from high foaming to low foaming. One interesting property of APGs is there solubility and stability in concentrated alkaline solutions (up to 40 wt% sodium hydroxide). Because of this they are finding increasing use in textile finishing and industrial cleaning applications.

6.3.2   Polymers from Renewable Resources

The major arguments for increased use of polymers from renewable resources centre around their improved biodegradability compared to most petroleum based polymers and the CO_2 neutrality over their lifetime. Neither of these arguments should be taken completely at face value. To get a true picture of the 'greenness' a full LCA must be carried out. For instance the use of energy in sowing, growing, harvesting, transporting, extracting, and modifying renewable feedstocks may be high compared to production of a commodity polymer like polyethene. For example, some LCA studies comparing the energy involved in production and recycle of paper and plastic bags suggests that plastic bags are the preferred option, from an energy perspective. Although it is generally true that polymers from renewable resources are intrinsically biodegradable chemical modification can dramatically reduce this. Cellulose is, of course, very biodegradable but in most of the useful cellulose based polymers a high proportion of the hydroxyl groups have been acetylated, this modification very significantly reduces the biodegradability of commercial products. That said the increased use of polymers based on renewable resources does have an important role to play in sustainable development. Polymers, however, do need to be produced efficiently with derivatization kept to a minimum. Recent consumer preferences for natural materials are starting to have a positive effect on the production of polymers from renewable resources. The floor covering linoleum (originally called oilcloth because it was produced by polymerizing linseed oil) is seeing something of a renaissance with the requirement for linseed oil now exceeding 100 000 tpa. Although this is more than double the amount used in 1975 it is still lower that the requirement in the late 1950s and early 1960s.

6.3.2.1 Polyhydroxyalkanoates (PHAs). This class of polymer was first launched commercially by ICI in 1990 under the trade name Biopol. Despite high hopes for mass commercial production of this material it has been slow to take off as a commercial polymer. In 1996 the business was sold to Monsanto who later sold it to Metabolix, who also had a small business producing PHAs.

PHAs are produced by certain classes of bacteria when they are grown on specific substrates, usually in the absence specific nutrients, especially nitrogen. Under these conditions the bacteria lay down an energy store for the future, in a similar way to the lay down of fat in animals about to hibernate. The most common method of production involves fermentation of a glucose solution (Scheme 6.9), containing propanoic acid, but in the absence of nitrogen with *Alcaligenes eutrophus*. The presence of the propanoic acid encourages the bacteria to produce copolymers of polyhydroxybutyrate containing 5–20 wt% of polyhydroxyvalerate.

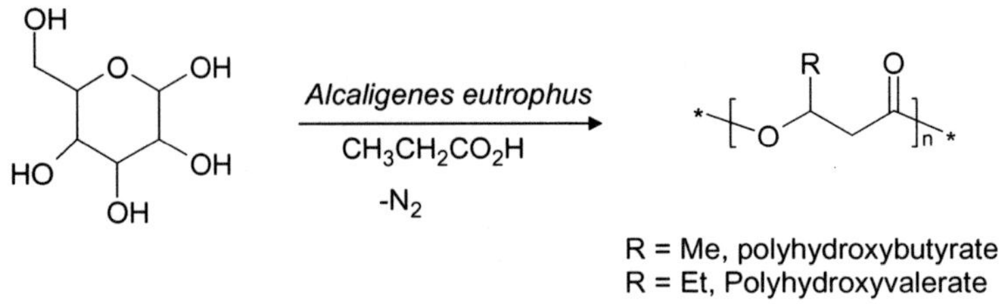

R = Me, polyhydroxybutyrate
R = Et, Polyhydroxyvalerate

Scheme 6.9   Production of polyhydroxyalkanoates.

This copolymer is more workable and flexible than pure poly-hydroxybutyrate, which is produced in the absence of propanoic acid. The actual production process is thought to involve three enzymes: initially two molecules of acetyl coenzyme A are linked, one of the ketone functions is then reduced by NADPH into the basic hydroxyalkanoate building block, which is polymerized by a third enzyme. PHAs are deposited in intracellular inclusions and are recovered by centrifuging the biomass, disrupting the cell walls, and recovering the PHA as a white powder, after washing and drying. Even though, under favourable conditions, PHAs can make up over 80% of the dry mass of the cell the polymer is still expensive to produce. This high cost has prevented rapid commercialization of the product, which at its peak was only produced at around 1000 tpa. Attempts to reduce costs have centred on using lower cost substrates such as palm oil waste and wastewater from distilleries.

Recently, cost reduction methods have focussed on genetic engineering techniques to produce PHAs directly in cells of fast growing crops such as sugar cane and corn and in poplar trees. Although at an early stage, genetic engineering technology has been successful and produces PHAs identical to the fermentation process; the methods of extracting the PHAs are still relatively costly, however, and often require use of noxious solvents like chloroform.

The huge interest in PHAs is largely a result of the polymers properties. Polymers can be produced, by the addition of tougheners, that have similar properties to polyethene and polypropene; hence if the price were right there would be a large established market for these materials. In particular, PHAs can be produced to match polyolefins in terms of tensile strength, melting point, and glass transition temperatures. This enables PHAs to be processed on existing equipment, which is almost essential for successful commercialization. The biodegradability of PHAs is also an important commercial driver, with Greenpeace effectively using this to market the first biodegradable credit card.

Table 6.4 Comparison of renewable and non-renewable routes to lactic acid.

Parameter	Renewable	Non-renewable
Energy use	High	Lower
Hazard potential	Low	High
Waste generation	High	Low
Nature of waste	Benign	Non-benign contamination?
Feedstock	Renewable	Non-renewable
Plant size	Larger	Smaller

6.3.2.2 Polylactates. Polylactates are an interesting class of bio-degradable polymers that may be made from either renewable or petroleum feedstocks. The synthesis of lactic acid raises real issues concerning the relative greenness of the renewable and non-renewable (HCN) route, as discussed in Chapter 2. Table 6.4 summarizes a comparison of the 'greenness' of both routes. Without a full life cycle assessment the choice of route on environmental grounds is not easy and at least partly depends on plant location and raw material availability.

Developments by Cargill have made the renewable route, based on fermentation of corn starch with *Lactobacillus acidophilus*, more competitive. In this ingenious process intermediate calcium lactate is reacted with a trialkyl amine and carbon dioxide. This produces amine lactate and regenerates calcium carbonate for reuse. When heated in hot water the amine lactate decomposes to high purity lactic acid and releases the amine for reuse. Overall, this route avoids the requirement for an esterification purification step and avoids the problem of producing significant amounts of calcium sulfate based waste. By far the major worldwide production of lactic acid now comes from renewable resources. Other than production of polylactic acid the main industrial outlet is in production of ethyl lactate, an increasingly common non-toxic degradable solvent. Applications are varied and include paints and automotive finishes as well as industrial solvent applications.

Despite high production costs polylactic acid (PLA) has been produced for many years as a high value material for use in medical applications such as dissolvable stitches and controlled release devices. The very low toxicity and biodegradability within the body made PLA the polymer of choice for such applications. In theory PLA should be relatively simple to produce by simple condensation polymerization of lactic acid. Unfortunately, in practice, a competing depolymerization process takes place to produce the cyclic lactide (Scheme 6.10). As the degree of polymerization increases the rate slows down until the rate of depolymerization and polymerization are the same. This equilibrium is

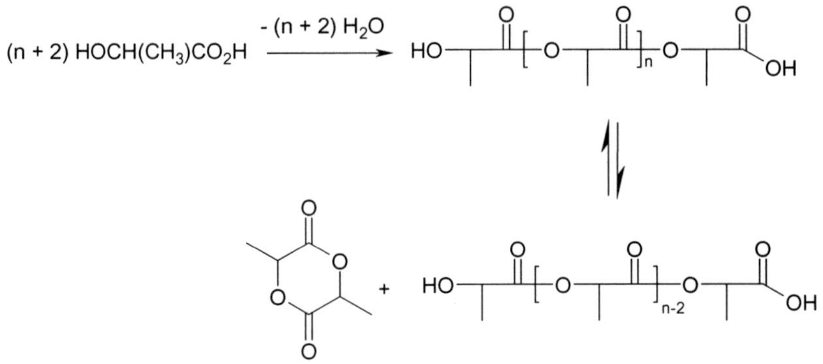

Scheme 6.10   Polymerization of lactic acid.

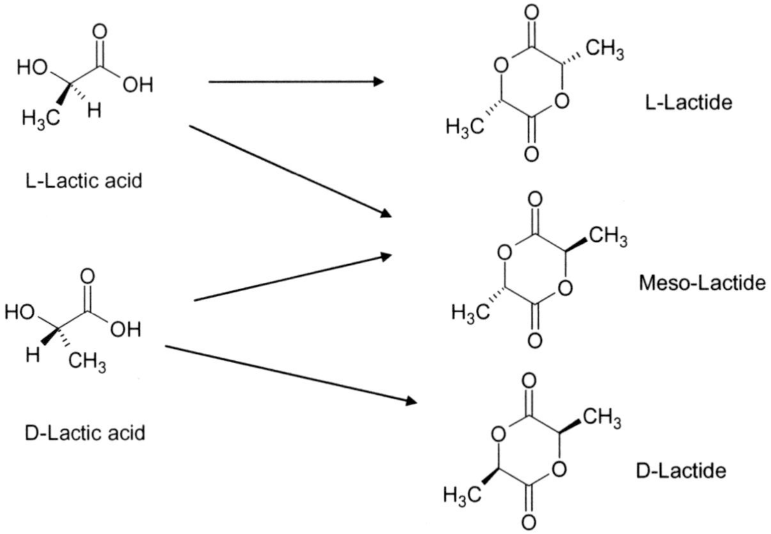

Scheme 6.11   Lactide optical isomers.

achieved before commercially useful molecular weights of PLA have been formed.

The solution to this problem has been to isolate the lactide and to polymerize this directly using a tin(II) 2-(ethyl)hexanoate catalyst at temperatures between 140 and 160 °C. By controlling the amount of water and lactic acid in the polymerization reactor the molecular weight of the polymer can be controlled. Since lactic acid exists as D- and L-optical isomers, three lactides are produced: D, L, and meso (Scheme 6.11). The properties of the final polymer depend not only on the molecular weight but vary significantly with the optical ratios of the lactides used. To get specific polymers for medical use the crude lactide mix is extensively recrystallized, to remove the meso isomer, leaving the required D, L mix. This recrystallization process results in considerable waste, with only a small fraction of the lactide produced being used in the final polymerization step. Hence PLA has been too costly to use as a commodity polymer.

A more cost-effective process for producing PLA at a scale of 140 000 tpa has now been developed. In essence the process relies on partially controlling the lactide isomer ratio by using a catalyst of tin oxide to enhance conversion of lactic acid oligomer into lactide and then using sophisticated distillation technology to isolate the required ratio of lactide isomers. Unwanted lactide is recycled back into the lactic acid feed, resulting in racemization. By careful control of the process PLAs with varying ratios of D- and L-lactic groups can be produced. This enables a variety of polymers, for various applications, to be produced from a single plant. The process has better than 95% yield, requires 30–50% fewer fossil resources than conventional plastics, and results in a 30–60% reduction in greenhouse gas emissions. Using this technology it is possible to produce PLA fibres for clothing, films for food packaging and agrochemical use, and bottles. PLA has particularly good mechanical properties, which could make it a suitable substitute for polystyrene.

6.3.2.3 Other Polymers from Renewable Resources. There are many other polymers that have been, are, or could be produced from renewable feedstocks. There are an even larger number of polymers that can be partially produced from renewable feedstocks. In terms of commercial production, starch based polymers are by far the largest. The major use of starch based polymers is in the paper industry but other applications include agricultural film, bags, and nappy back-sheets. Starch is also incorporated at small levels (6–15%) into some polyolefin materials to increase biodegradability. The success of this approach is still doubtful and largely depends on the degradation conditions. For many years cellophane was a popular packaging material made from cellulose. The manufacturing process involved treating the cellulose with sodium hydroxide followed by carbon disulfide. The resulting xanthate was then spun into a cellulose fibre or cast into a film by passage through an acid bath. Largely due to the environmental hazards of handling large amounts of CS_2 the product was superseded by petroleum based polymers.

The liquid obtained from roasted cashew nut shells (CNSLs) contains a high proportion of cardinol (Scheme 6.12) that consists of a mixture of

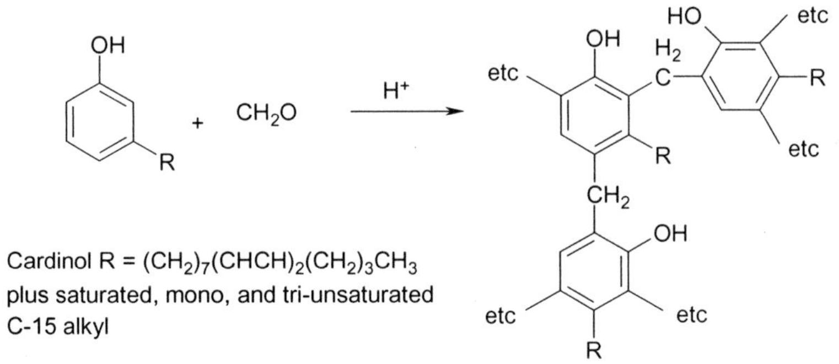

Cardinol R = (CH₂)₇(CHCH)₂(CH₂)₃CH₃
plus saturated, mono, and tri-unsaturated
C-15 alkyl

Scheme 6.12   Phenol resin friction materials from CNSLs (cashew nut shells).

saturated and (mainly) unsaturated C-15 *meta*-alkylphenols. CNSL–formaldehyde resins (sometimes incorporating further phenol) have long been used in car brake-linings. The resins, incorporating inorganic fillers have very good friction properties, particularly for low temperature braking. They also have good thermal resistance as well as producing less noise than may alternatives.

   The other major outlet for CNSL based polymers is in surface coatings. Generally, these polymers consist of low molecular weight resins made by thermally treating CNSL at around 100 °C, which causes oligomerization through the carbon–carbon double bonds. Varnishes made with CNSL resins have very good toughness, gloss, and adhesive properties; because of the very dark colour of CNSL applications are limited to black varnishes and primers. More recently CNSL–formaldehyde resins have been used to make waterproof roofing materials by impregnating natural fibres such as sisal. This low cost option may offer significant opportunities in developing countries such as Brazil and India, which produce both raw materials.

6.3.3   Some Other Chemicals from Natural Resources

Some of the potential uses of the fats and oils found in plants have been reviewed and briefly some uses of the carbohydrate based polymers have been discussed. Plants contain a whole variety of other chemicals, including amino acids, terpenes, flavonoids, alkaloids, *etc*. When the potential for these naturally occurring materials is combined with the secondary products that can be obtained by fermentation or other microbial processes or traditional chemical transformations the array of chemicals that can simply be created from renewable resources is huge. This section considers a few of the more interesting examples.

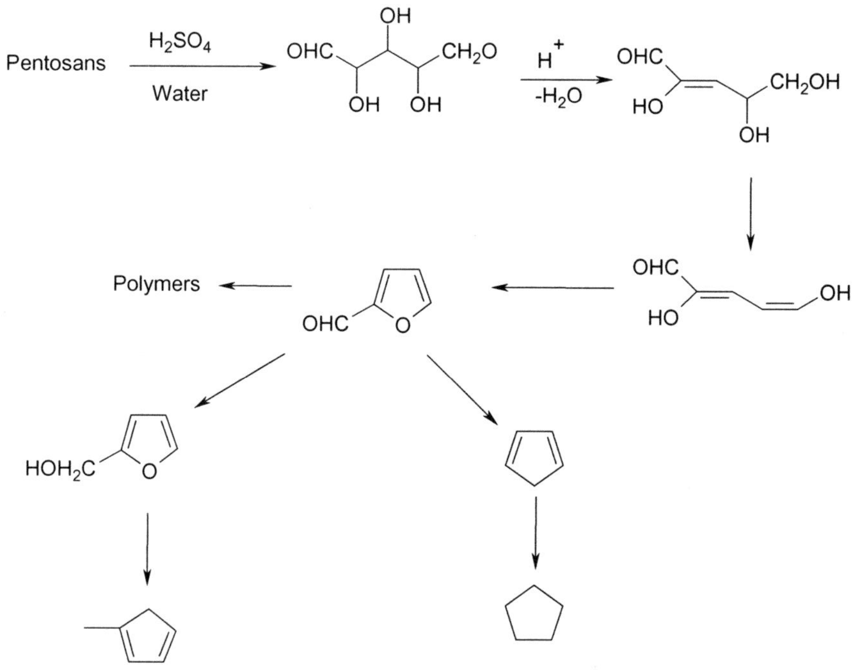

Scheme 6.13 Furfural production and uses.

6.3.3.1 Furans. Production of furfural (Scheme 6.13) by dilute sulfuric acid hydrolysis of pentosan-containing hemicellulose material has been carried out for almost 100 years, with annual production rates approaching 300 000 tonnes. Although the process is commercially viable the furfural is expensive, limiting its use to speciality markets. There are two major inefficiencies in the process that increase the cost of the final product. First, although the initial hydrolysis of pentosans to pentose is quite rapid the subsequent dehydration and cyclization steps are slow. During this period breakdown of the cellulose component occurs, leaving a product with only fuel value. Second, only around 50% of the available pentose is converted into furfural, with many side reactions taking place. There is currently much research taking place aimed at both improving the efficiency by using strong acid catalysis and by trying to integrate furfural production with other processes such as bioethanol or acetic acid production.

There are many current and potential uses for furfural. Furfural itself is used as solvent for selectively extracting aromatic and alkene components from lubrication oils and is also used to a smaller extent in as a pharmaceutical and flavour and fragrance intermediate. Catalytic hydrogenation gives furfuryl alcohol and is the key monomer for preparing furan resins, which compete with phenolic resins in foundry applications. Their good thermal resistance, non-burning, and low smoke emission properties make them ideal for uses in moulds and foundry cores. Further catalytic reduction of the side group gives 2-methylfuran; although current commercial applications for this are limited (it is a useful dienophile) it is interesting for the fact that it has one of the highest octane values for an organic chemical. Although the current cost is far too high to use this material as a gasoline additive it was used as an additive in military applications in World War II. 2-Methyltetrahydrofuran formed *via* hydrogenation is also a useful fuel additive. Decarbonylation of furfuryl alcohol gives furan, which can be hydrogenated to the important solvent tetrahydrofuran (THF). The flammability and propensity to form peroxides will limit widespread use of THF on environmental and safety grounds.

6.3.3.2 Levulinic Acid. Levulinic acid is sometimes referred to as a platform chemical, *i.e.* one from which a whole product range can be built through relatively simple chemical transformations, providing the raw material price is low enough. Levulinic acid is produced by high temperature acid hydrolysis of cellulose to glucose followed by a further controlled high temperature ring-opening dehydration step (Scheme 6.14). Although this technology is well known, yields of levulinic acid are low due to formation of tars. Recent innovations by Biofine have resulted in a more economical process in which waste material from paper pulp mills can be can be converted in well over 70% yield from the available glucose. Crucially the product is removed from the acidic medium as soon at it is formed, thereby preventing build up of tars. Levulinic acid may be readily cyclized under acid conditions to furfural or by high temperature catalytic hydrogenation into 2-methytetrahydrofuran. Hence levulinic acid provides an alternative and perhaps more cost-effective platform to the range of furan compounds discussed above.

5-Aminolevulinic acid (DALA) is a useful biodegradable herbicide, triggered by sunlight, which only kills dicotyledons and hence is potentially a useful weed killer for use with grass, corn, wheat, *etc.* The high cost of production from petrochemical sources has precluded its use; however, it may now be produced more competitively from levulinic acid. Levulinic acid may be condensed with phenol under acid conditions to give the diphenolic acid, which has potential applications in waterborne coatings. Succinic acid may also be produced from levulinic acid; however, recent studies have shown that it may be more efficiently produced from fermentation of glucose using *E. coli* (AFP111). This route produces succinic acid at up to $50 \, \text{g} \, \text{L}^{-1}$, which is relatively high

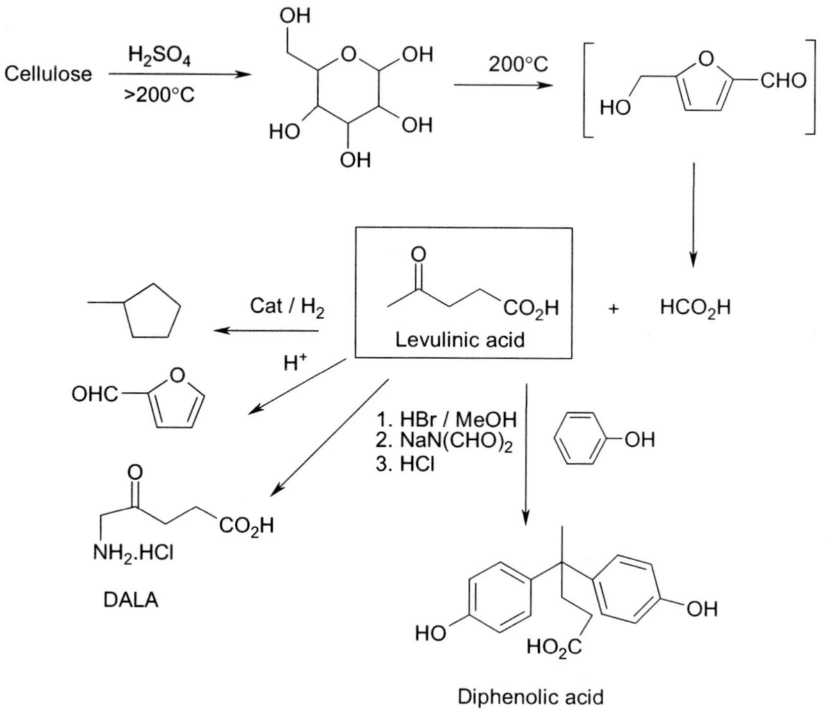

Scheme 6.14 Manufacture and uses of levulinic acid.

for a fermentation process. A commercial plant using this technology is planned and is expected to compete with the current commercial route involving hydrolysis and hydrogenation of maleic anhydride.

6.3.3.3 Adipic Acid. Adipic acid is currently produced at around 9 000 000 tpa from benzene. The initial step involves hydrogenation to cyclohexane followed by an inefficient oxidation step to give a mixture of cyclohexanol and cyclohexanone (KA mix). This is then oxidized to adipic using nitric acid. There are many problems with this process, discussed in elsewhere in this book, but one of the major ones concerns the production of stoichiometric amounts of nitrous oxide, a potent greenhouse and ozone-depleting gas. Alternatively, adipic acid may be produced by simple catalytic hydrogenation of *cis,cis*-muconic acid, which can be directly synthesized from glucose using genetically modified *E. coli* (Scheme 6.15).

The shikimic acid pathway is the well-known natural route for synthesis of the important aromatic amino acids tryptophan, tyrosine, and phenylalanine from glucose. A key intermediate in this process is

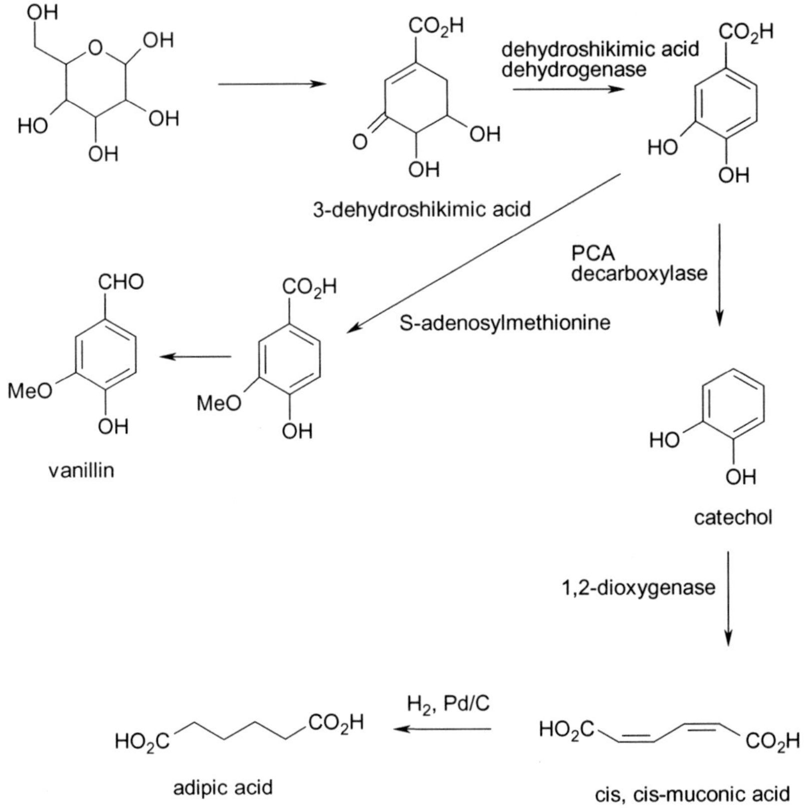

Scheme 6.15 Synthesis of adipic acid from glucose using *Escherichia coli*.

3-dehydroshikimic acid; following production of this material the natural pathway can be diverted in several directions using genetically
modified *E. coli*. By inserting genes that produce dehydroxyshikimic
acid dehydrogenase and PCA decarboxylase from *Klebsiella pneumoniae*
and 1,2-dioxygenase from *Acinetobacter calcoaceticus* the pathway can
be altered to produce muconic acid in high yield. As is evident from
Scheme 6.15, catechol is an intermediate in the synthesis of muconic
acid; consequently, by removing the 1,2-dioxygenase gene the process
can be altered to produce catechol. Catechol is an important speciality
chemical used in the flavour and fragrance industry in the production of
vanillin; it may also be employed in the production of L-Dopa used in
the treatment of Parkinson's disease. Vanillin may be produced directly
without the need to isolate catechol; the key step in this sequence is
methylation of the hydroxy group by (*S*)-adenosylmethionine. At present, the economics of the process do not compare favourably with
petrochemical routes.

6.3.3.4 1,3-Propanediol. 1,3-Propanediol is conventionally made by the hydration of acrolein, or by the hydroformylation of ethylene oxide to afford 3-hydroxypropionaldehyde. The aldehyde is hydrogenated to give 1,3-propanediol. Dupont have now developed a commercial process using a genetically modified strain of *E. coli* in a fermentation process based on simple sugars such as corn starch. The separation process is relatively complex requiring:

- high temperature microfiltration and ultra-filtration to remove the cell walls and proteins from the reactor liquid;
- ion exchange to remove impurities;
- flash evaporation to increase propane diol concentration from 10% to 80%;
- distillation to achieve the required purity.

Despite these relatively energy intensive processes, the biotechnology route is claimed to use 40% less energy and produce 20% less greenhouse gases than conventional processes. 1,3-Propanediol is used in various applications such as solvent adhesives and production of PTT (polytrimethylene-terephthalate), a new polymer that is used for the production of high quality fibres.

6.4 ALTERNATIVE ECONOMIES

As will be evident from the above discussion many valuable chemicals can be made from renewable resources. In many cases current production methods fail to compete effectively with routes from fossil sources. With advances in biotechnology and increasing oil prices renewable feedstocks will become more commercially attractive, especially for fine, speciality, and pharmaceutical chemicals. If future bulk chemical production were to be based on renewable feedstocks then it would be highly desirable, by analogy with petroleum refining, to have a simple system that allows production of most of our chemical requirements from a single starting point. It would also be desirable if thought processes could be changed such that the use of alternative starting materials to make products with the functions required, but not necessarily the chemistry currently used, are embraced. For example, it is inefficient in trying to produce benzene in huge amounts for use as a general feedstock from a renewable resource. It takes considerable energy, effort, and cost to produce an unsaturated hydrocarbon from a natural resource containing large amounts of oxygen. The main advantages of this common feedstock approach are economy of scale and maximizing the usefulness of the whole resource. The concept of platform chemicals goes someway towards achieving this.

6.4.1 Syngas Economy

As outlined earlier one of the economic issues concerning increased use of biomass is the expense and energy consumed in transporting vast quantities of solids with a high water content. When compared with the highly developed pipe network used for transporting oil and gas to chemical factories the disadvantage is obvious. One way to overcome this disadvantage is to convert the biomass feedstock into a liquid or gas *in situ* for transportation within the existing infrastructure. Synthesis gas would be an obvious choice on which to base this new economy.

The production of synthesis gas from natural gas and coal is the basis of the 33 000 000 tpa methanol production and is also used in the production of ammonia. After removal of sulfur impurities methane and water are reacted over a nickel oxide on calcium aluminate catalyst at 730 °C and 30 atm pressure, the reaction is highly endothermic (210 kJ mol^{-1}) [Equation (6.6)]:

$$CH_4 + H_2O \leftrightarrow CO + 3H_2 \qquad (6.6)$$

Methanol production is carried out over a catalyst containing Cu and Zn oxide on alumina at around 300 °C and 100 atm according to Equation (6.7). It is actually thought that most of the methanol is produced from CO_2 formed from reaction of CO with steam (6.8):

$$CO + 2H_2 \leftrightarrow CH_3OH \qquad (6.7)$$

$$CO_2 + 3H_2 \leftrightarrow CH_3OH + H_2O \qquad (6.8)$$

Whilst the basic process for generation and conversion of syngas is well established, production from biomass poses several challenges. These centre on the co-production of tars and hydrocarbons during the biomass gasification process, which is typically carried out at 800 °C. Recent advances in production of more robust catalysts and catalytic membrane reactors should overcome many of these challenges.

Once syngas and methanol can be produced viably from renewable resources then established synthetic pathways (as there is insufficient space to discuss the details here readers are invited to consult a textbook of industrial chemistry) can be used to produce a whole variety of bulk

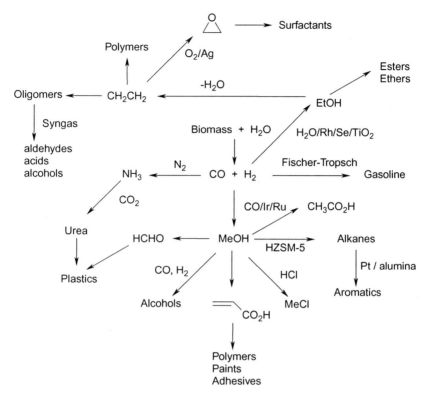

Scheme 6.16 The syngas economy – some examples.

chemical feedstocks (Scheme 6.16). By analogy, syngas and/or methanol will become the petroleum feedstock of the future.

6.4.2 Hydrogen Economy

Hydrogen has been mentioned already as the key fuel for fuel cells and an integral component of synthesis gas. If challenges of transportation, storage, and cost of production can be overcome it is likely that hydrogen will play a larger part in our lives as fossil fuels become scarce. This will be largely for energy but hydrogenation is likely to become an even more important reaction as highly oxygenated biomaterials become an important raw material for the chemical industry.

Hydrogen may be produced from any form of biomass, including waste paper and municipal waste, the most suitable ones containing 5–30% water, using gasification technology. The latest technology uses fluidized beds and as part of a combined heat and power plant can achieve thermal efficiencies over 35%. A major attraction of preparing hydrogen from biomass is that the product is classed as renewable for tax purposes. At present, the costs are still several times that of producing hydrogen from large-scale steam reforming of natural gas.

Water is a potential valuable source of hydrogen, but it is extremely thermally stable, with temperatures of over 2000 °C being needed to cause decomposition. Methods of reducing the energy input through uses of catalysts continue to be studied but many obstacles remain.

Presently, electrolysis of water is not commercially attractive for large-scale production, but may become so in the medium to longer term; however, a few plants have been built in countries like Norway and Canada that have a surplus of hydroelectric energy capacity. To maximize conductivity and therefore minimize electricity use a concentrated solution of potassium hydroxide, in purified water, is normally used as electrolyte.

A few natural bacteria in blue–green algae have the ability to use sunlight to decompose water – biophotolysis. The process is not very efficient due to the co-production of oxygen which inhibits the hydrogen-producing enzyme. Research is underway to overcome these problems and develop commercially viable systems.

6.5 BIOREFINERY

The biorefinery concept would contain many of the elements discussed already and would make maximum use of a renewable feedstock by using different technologies, such as supercritical fluid extraction, gasification, fermentation as well as standard chemical transformations to give fuels, platform chemicals, and specialties.

The first stage in a biorefinery is likely to involve some kind of extraction process, ideally using non-organic solvents, *e.g.* supercritical CO_2. This could be used to obtain easily extracted materials such as terpenoids and waxes without destroying the chemical structure of the biomass.

Most biorefinery concepts involve gasification technology to produce syngas. This may then be converted into hydrocarbon fuels using Fischer–Tropsch technology according to Equation (6.9):

$$(2n + 1)H_2 + nCO \rightarrow CnH(2n + 2) + nH_2O \qquad (6.9)$$

The reaction is carried out at high pressure, to favour formation of long-chain hydrocarbons, and moderate temperature (around 200 °C) to limit methane formation. Cobalt, ruthenium, or iron catalysts are commonly used but the presence of even small amounts of sulfur in the feed will poison the catalyst. Ideally, biomass waste that cannot be converted into higher value products would undergo this process.

Using the latest fermentation technology involving genetically modified yeast or bacteria a whole range of platform chemicals can be derived from sugars found in biomass. These include ethanol, 1,4-diacids, 2,5-hydroxymethylfurfural, and 1,3-propanediol, all of which can be converted into a range of valuable material, using conventional chemistry. Fatty acids, as described above, can also be extracted to lead to further platform chemicals. Glycerol, the by-product of biodiesel and fatty acid production, is probably becoming the most important platform chemical. Useful derivatives include 1,2-propanediol (used in antifreeze), polyglycerol (used in surfactants), acrolein, and 1,3-propanediol discussed earlier. Further examples of the use of glycerol are discussed in Chapter 10.

6.6 CONCLUSIONS

It is now well recognized by governments that, for long-term sustainable development (or even survival) and the future stability of the economy, the worlds' dependence on fossil resources must be reduced. The first priority in this area is undoubtedly to reduce the amount of fossil fuels used for energy. Alternative energy sources like solar and wind are growing rapidly but are unlikely to have a huge impact within the next 20 years; however, when coupled with other initiatives such as wood burning power stations and a resurgence in carbon-neutral nuclear power, they will help reduce our dependence on oil and gas. In the longer term perhaps the best hope for low carbon transport fuel lies with fuel cells. Initially these will use fossil based hydrogen or methanol but at a reduced rate with reduced pollution. In the short to medium term the amount of biofuels blended into conventional feedstocks is likely to increase. But tensions between use of land for food and energy crops will need to be carefully managed in many parts of the world.

The use of renewable resources for manufacturing specific performance and speciality chemicals and for fibres to replace synthetic ones is growing. The driver for this is improved cost/performance. To have a major impact on the amount of oil and gas used there is a need to convert biomass into new, large-scale basic feedstocks such as synthesis gas or methanol. Many technical developments in separation science as well as improving the overall 'yield' of chemicals are required before renewable feedstocks can compete effectively with oil and gas (at current prices), but the gap will continue to narrow.

REVIEW QUESTIONS

1. Discuss the contribution of chemicals to climate change. Describe the positive roles chemists and chemical engineers may play in helping meet the challenges posed by climate change.
2. Discuss the reasons why fuel cells are considered to be a significant advance towards development of a sustainable energy source. Assess the potential usefulness of the various types of fuel cell as alternatives for internal combustion.
3. Starting from the appropriate naturally occurring fatty acid devise synthetic strategies for the following:
   (a) *N*-stearylpyrrolidine,
   (b) oleyl(trimethyl)ammonium chloride,
   (c) the sulfosuccinate of stearyl alcohol 9 mole ethoxylates.
4. Using corn as your starting material describe synthetic strategies for the following chemicals:
   (a) phenyl methyl ether,
   (b) 1,2-propanediol,
   (c) formaldehyde,

   (d)

FURTHER READING

J. Houghton, *Global Warming: The Complete Briefing*, Cambridge University Press, Cambridge, 2004.

G. Olah, A. Goeppert, and G. K Surya Prakash, *Beyond Oil and Gas: The Methanol Economy Second edition*, Wiley-VCH Verlag, Weinheim, 2009.

D. A. J. Rand and R. M. Dell, *Hydrogen Energy: Challenges and Prospects*, RSC Publishing, Cambridge, 2008.

E. S. Stevens, *Green Plastics*, Princeton University Press, Princeton, 2002.

CHAPTER 7

Emerging Greener Technologies and Alternative Energy Sources

7.1 DESIGN FOR ENERGY EFFICIENCY

The preceding chapters discussed some of the main chemical developments that can help reduce waste, lower harmful emissions, improve process efficiency, and generally aid development of more sustainable products and processes. In this and the following chapter the focus will be more on the technology aspects which can lead to improved process and energy efficiency as well as process cost reduction.

Most chemical processes use thermal sources of energy originating from fossil (or nuclear) fuels. The energy input to the process is non-specific, *i.e.* it is not directly targeted at the chemical bond or even the molecules undergoing reaction. Much of the energy is 'wasted' in heating up reactors, solvent, and even the general environment. For some processes alternative, more specific forms of energy, *e.g.* photochemical, and microwave energy, may be beneficially applied. The use of such alternative forms of energy is not new but they are being taken more seriously by manufacturing industries and hence can be viewed as emerging technologies. Even with thermal sources of energy, conservation measures can be applied to reduce cost and environmental impact; some such measures are exemplified below:

- *Monitoring, control, and maintenance:* Many pieces of equipment operate below design specification, and temperature control of many processes is far from optimum. By use of modern

Green Chemistry: An Introductory Text, 2nd Edition
By Mike Lancaster
© Mike Lancaster 2010
Published by the Royal Society of Chemistry, www.rsc.org

computerized digital monitoring and control systems optimum system performance is more easily achieved. Regular maintenance of equipment, including replacement of filters on pumps and regular lubrication, can improve energy efficiency and extend equipment life.

- *Loss prevention:* Walking around many chemical factories will reveal obvious sources of energy loss, probably the most noticeable being leaky steam valves. Good housekeeping can do much to conserve energy.
- *Waste heat recovery:* Many processes, and whole factories, now employ waste heat recovery, *e.g.* hot flue gasses pass through a heat exchanger before being emitted, the resulting hot water being put to beneficial use. Hot liquid streams from one process are also used to directly heat incoming streams from another.
- *Matching energy sources to requirements:* Most householders realize that heating a whole house to, say 20 °C, using electrical resistance heaters (electric fires) is more costly and uses more energy than a modern gas central heating system. The same principle applies to the process industries, *e.g.* using high pressure steam to heat a solution to 35°C is highly inefficient. Hence most industrial sites have various energy sources to meet specific process demands. Older chemical plants often have over-sized boilers for modern needs, replacing these with smaller more efficient equipment, although costly, can reduce energy consumption considerably.

By using techniques such as the above, together with sophisticated analysis programmes, coupled with improvements in the energy efficiency of equipment as well as improved process chemistry, the relative energy consumption per tonne of product has been declining for the past 50 years (Figure 7.1). Recent figures from the USA indicate that energy reduction per tonne of product is becoming harder to achieve. This is probably because simple cost-effective measures have already been carried out; future large gains will probably require significant technology developments. Despite these energy savings the chemicals and petroleum refining industries are major energy users; in the USA it is estimated that these two sectors account for 50% of industrial energy usage; the situation is not likely to be very different in many other developed countries. Looking more closely at the chemical industry, roughly a quarter of the energy used is done so in distillation and drying processes. This highlights the necessity for considering the whole process, not just the reaction stage, when undertaking research and development work. During the period 1990–2006 chemicals production in

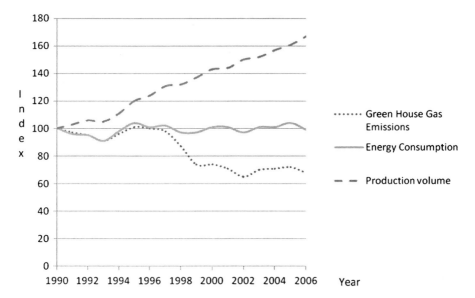

Figure 7.1 Relative decline in energy used in chemicals production (EU data).

Europe rose by 67%, but energy consumption remained roughly constant. Greenhouse gas emissions were reduced by around one-third.

Before discussing 'emerging technologies' a few examples of how some industrial energy intensive processes have been made more efficient will be given. As discussed in Chapter 4, development of the Haber process for ammonia made large-scale commercial production viable by lowering the activation energy required, a large energy saving, compared to using temperatures over 3000 °C required by the electric arc process. Although the underlying principles of the Haber process have changed little, the energy consumption of the process in now less than 40% of the original process. This has been brought about by using gas in place of coal as a feedstock as well as optimization of turbine equipment and steam distribution networks. Development of radical flow converters using small sized catalyst particles as well as the latest ruthenium based catalysts have also made significant contributions to energy efficiency improvements. In fact the ammonia process is now so energy efficient that less process energy is used in ammonia production than in production of acetone, butanol, nylon 6,6, and interestingly extraction of rapeseed oil. Some of the more energy-intensive organic bulk chemical processes currently used include adipic acid and propylene oxide manufacture.

Detailed information of the energy requirements of competing processes is rarely available in the open literature. However, particularly for bulk chemicals, development of a much more energy efficient process may be expected to provide significant commercial benefits resulting in the abandonment of older energy intensive processes. An excellent example of the differing energy requirements of competing processes is provided by titanium dioxide production. There are two competing processes:

The sulfate process essentially involves three energy intensive stages:

1. dissolution of the ore, ilmenite, in sulfuric acid and removal of iron impurities;
2. formation of hydrated TiO_2 by treatment of the sulfate with base;
3. dehydration in a calciner.

By contrast the chloride process can, for simplicity, be broken down into two relatively energy efficient steps:

1. chlorination of the ore with Cl_2 and purification of $TiCl_4$ by distillation;
2. oxidation by burning.

There is a difference of a factor of five in energy consumption between the two processes, largely due to avoidance of evaporation of large amounts of water in the latter process. Despite this both processes still operate, although the chloride process does dominate. There are two main reasons for this: firstly the sulfate process can use lower grade and therefore less expensive ores and secondly it produces anatase pigments as well as rutile, which is the sole product of the chloride process.

7.2 PHOTOCHEMICAL REACTIONS

A fairly obvious prerequisite for a photochemical reaction to occur is that the atom or molecule must absorb light (Grotthuss–Draper law). Furthermore, one photon of light can only activate one molecule (Stark–Einstein law). At face value this law would suggest that the quantum yield for a reaction (the number of molecules of reactant consumed per photon adsorbed) should have a maximum value of 1; this of course, has implications for the energy required, since photons are usually generated from electrical energy. Many photochemical reactions do, however, have quantum yields of several thousand, implying that the photon is simply being used to initiate a chain reaction. The current cost of generating photons (unless visible light can be used) is generally such that, unless a reaction has a quantum yield far higher than unity, it is unlikely to be commercially attractive unless there is no viable alternative. In practice quantum yields are not that easy to measure and it is more useful to measure total energy consumption.

Table 7.1 Some mechanisms for loss of excitation energy.

Processes leading to reaction	Processes that may lead to reaction	Processes that do not lead to reaction
Dissociation of the molecule, usually into radicals.	Intermolecular energy transfer giving another electronically exited species, which may undergo reaction.	Luminescence, including phosphorescence & fluorescence.
Extrusion or elimination of small stable species such as CO_2.	Intramolecular energy transfer.	Quenching involving translational or vibrational excitation of another molecule.
Isomerization.	Ionization by loss of an electron – this usually results in reaction of the cationic species.	
Direct reaction with another molecule involving addition or elimination.		

The reader is referred to the Further Reading section for a full treatment of the theory of photochemical reactions. When a photon is absorbed it must transfer all its energy to the absorbing molecule, the molecule being promoted to a higher energy state. For many molecules the energy required for promotion from the electronic ground state to the lowest excited state falls in the visible and UV regions of the electromagnetic spectrum. For most molecules the ground state is a singlet in which the two residing electron spins are paired. On absorption of the photon one electron is promoted to an unfilled orbital, the electron spin initially being retained to produce a singlet exited species. Although formally forbidden by selection rules, spin inversion to a lower energy triplet state, *via* a process called intersystem crossing, does frequently occur. The electronically excited molecule may then undergo several processes by which the excess energy is lost (Table 7.1); some of these do not result in chemical reaction, reducing the quantum yield.

7.2.1 Advantages of and Challenges Faced by Photochemical Processes

Other than directly targeting energy at specific molecules, hence reducing energy consumption, photochemical processes have several other 'green' credentials:

- Photons are very clean reagents, leaving no residues. A photo-initiated process therefore has potential advantages even when compared to reactions initiated by use of catalysts. Such processes may

use fewer raw materials compared to non-photochemical alternatives, *e.g.* comparison of photo- and radical-initiated halogenation reactions.
- Since the energy is more directed reaction temperatures are generally low; this may give higher selectivities, reducing by-product
formation from competing reactions.
- As will be discussed more fully below some reaction pathways are
more readily available *via* photochemical processes, leading to
products that would be difficult to make by other routes.

Although academic research on photochemistry dates back many
years its uptake by industry has been limited, this is, in part, a result of
there being significant unsolved inherent problems:

- Reactor fouling is a fairly common problem in chemical manufacturing. Unless very severe it is often only a minor inconvenience
in thermal processes, the main effect being a slight reduction in heat
transfer efficiency. In photochemical processes even small amounts
of fouling on the photochemical window or reactor wall, through
which the light has to pass, can completely prevent reaction
occurring. In thermal processes fouling can often be minimized
through selection of reactor materials, *e.g.* glass, stainless steel,
Inconel, PTFE-lined, *etc.*; in photochemical processes there is little
flexibility in choice of 'window' material due to the requirement for
transparency at a particular wavelength.
- Radiation of a particular wavelength (monochromatic) is required
to initiate a specific electronic transition; however, most UV and
visible light sources are polychromatic. For example, common
mercury arc lamps emit around 50% of their energy in the range
405–578 nm. Hence for most processes well over half of the electrical energy supplied to the lamp is wasted, reducing the overall
energy efficiency and increasing process costs. In addition some
lamps also emit some of their energy as heat; again this is wasteful
of energy and necessitates installation of a cooling device (frequently circulating cold water).
- Light sources are often expensive, especially if made of thick-walled
quartz as in high-pressure mercury lamps, and delicate, hence
equipment costs may be high compared to thermal processes.
- Since the power of transmitted light drops off as the square of the
distance from the light, for efficient reaction and energy usage the
reactants must be as close as possible to the light source. This has
practical implications for design of industrial reactors.

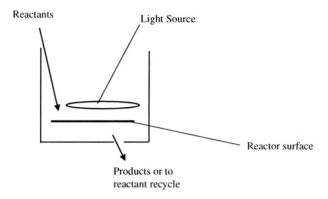

Figure 7.2　Schematic illustration of a 'non-contact' type photochemical reactor.

In recent years significant research effort has gone into meeting these challenges, with particular focus on development of reactor technology. One obvious way of preventing fouling is to avoid direct contact between reactants and any window or wall (see Figure 7.2 for an illustration). Some success has been achieved but the need to prevent power loss means that the light source still needs to be close to the reactors and hence splashing of reactants onto the window is a problem. Some other reactor types are discussed below for particular reactions.

7.2.2　Examples of Photochemical Reactions

Two commercial photochemical processes quoted in most texts are the synthesis of caprolactam and vitamin D_3. The caprolactam process, operated at well over 100 000 tpa by Toray for many years, is an excellent example of the scale at which such reactions can be successfully run, even though the quantum yield is only 80%. The chemistry (Scheme 7.1) involves a radical process through production of $NO^{\cdot}$ and $Cl^{\cdot}$ radicals from NOCl by light of 535 nm wavelength. The commercial success of this process is partially due to the relative low cost of producing this 'visible region' light using a low-pressure mercury lamp doped with thallium iodide. The thallium iodide both increases the intensity of the required wavelength and reduces the intensity of lower wavelengths, which initiate polymer formation; hence fouling is reduced to acceptable levels. As may be expected from the scale of the reaction, a continuous counter-current flow reactor is used, the inside of which is made of glass. The reactants do not come into direct contact with the multi-lamp light source but pass through a thin adjacent glass vessel, which can de cleaned as required. The oxime salt passes out through the bottom of the reactor with unused gaseous reactants being recycled from the top of the reactor.

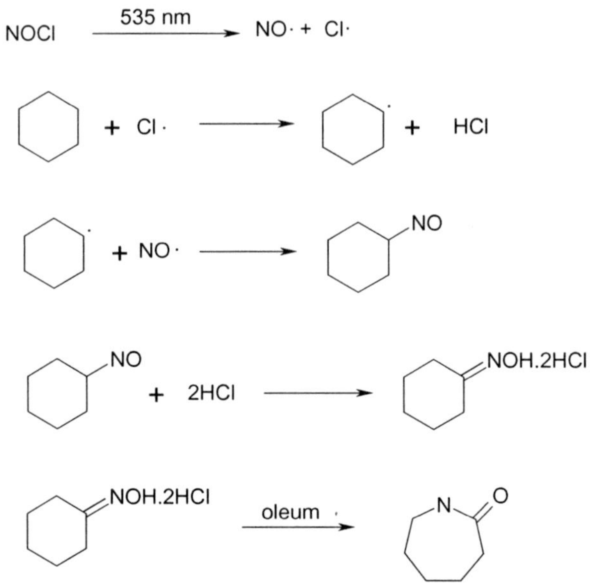

$$\text{NOCl} \xrightarrow{\text{535 nm}} \text{NO}\cdot + \text{Cl}\cdot$$

Scheme 7.1   Photochemical synthesis of caprolactam.

Without knowing all the process details, including actual energy consumption as well all reagent manufacturing processes, it is not really possible to assess if this process is 'greener' than the competing one based on air oxidation of cyclohexane followed by reaction with hydroxylamine hydrogen sulfate. However, the number of process steps has been reduced. Recent laboratory work has shown that cyclohexane can be oxidized in a high quantum yield reaction using photocatalysts based on iron porphyrins or tungstates. If this finding proves commercially viable then many of the safety issues of the current inefficient oxidation process may be overcome.

The other commonly quoted industrial photochemical process is the production of vitamin D_3 involving a photochemical electrocyclic ring opening followed by a thermal 1,7-hydride shift (Scheme 7.2). This is a further example of a successful low quantum yield process; in this case there is no viable thermal alternative. Vitamin A acetate has also been produced commercially using a photochemical isomerization process to convert a mixed terra-alkene precursor into the all-trans form.

Photochemistry offers the possibility of simple synthesis of some materials that would be very difficult to synthesize by other means. Stereoselective synthesis of four-membered rings is an excellent example (Scheme 7.3). As discussed previously, pericyclic reactions, such as the disrotatory photochemical ring closure of dienes, governed by the Woodward–Hoffman rules are inherently green.

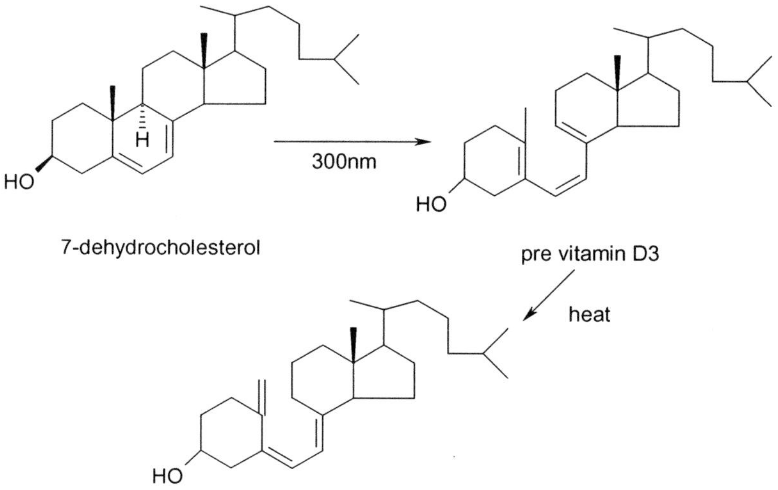

Scheme 7.2 Vitamin D$_3$ synthesis.

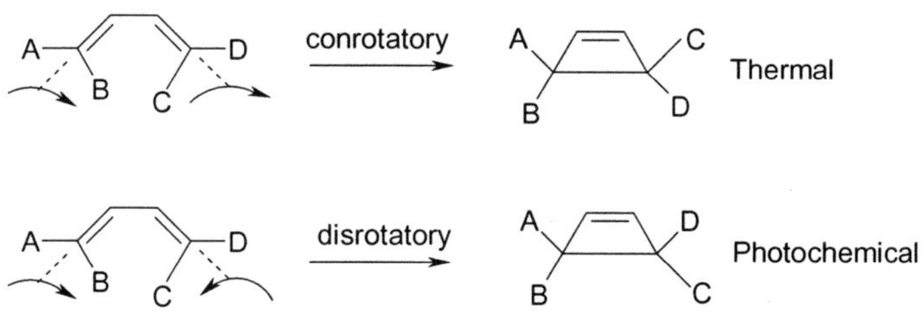

Scheme 7.3 Photochemical ring closure of dienes.

Photochemical free-radical reactions, exemplified by halogenation reactions, usually have very high quantum yields and proceed at low temperatures, thereby limiting side reactions. Chlorination of methane and synthesis of the degreasing solvent trichloroethane are commercial processes. Whilst the process may have some green elements products of these reactions are being replaced by more environmentally friendly ones. An interesting free radical bromination has been carried out at the 20 kg scale (Scheme 7.4). The choice of the unusual brominating agent was made on safety grounds since the more common, but highly re-active, *N*-bromosuccinamide produced a very large exotherm. The pilot reactor simply consisted of a sunlamp around which the reactants flo-wed in a PTFE tube; again the process economics are helped by the fact that visible light could be used.

Other than energy considerations, on which there is little comparative data, the most important 'green' role for photochemistry is in improving

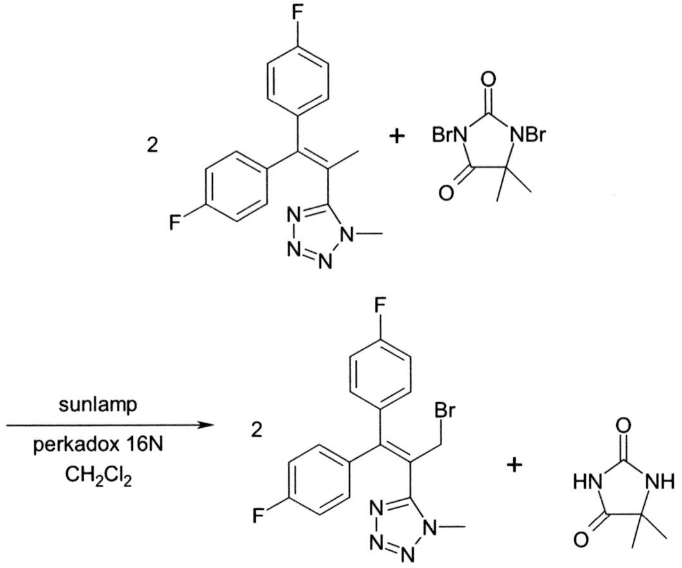

Scheme 7.4 Pilot-scale example of a photochemical bromination.

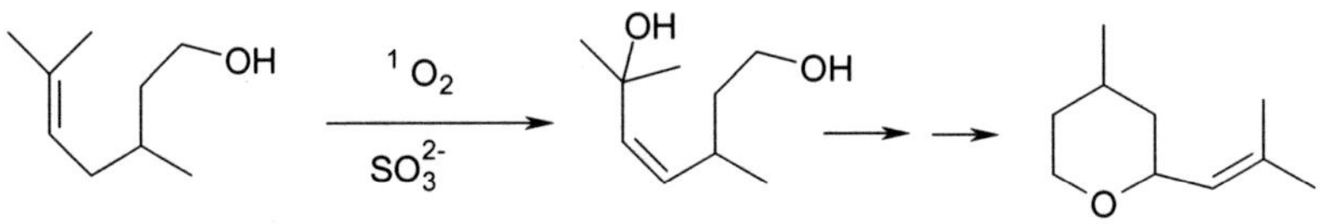

Scheme 7.5 Synthesis of Rose Oxide involving $^1O_2$.

atom economy. Although only a preliminary research result, an excellent example of this is the avoidance of the need for stoichiometric amounts of Lewis acid catalysts in the synthesis of some acylated aromatic compounds. Benzoquinone can be reacted with an aldehyde under a sunlamp to an acylhydroquinone in up to 88% yield. The alternative procedure would involve reaction of an acyl chloride with hydroquinone and a stoichiometric amount of aluminium chloride. Oxidation reactions are also possible using singlet oxygen, which is generated using a sensitizer such as Rose Bengal; potentially such a process could be used to replace ones involving stoichiometric oxidants such as chromium oxide. The fragrance chemical Rose Oxide has been made by a route involving singlet-oxygen generation (Scheme 7.5).

Like many of the topics discussed in this book, photochemical reactions are likely to be used in niche applications for commercial and environmental reasons. Unless there is a major a breakthrough in reactor and lamp design widespread use of this technology is unlikely. Perhaps the best hope of producing high intensity monochromatic sources of radiation rests with lasers, but currently equipment costs are too high to justify their use for commercial chemical production.

Systematic work has been carried out to compare thermal and photochemical carbon–carbon bond forming reactions, using a range of environmental measures, discussed in Chapter 3 to assess the relative greenness. Results were mixed; one of the key points noted was that solvents played a role in the greenness of reactions and in many cases outweighed the benefits of the lower temperatures of the photochemical compared to the thermal processes.

7.3 CHEMISTRY USING MICROWAVES

Photochemistry, at least in principle, offers the possibility of targeting energy at a particular bond in a molecule; other sources of energy have less specific targeting but go someway to achieving this goal. As anyone who has used a microwave oven will be aware certain substances heat up extremely quickly whilst others take much longer. For example, microwaving a mince pie for a few seconds will leave the pastry quite cool but the inside will quickly become too hot to eat, a clear example of energy being targeted.

7.3.1 Microwave Heating

Microwaves have wavelengths between 1 mm and 1 m and hence have similar frequencies to RADAR and telecommunication devices. So as not to cause interference with these systems the frequency of radiation that can be emitted by household and industrial appliances is strictly regulated, with most appliances operating at a fixed frequency of 2.45 GHz. To some extent this reduces the flexibility of such equipment.

The overall mechanism of how energy is imparted to a substance under microwave irradiation is complex, consisting of several different aspects. One process, which explains why microwaves heat certain substances and not others, is termed dipolar polarization. When a substance possessing a dipole moment (water is the obvious example) is subject to electromagnetic radiation it will attempt to align itself to the electromagnetic field by rotation. In liquids this rotation causes friction between adjacent molecules, which in turn causes a temperature rise. Of course the rates of rotation will influence the heating rate and are related to the radiation frequency. Although this effect is produced by all electromagnetic radiation, at high frequencies the change in direction of the field is too rapid to allow rotation to occur hence there is no heating effect; but at low frequencies the rate of rotation is slow, having minimal heating effect. In the microwave region of the spectrum rotation rates are high enough to produce rapid temperature rises in dipolar substances. Liquids that do not have a dipole moment (or in which one can not be induced) are not directly heated by microwaves. By adding a small amount of a dipolar liquid to a miscible non-dipolar liquid, the mixture will rapidly achieve a uniform temperature under irradiation. Because of the large distance between molecules, gases are not heated by microwave irradiation.

For substances containing ions the most effective heating mechanism is conduction. The ions will move through the solution under the influence of the electric field, undergoing frequent collisions, with this kinetic energy being converted into heat. One other significant mechanism of energy transfer is called 'dielectric loss'. Dielectric loss is used to account for the fact that materials of similar polarity heat up at different rates. The theory is rather complex but, in essence, it is a measure of the efficiency of a substance in converting absorbed radiation into heat. The higher the dielectric loss (determined experimentally) the more rapid will be the rate of heating of materials with similar dipoles.

From the above discussion it will be evident that although certain materials can be heated selectively the energy will soon be uniformly distributed throughout a homogeneous reaction medium. Microwaves may be considered a more efficient source of heating than conventional steam or oil heated vessels since the energy is directly imparted to the reaction medium rather than through the walls of a reaction vessel.

7.3.2 Microwave-assisted Reactions

One of the reasons that have sparked the phenomenal growth in research in microwave chemistry in the last 15 years is the realization that it can provide a rapid method for screening reactions. With a heating rate of $10\,^{\circ}\mathrm{C\,s^{-1}}$ being achievable it is easy to see how the overall reaction time can be considerably shortened. Although there are examples of improved reaction selectivities and yields using microwave heating, any specific microwave effects other than can be obtained by rapid heating have not been conclusively proven, although it has been postulated that microwave-induced reactions follow the most polar pathway possible.

Even though well-designed industrial single-mode ovens are now available most microwave experiments have been carried out in domestic multi-mode ovens, which have poor control. This has led to several explosions occurring due to overheating of organic solvents. Safety aspects, coupled with the natural limitations on solvents imposed by microwave heating, have led to many reactions being carried out in water or, more commonly, under solvent-free conditions. This is often quoted as a major 'green' advantage of microwave chemistry – although care needs to be taken that the 'solvent problem' is not simply being moved pre- or post-reaction. Discussion will be limited to these two systems, but it should be noted that, due to the high polarity and non-volatility, ionic liquids might be ideal for efficiently carrying out high temperature reactions, since temperatures of over 200 °C should be rapidly attainable. As with the other 'emerging technologies' discussed in this chapter there are virtually no reports of comparative energy usage for particular reactions.

7.3.2.1 Microwave-assisted Reactions in Water. It was noted earlier that despite microwave energy being targeted at polar molecules a reaction mixture usually rapidly reaches temperature equilibrium through molecular collisions *etc.* An exception to this rule has led to a high-yielding synthesis of a thermally unstable Hoffmann elimination product (Scheme 7.6). In this example a poorly mixed two-phase water–chloroform system was used. Being polar the starting quaternary ammonium compound was water-soluble. Microwave irradiation quickly heated the water phase to over 100 °C, causing rapid elimination (reaction time 1 min). The less polar product rapidly partitioned into the chloroform phase, which being less polar had only reached a temperature of 48 °C. This low temperature enabled the product to be isolated in 97% yield – twice that using conventional heating.

Water-based microwave reactions have been relatively well studied for hydrolysis and hydrogen peroxide oxidation reactions were the natural solvent of choice would be water (Scheme 7.7). In the hydrolysis of benzamide with sulfuric acid quantitative conversion was achieved in 7 min, under microwave irradiation at 140 °C compared to a 90%

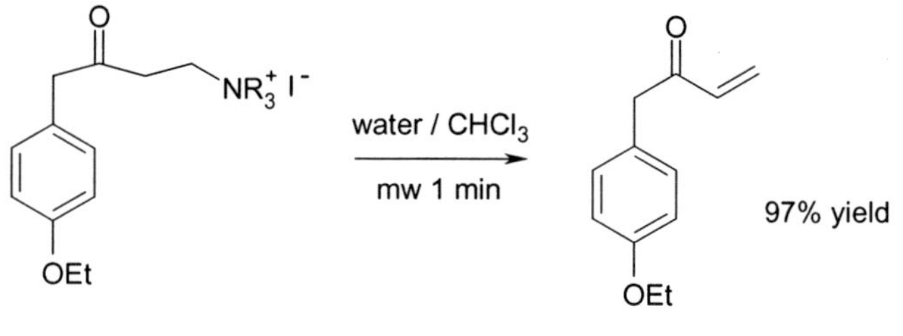

Scheme 7.6 Microwave-assisted Hoffmann elimination.

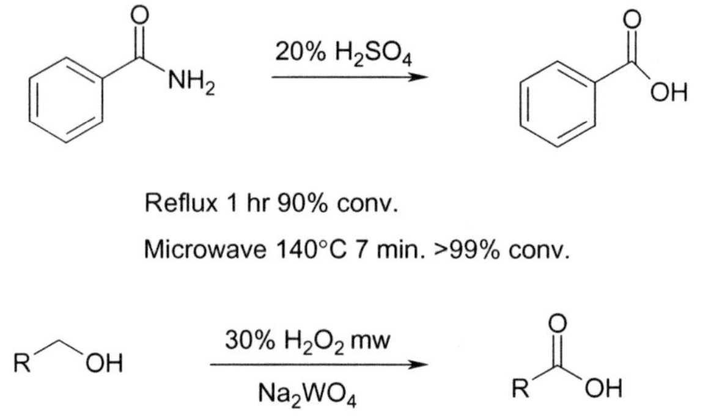

Scheme 7.7 Water-based microwave-assisted reactions.

conversion after a 1-hour reflux using a conventional heating source. The difference in reaction time can be accounted for by the difference in reaction temperature coupled with the very rapid heating of the microwave reactor. A range of primary alcohols have been oxidized to the corresponding carboxylic acid using sodium tungstate as catalyst in 30% aqueous hydrogen peroxide. Yields, although variable, of up to 85% have been achieved in a rapid, clean, safe, and atom efficient reaction.

7.3.2.2 Solvent-free Reactions. Solvent-free reactions are especially suitable for microwave heating. Since energy conduction is not required (unlike when using more conventional heat sources). In the absence of solvent, the radiation is directly absorbed by the reactants, giving enhanced energy efficiency. In addition many non-crystalline solid supports absorb microwave energy efficiently but are rather poor (*e.g.* some metal oxides) at conducting heat. Rate enhancement of Diels–Alder reactions by Lewis acids has been exploited by using solid catalysts such as K10 montmorillonite in solvent-free systems; *endo/exo* ratios are very similar to those obtained using non-microwave heating sources.

There are several examples of N-alkylation and acylation being successfully carried out, sometimes using a solid catalyst. Whist most reactions proceed as expected, where different isomeric products are obtainable significantly different ratios have sometimes been noted compared to when conventional heating sources are used, even when reaction temperatures are nominally the same (Scheme 7.8). The reasons for this are not understood but may be connected to the thermal profile of the reaction.

The rapidity of microwave-assisted reactions is well exemplified in the field of oxidation chemistry (Scheme 7.9). By simply mixing the solid

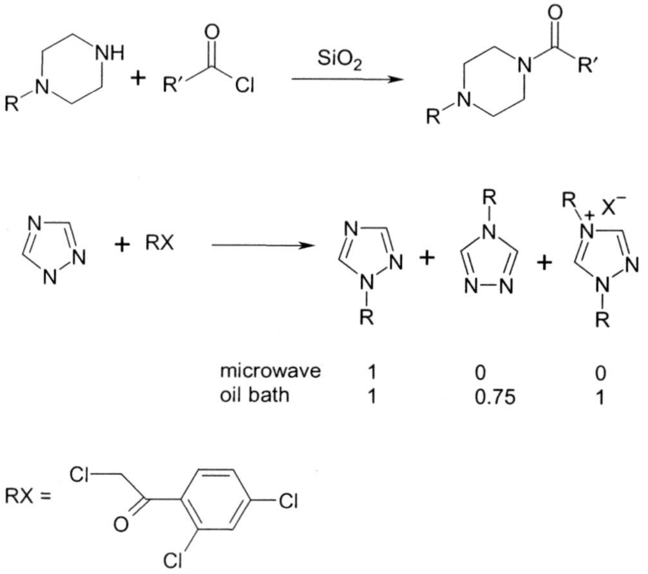

Scheme 7.8   Microwave-assisted N-alkylation and acylation reactions.

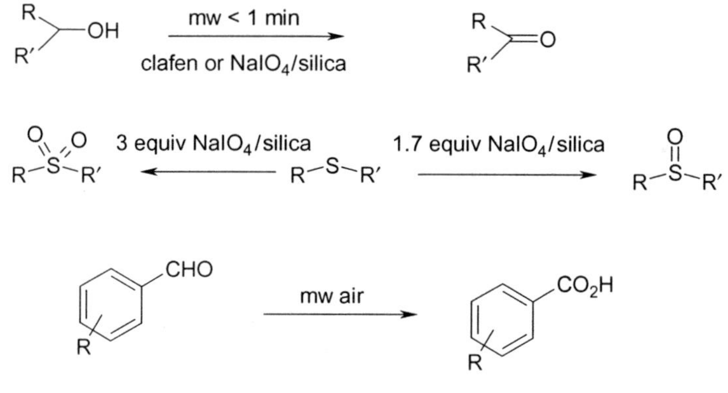

Scheme 7.9   Some rapid oxidation reactions.

oxidizing agent (clafen and MnO_2 impregnated on silica have been widely used) with a range of secondary alcohols and irradiating for periods of less than 1 min yields of ketones in excess of 90% are often obtained. An interesting selective oxidation of thiols has been achieved using sodium periodate on silica. Depending on the amount of oxidant used either the sulfoxide or sulfone can be obtained in very high selectivity in less than 3 min. A final example of the simplicity of microwave reactions is the simple irradiation of aromatic aldehydes in air to give benzoic acids. Although yields are only moderate it is a useful indication of the potential simplicity and convenience of what can be achieved using this technology.

Chemistry related industrial applications of this technology are limited, but are starting to be developed, as equipment becomes available. The first such application was the devulcanization of rubber in which microwaves efficiently break the C–S bonds, causing depolymerization. Synthetically, a small-scale *in situ* process for producing HCN has been patented by Dupont. The process involves passing a mixture of gaseous ammonia and methane over a Pt on alumina catalyst in a continuous flow system (7.1). Since the reactants are in the gaseous phase it is only the catalyst that absorbs the microwaves, with some reports suggesting that local catalyst temperatures of around 1200 °C are achieved. The concept behind this process is that the highly toxic HCN could be made 'just in time' as required (*cf.* phosgene manufacture). Owing to the rapid heating effects of microwave irradiation it readily lends itself to such small but productive intensified process plants (Chapters 8 and 9):

$$CH_4 + NH_3 \xrightarrow{\text{Pt/alumina}} HCN + 3H_2 \qquad (7.1)$$

7.4 SONOCHEMISTRY

Ultrasound refers to sound waves with frequencies higher than those detectable by the human ear, *i.e.* around 18 kHz. The terminology has become common knowledge due to widespread use of ultrasound scanning equipment in medical applications. Here high frequency waves in the range 3–10 MHz are pulsed at low power levels and the scattering pattern as they 'hit' various body parts and echo back are analysed and reconstructed into an image. The ultrasound frequencies of interest for chemical reactions (typically 20–100 kHz) are much lower than used for medical applications, but the power used is higher; the two should not be confused. Application of the devices used for chemical reactions to the body could result in tissue damage whilst trying to do a chemical reaction with a medical scanner would be futile.

When a sound wave, propagated by a series of compression and rarefaction cycles, passes through a liquid medium it causes the molecules to oscillate around their mean position. During the compression cycle the average distance between molecules is reduced and, conversely, it is increased during rarefaction. Under appropriate conditions in the rarefaction cycle the attractive forces of the molecules of the liquid may be overcome, causing bubbles to form. If the internal forces are great enough to cause collapse of these bubbles very high local temperatures (around 5000 °C) and pressures (over 1000 atm.) may be created. It is these very high temperatures and pressures that initiate chemical reaction. For there to be sufficient time for bubble collapse a wave of an appropriate frequency and power must be used. At the high frequencies used in medical applications the compression cycle follows the rarefaction cycle too quickly to allow bubble collapse. Hence no hot spots are created and therefore this frequency of ultrasound is quite safe to use on the body.

The sound waves are generated by converting electrical energy using a transducer. While there are several different types available, only the most common, piezoelectric transducers, will be described here. When equal and opposite electrical charges are applied to opposite faces of a crystal of quartz expansion or contraction occurs. Application of rapidly reversing charges sets up a vibration that emits ultrasonic waves. This is called the piezoelectric effect; whist quartz crystals are still used more modern piezoelectric transducers are made from ceramic impregnated barium titanate. Such transducers are expensive and fragile and a disc of the material is usually placed between protective metal layers. Modern devices convert over 95% of the electrical energy into ultrasound. To amplify the waves the transducers are often incorporated into a horn-like device with a titanium tip; using such devices, which may be inserted into a reactor, a power of a few hundred $W\,cm^{-2}$ can be obtained. Such are the forces exerted at the tip that erosion of the titanium may occur. In principle the frequency can be varied by altering the current, in practice devices only have one optimum operating frequency, depending on its dimensions. The remainder of our discussion will focus on the use of ultrasound in synthetic chemistry; the same 'power ultrasound' has, however, found many applications outside the synthesis field, some of which are highlighted in Table 7.2.

7.4.1 Sonochemistry and Green Chemistry

The term sonochemistry is used to denote reactions initiated by ultrasound. As with microwaves many studies have tried to discover 'special

Table 7.2 Some other uses of power ultrasound.

Application	Examples
Cleaning	Laboratory glassware; jewellery; computer components; large delicate archaeological items
Engineering	Welding and riveting of plastics, ceramic processing; drilling aid for hard, brittle materials; filtration; degassing; pigment dispersal
Biology	Disruption of cell membrane to allow extraction of contents

effects' but in most cases the reaction follows the predicted pathway, but rates are often higher, while bulk reaction temperature remains low, which is indicative of good energy efficiency. There are cases in which different pathways are followed to thermal reactions; for example, radicals are sometimes formed in ultrasonic reactions due to the high local temperatures. An interesting example of 'sonochemical switching pathways' is the reaction between benzyl bromide and potassium cyanide with an alumina catalyst in toluene. In the absence of ultrasound, alkylation is the preferred pathway (Scheme 7.10) but when ultrasound was applied benzyl cyanide was produced in 76% yield. This difference has been explained in terms of the ultrasound forcing the cyanide onto the surface of the alumina, thereby enhancing cyanide nucleophilicity and reducing the Lewis acid character of the alumina.

Owing to the widespread use of ultrasonic cleaning baths it is not surprising that many early sonochemical experiments were directed at reactions were clean metal surfaces were thought to be the cause of inefficiencies. Reactions typified by Grignard and Simmons–Smith (Scheme 7.11) are often not predictable, sometimes having long induction periods followed by violent exotherms. Frequently, small amounts of iodine are added to such reactions to 'clean' the metal

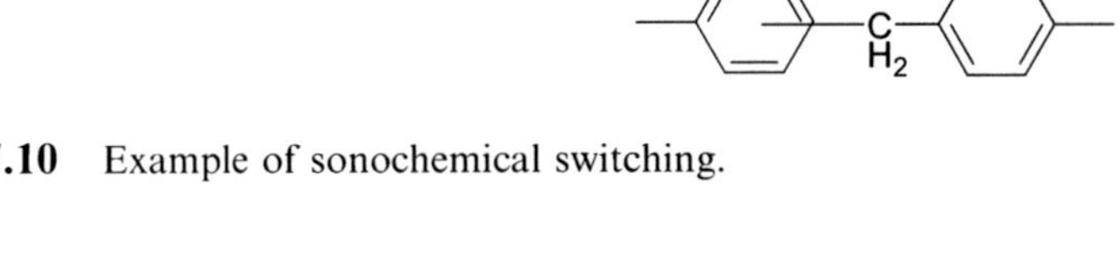

Scheme 7.10 Example of sonochemical switching.

Ultrasound $\Longrightarrow$ 91% yield, no exotherm, no iodine

No ultrasound $\Longrightarrow$ 51% yield, unpredictable exotherm, iodine

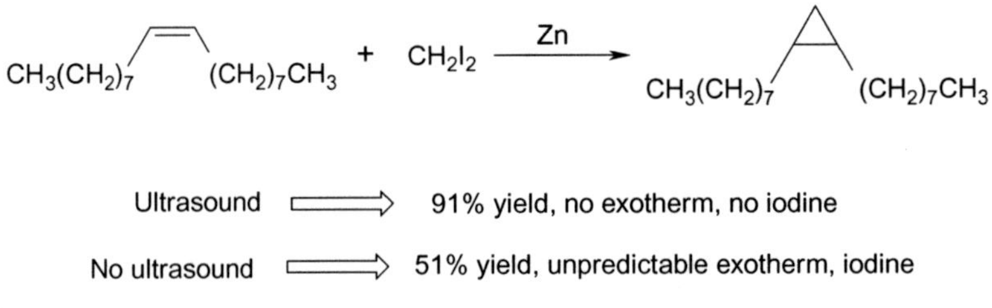

Scheme 7.11 Sonochemical Simmons–Smith reaction.

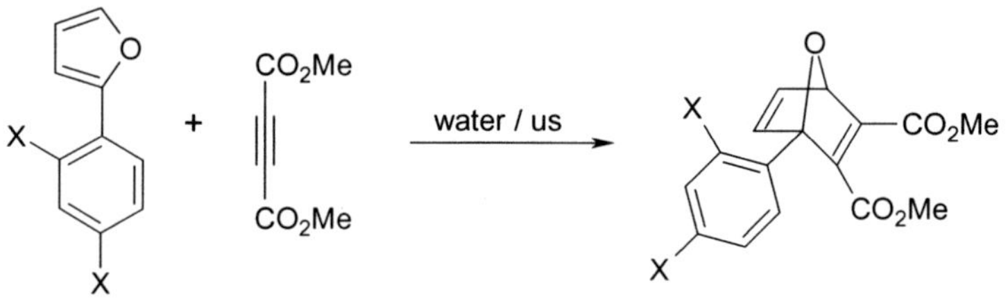

Scheme 7.12 High yielding Diels–Alder reaction initiation by ultrasound.

surface. Ultrasound has proved a very effective alternative to iodine, giving smooth reactions and high yields.

The range of reactions carried out under sonochemical conditions is large and growing rapidly; in many cases some 'green' benefits are obvious. Typical reaction types assessed include (i) oxidation, which can often be carried our more rapidly at lower temperatures, (ii) radical reactions, with the radicals being generated under mild conditions, and (iii) synthesis of nanoparticles. Here the high temperatures generated, followed by rapid cooling, are favourable to the formation of these particles. One reaction type, which has not been particularly well reported, is the Diels–Alder reaction. One of the few particularly successful examples involves addition of a dimethyl acetylenedicarboxylate to a furan (Scheme 7.12); sonication in water at temperatures in the range 22–45 °C gave virtually quantitative yields. The high yields have been attributed more to the solvent than to the sonication but it is a good example of what can be achieved by combining green technologies.

An interesting example of unexpected product formation is the ultrasonic ring opening of 1,2-benzocyclobutenes with poly(ethylene glycol) substituents on the four-membered ring. Both cis and trans isomers produced the same product – the one expected from the trans isomer according to Woodward–Hoffmann rules. Although not proven the explanation for this has been put down to the idea that thermal energy subjects reactants to random Brownian motion, whereas mechanical energy provides a direction to atomic motions. Therefore, forces from ultrasound efficiently direct the energy by straining the molecule, and thus reshaping the potential energy surface.

7.5 ELECTROCHEMICAL SYNTHESIS

Electrochemistry is widely used in industry, *e.g.* in effluent treatment, corrosion prevention, and electroplating as well as electrochemical synthesis. Electrochemical synthesis is a well-established technology for major processes such as aluminium and chlorine production; there is, however, increased interest in the use of electrochemistry for clean synthesis of fine chemicals. The possible 'green' benefits of using electrochemical synthesis include:

- often water-based processes;
- usually mild operating conditions (relatively energy efficient);
- atom efficient – replacement of reagents by electrons;
- novel chemistry possible.

The basis of electrochemical synthesis is the electrochemical cell, of which there are many types, both batch and continuous flow, with a multitude of electrode variations. A common type of cell used in synthesis is shown in Figure 7.3 – exemplified by the membrane cell for the production of chlorine. In addition to the basic features shown many cells have some method of improving mass transport. Electrochemical cells are generally more expensive than the average and although chemical reagent costs are low, electron costs are relatively high at approximately £10 000 per tonne equivalent. Careful electrode choice is essential for the efficient, cost-effective operation of electrochemical cells. A range of factors need to be considered, including operational stability, especially resistance to passivation and corrosion, energy consumption, and cost. In most cases production of oxygen and/or hydrogen also needs to be

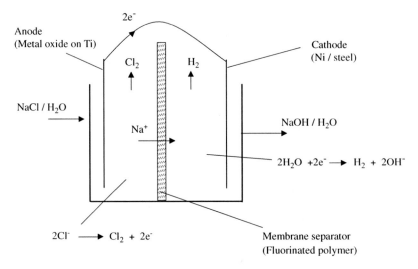

Figure 7.3 Schematic of an electrochemical membrane cell for production of Cl_2.

minimized. For these reasons lead, cadmium, zinc, and carbon electrodes are quite often used.

7.5.1 Examples of Electrochemical Synthesis

The membrane cell for production of chlorine is the most recent of three competing processes for chlorine production, the others being the mercury and diaphragm cell processes. In the flowing mercury cell sodium ions are discharged in preference to hydrogen ions due to the high overvoltage of hydrogen at the mercury cathode. The amalgam formed is subsequently decomposed with water outside the main cell (in another electrochemical cell containing graphite electrodes), regenerating mercury for return to the process. Although the cells are expensive and energy consumption is high the process does produce high purity sodium hydroxide at the required concentration of 50%. Environmentally, the process is not good, with mercury discharges being a cause for concern.

The diaphragm cell uses an asbestos diaphragm to separate anode and cathode compartments; however, due to the higher pressure in the anode cell (to prevent backflow of hydroxide) brine flows through into the cathode compartment, leaving the sodium hydroxide contaminated with sodium chloride. Evaporation of the product to 50% caustic concentration reduces the sodium chloride content to less than 1%, which is acceptable for most uses. Although cell costs are relatively low energy consumption is similar to that used in the mercury cell.

Environmentally, the preferred process is the membrane cell, since there is no toxic effluent and, at a consumption of less than 3 MW per tonne of Cl_2, energy use is over 10% lower than competing processes. The membrane is very selective, only allowing cations to pass; however, the sodium hydroxide is produced at 35% concentration, which is less than optimum. Cell costs are lower than the mercury cell. During the last 25 years, however, mercury discharges have been reduced by some 95% and now stand at around 1.2-g $Hg/teCl_2$. In Japan the mercury process has been phased out, largely due to intense public concern regarding mercury discharges following the Minamata Bay incident in 1965.

Adiponitrile is produced in over 1 million tpa, being used in the manufacture of hexamethylene diamine and (to a small extent) adipic acid; it is by far the highest volume organic material that is produced electrochemically. The mechanism (Scheme 7.13) involves electrolytic reduction of acrylonitrile followed by protonation, further reduction, Michael addition, and a final protonation step.

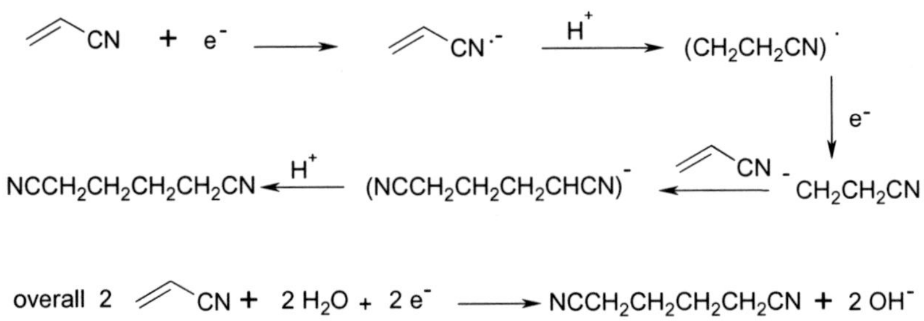

Scheme 7.13 Electrochemical adiponitrile synthesis.

Although superficially simple, to obtain high yields and an efficient process the design of cell and electrolyte is quite complex. One of the main problems is minimizing the protonation of the $^-CH_2CH_2CN$ anion to give the main by-product, propionitrile. This is achieved through addition of quaternary ammonium salts to the electrolyte; this produces a layer adjacent to the electrode, which is deficient in proton donors due to adsorption of lyophilic cations. By selection of conditions the required protonation steps are not prevented. Early processes used membrane type cells with a homogeneous electrolyte in the cathode compartment; this resulted in low conductivities (due to low water content), high energy use, and a difficult product separation. Modern equipment employs an undivided cell arrangement with cadmium cathodes and carbon steel anodes. The electrolyte is an emulsion containing sodium phosphate, borax, sodium EDTA, reactant, product, and quaternary ammonium salt. Although selectivities, at just under 90%, are a little lower the product is much easier to separate and energy consumption is less than 50% of the original process. The issue of ensuring high conductivity whilst maintaining a high organic reactant concentration is a general one. Several techniques are employed to overcome the problem, including addition of water miscible organic solvents, control of pH, and use of phase transfer catalysts.

Several electrochemical syntheses compete effectively with thermal routes. In Chapter 6 routes to sebacic acid were briefly described, one of which is the high-temperature hydrolysis of the renewable resource castor oil. Asahi and others have reported electrochemical routes based on methyl adipate. The key step is anodic coupling of the sodium salt of the ester using a platinum-coated titanium anode at 55 °C, which proceeds with current efficiencies as high as 90% and yields of over 90%. The resulting ester is hydrolysed to give sebacic acid. Despite high cell

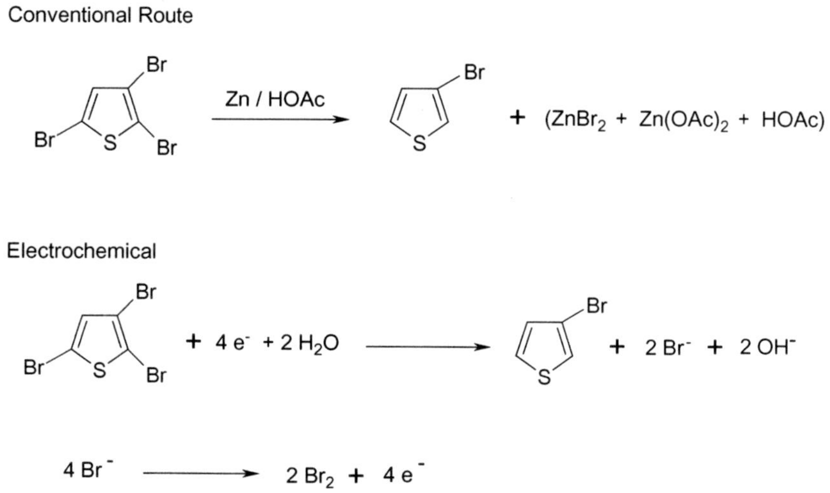

Conventional Route

Electrochemical

$$4\,Br^- \longrightarrow 2\,Br_2 + 4\,e^-$$

Scheme 7.14 Competing routes to 3-bromothiophene.

costs the process can compete due to lower raw material costs. Comparative data on energy use is not available but the electrochemical process is likely to be more efficient. Sebacic acid production provides yet another example where it is difficult to assess the relative 'green' merits of competing technologies (*i.e.* use of a renewable feedstock *versus* a more efficient process).

Synthesis of 3-bromothiophene provides an example of the obvious environmental benefits of the electrochemical route compared to a conventional process (Scheme 7.14). Both routes start from 2,3,5-tribromothiophene, obtained *via* bromination of thiophene. The conventional route uses an excess of zinc and acetic acid as the reducing agent, producing a large waste stream containing zinc bromide and waste acid. In the electrochemical route the use of a metal reducing agent is avoided and no bromine is wasted since the resulting bromide is oxidized to bromine, for recovery, at the anode. As will be evident from Scheme 7.14 additional bromide is required; this is added to the electrolyte as sodium bromide. To obtain adequate solubility of the starting materials dioxane is used as a co-solvent.

The scope of organic and inorganic syntheses that can be carried out electrochemically is very large. In particular, reduction of carboxylic acids, nitro compounds, and nitriles have been widely reported. Oxidation of aromatics and methyl aromatics has also been well studied. Other than chlorine and sodium hydroxide several other important inorganic compounds are manufactured using electrochemical technology; these include persulfate, permanganate, and perchlorate.

7.6 CONCLUSIONS

Energy consumption, particularly for large volume chemicals, has been taken seriously by industry for many years. Through a combination of new processes, particularly catalytic ones, improved engineering designs, and 'good housekeeping', energy use continues to fall. Unfortunately, energy consumption is still not considered particularly important by the average research chemist. The recent emergence of technologies that input energy in an alternative forms (microwave and ultrasound) together with renewed interest in photochemistry and electrochemistry for green chemistry will lead to selected future processes being more energy efficient as well as cleaner. Ideally, each research laboratory should be equipped with photochemical, microwave, ultrasonic, and electro-chemical reactors as well as heating mantles and oil baths; only then will these technologies become mainstream.

REVIEW QUESTIONS

1. Starting from cyclohexane compare the 'greenness' of the photo-chemical and non-photochemical industrial routes to caprolactam. Draw process flow sheets making assumptions for the waste materials generated where necessary. Discuss the volume and nature of waste generated from each process (including those from non-common starting materials) as well as the likely energy requirements.
2. Suggest synthetic methods for compounds **7.1**–**7.3**. At least one of the steps in each method should involve energy input from a non-thermal source. Discuss the benefits of using this non-thermal energy source for the particular reaction.

| 7.1 | 7.2 | 7.3 |

3. Discuss and compare the mechanisms of energy transfer using high-pressure steam, microwaves, and ultrasound. Discuss the role and limitations of solvents for carrying out a chemical reaction using these energy sources.
4. Industrial electrochemical reduction processes exist for the con-version of 3-hydroxybenzoic acid into 3-hydroxybenzyl alcohol and

of 4-nitrobenzoic acid into 4-aminobenzoic acid. How may these processes be carried out? Compare these processes in terms of the 'Principles of Green Chemistry' with alternative non-electrochemical methods.

FURTHER READING

B. B. Hamel, B. A. Hedman, M. Koluch, B. C. Gajanana, and H. L. Brown, *Energy Analysis of 108 Industrial Processes*, Fairmont Press, New York, 1996.

P. Lindstrom, J. Tierney, B. Wathey, and J. Westerman, Microwave assisted organic synthesis – a review, *Tetrahedron*, 2001, **57**, 9222–9283.

T. J. Mason and P. J. Lorimer, *Applied Sonochemistry. The Uses of Power Ultrasound in Chemistry and Processing*, Wiley-VCH Verlag, Weinheim, 2002.

S. Protti, D. Dondi, M. Fagnoni, and A. Albini, Assessing photochemistry as a green synthetic method, *Green Chemistry*, 2009, **11**, 239–249.

CHAPTER 8

Designing Greener Processes

8.1 INTRODUCTION

Previous chapters have discussed aspects of the chemistry involved in the design of 'greener' processes and products along with the techniques for minimizing and controlling waste. Most of the discussion has been centred on what chemists can do but, as highlighted in Chapter 2, engineering has a vital role to play. This chapter examines some engineering aspects of green process development; the use of these technologies can lead to safer, cleaner, and more cost-effective processes.

Scale-up from the laboratory is not always straightforward and what may appear to be a green efficient reaction on the bench may not result in a green process without appropriate development, including the correct choice of reactor and other plant equipment. Chemical processes can be divided into two main types, batch and continuous; most fine chemicals and pharmaceuticals are made in batch reactors whilst the majority of bulk chemicals are made in continuous plants. Semi-batch plants are used in some cases; these processes involve additional ingredients being added at certain stages into an otherwise batch process. Multi-purpose plants tend to go hand-in-hand with batch processing; these are relatively common in the fine chemicals industry, while most bulk processes use dedicated plant. Reactor choice and operation can have a dramatic effect on the overall eco-efficiency of a process.

Green Chemistry: An Introductory Text, 2nd Edition
By Mike Lancaster
© Mike Lancaster 2010
Published by the Royal Society of Chemistry, www.rsc.org

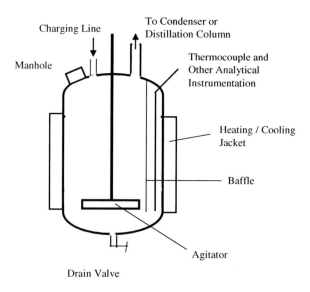

Figure 8.1 Typical batch reactor.

8.2 CONVENTIONAL REACTORS

8.2.1 Batch Reactors

A typical batch reactor (Figure 8.1) will be made of stainless steel or glass-lined steel. Stainless steel vessels have better heat transfer characteristics than their glass-lined counterparts and a wider array of reactor internals are available. Glass-lined vessels on the other hand can offer superior fouling or corrosion properties and are often used for reactions involving strong acids. Mixing will be *via* an internal agitator, possibly enhanced by internal baffles. The top of the reactor will be fitted with a variety of ports that will be used for sampling, instrumentation, reactant/product inlet/outlet, a venting line, *etc*. Often there will be a line from the head of the reactor to a reflux condenser, for additional heat removal, or a distillation column. At the bottom of the reactor is the drain valve, which often leads to a filter.

Batch reactors are often viewed as just large versions of laboratory equipment and it is therefore often assumed that scale-up will be straightforward. This assumption may prove to be untrue, and result in scale-up problems, many of which may be connected with differences in heat and mass transfer. Most of the following discussion will focus on heat transfer but it is worth pointing out some of the consequences of poor mass transfer. A very important possible consequence of poor mass transfer is a delay in mixing, followed by a generation of large and dangerous exothermic reactions. This can occur when a reagent has been added faster than efficient mixing occurs, allowing relatively high concentrations to build up that then start to react rapidly. Another effect of poor macro-mixing is the build up of local high concentrations of added neutralizing agents, at the end of a batch; this may result in excessive hydrolysis, for example. Some of these issues can be avoided by monitoring the rate of the laboratory reaction as a function of agitator speed and ensuring that work is carried out in a region in which rate is independent of speed.

Problems associated with heat transfer impact on the safety and efficiency of a process as well as the economics. Heat transfer characteristics vary with reactor design but can always be described by Equation (8.1):

$$Q = U \times A \times \Delta T \qquad (8.1)$$

where Q is the amount of heat transferred (W),
U is the heat transfer coefficient (W m^{-2} °C^{-1}),
A is the heat transfer area (m^2),

ΔT is the temperature difference between the reactants and the heating or cooling medium.

On moving from a laboratory flask to a large commercial reactor the volume of reactant rises much more rapidly than the external surface area of the reactor available for heat transfer. For example, a 1-litre laboratory flask has a heat transfer area of around 5×10^{-2} m^2, but a typical jacketed commercial 10 m^3 reactor will only have a heat transfer area of around 20 m^2, a relative reduction in surface area to volume ratio of 25. This reduction in surface area to volume ratio means that the heat transfer efficiency of a full-scale reactor is often lower than in a laboratory reactor, resulting in longer heating up and cooling down periods. In some instances prolonged heating or cooling periods can have adverse effects on the process. For example, slow hydrolysis reactions, which may not be detectable at laboratory scale, may become significant, reducing the yield and selectivity and making purification more difficult, thus leading to more waste, increased energy usage, and higher costs. Additionally, because of the large thermal mass and low heat transfer area, large volume reactors will be slow to respond to external temperature changes (thermal lag). In cases where rapid cooling of exothermic reactions is required to prevent further reaction of the product, it is likely that more by-products will be formed in a full-scale reactor than in a laboratory one. To some extent, this problem can be off set by the use of additional solvent as a heat sink, but this approach cannot be recommended to the green chemist. Utilizing the latent heat capacity of a boiling solvent is also a common method of controlling the temperature of exothermic processes in batch reactors. In this case the reactor is fitted with a condenser and a solvent with an appropriate boiling point chosen.

From examination of Equation (8.1), it can be seen that several things can be done to improve the heat transfer rate. Quite often the simplest approach is to increase the temperature differential, by using higher pressure steam or a hot oil supply. In some cases this may have adverse effects, *e.g.* a very hot wall temperature may lead to fouling, or worse initiate unwanted reactions; this is likely to be more pronounced in cases where mass transfer is poor. In some instances this practice may have safety implications, as in the case of Seveso discussed below. Another approach is to increase the area available for heat exchange, *e.g.* by the addition of heating/cooling coils inside the reactor. This may have additional advantages, such as increasing the turbulence in the reactor; it will, however, make reactor cleaning more difficult and can result in 'dead spots' or localized areas of poor mixing. Glass-coated coils (for use in glass-lined reactors) are very expensive, non-standard items.

An efficient alternative to having a jacketed reactor is to have an external, high surface area, heat exchanger through which the reaction medium passes; this type of reactor is often called a Buss loop reactor, after the company who developed the technology. The heat exchanger consists of many small diameter tubes or plates through which the heat transfer medium flows; the tubes are also often much thinner than the reactor walls, providing improved temperature response times. Separate mechanical stirring is not required in these reactors, adequate mixing being obtained by circulation through the heat exchanger. Such reactors are frequently used for hydrogenation processes and provide relatively efficient mass transfer as well good temperature control for an exothermic process.

One method of ensuring that exothermic reactions do not run away is attributed to Shinskey. The heat generated in an exothermic reaction is related to the reaction rate (r) and the heat of reaction ΔH. If the rate of increase in heat evaluation is less than the rate of increase in heat removal capacity (8.2) then, providing the cooling system is working, a runaway reaction will not occur:

$$\frac{d}{dT}(U \times A \times \Delta T) > \frac{d}{dT}(r \times \Delta H) \qquad (8.2)$$

In the pharmaceutical industry, and to some extent the fine chemicals industry, an important advantage of a batch reactor is traceability. The product from a particular batch will have a uniform consistency, can be uniquely labelled, and readily traced. In contrast product from a continuous process may change gradually over time and it is therefore more difficult to trace a particular impurity or fault in the material. Batch reactors are, however, rarely the most efficient in terms of throughput and energy use when the reaction kinetics are fast. Batch systems are also much more labour intensive than continuous processes.

8.2.2 Continuous Reactors

Although batch reactors offer some advantages that make them particularly attractive in some industry sectors, most bulk chemical production utilizes some form of continuous reactor, due to their overall greater efficiency. Continuous reactors fall into two categories: plug flow, which includes fixed bed reactors, and mixed flow, usually a continuous stirred tank reactor.

8.2.2.1 Plug Flow Reactors. Plug flow reactors are in some ways similar to batch reactors in that reaction is taking place under a range of conditions. In a batch reactor the concentration of the various components changes with time whilst in a plug flow reactor they change with distance along the reactor. For a given production rate of material the reactor volume required for either a continuous plug flow reactor or a batch reactor will therefore be the same. However, the slow turnaround times (charging, heating, cooling, discharging, and cleaning) associated with batch reactors mean that for a given annual production they will need to be larger, therefore having a larger inventory, and often more costly than a continuous plug flow reactor. Pressure drop will always be a significant consideration in the design of a plug flow reactor as it impacts on the dimensions of the reactor and the design of auxiliaries such as associated pumps and downstream treatment equipment. Any fouling caused by the reaction system will impact on both pressure drop and heat transfer and will also be of significant importance. Plug flow reactors tend not to be suitable for reactions requiring long residence times. Types of plug flow reactor in common use include tube-in-tube and fixed bed reactors.

8.2.2.2 Continuous Stirred Tank Reactors (CSTR). CSTRs share many physical similarities with batch reactors, tending to be large stainless steel vessels with an agitator, possibly baffles and either a jacket or internal coils for heating and cooling. These physical similarities may prove misleading, however, as the reaction regime in this kind of reactor is very different to that found in a batch unit; this is due to the

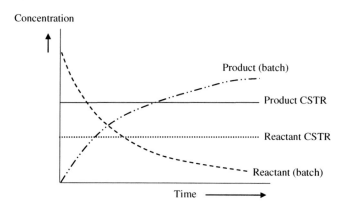

Figure 8.2 Reaction profile of batch and CSTR reactors.

CSTR having a constant in and out flow of materials. In a CSTR the concentration of all the components are constant (once equilibrium has been achieved) and the reaction proceeds under constant conditions (Figure 8.2).

The relative size of a CSTR compared to a batch or plug flow reactor is kinetic specific but, in general, CSTRs will have a larger volume. CSTRs offer advantages, however, in that the concentration of components can be optimized to minimize by-product formation; on the negative side there will be significant starting materials to separate from the product for recycle.

CSTRs may be used in series; in this case the conditions in each reactor will be different. The greater the number of CSTRs in series the closer the overall behaviour will be to a plug flow reactor. Although often more costly than a single batch reactor, a series of CSTRs may offer several advantages: safety advantages in terms of relatively low individual reactor inventory may be important as could the faster heat transfer rates associated with smaller reactors. Flexibility is another great advantage, *e.g.* additional reagents that would react adversely with one starting material can be added further downstream. Again knowledge of the kinetics is needed for a detailed comparison of the reactor volumes required for each specific reactor type.

For reactions with fast kinetics, continuous processes carried out in a simple tubular reactor will usually be more efficient but are rarely used outside the bulk chemicals industry. One reason for this is their lack of applicability to a wide range of processes, particularly ones requiring long residence times; another is that they can be more difficult to clean between different processes. For reactions requiring rapid heat transfer or very short reaction times, *e.g.* isolation of a reaction intermediate, they are the reactors of choice. One fairly common source of problems in scale up of continuous processes is that of sampling. Process development work usually involves frequent reactor sampling; at full scale samples are taken less frequently, and are much smaller relative to the reactor volume. Frequent sampling has the effect of turning over the reactor contents in a very short space of time at laboratory scale. One consequence of this is that trace by-products are being removed from the reactor, which would build up at full scale, possibly causing separation problems. The solution to this is to return all unused samples to the reactor, preventing loss of by-products.

As can be seen from the brief preceding discussion the choice of reactor is a complex decision, which relies heavily on knowledge of the reaction kinetics. These can be obtained from the rate of formation of products and by-products under carefully controlled conditions. Reaction kinetic data is independent of scale but the physical effects of scale-up still need to be addressed. By using a reaction calorimeter, heat evolution and thermal resistance data can also be obtained that can be extremely useful in the ensuing design process. The design of an appropriate reactor system that results in minimal by-product formation, optimal energy usage, and where safety hazards are minimized is the cornerstone of a green approach to process development. An effective reactor design can dramatically reduce downstream processing requirements and hence costs and waste.

8.3   INHERENTLY SAFER DESIGN

Although stringent safety standards are adhered to in designing chemical plants and Hazard and Operability (HAZOP) studies are performed well before the plant is built, most processes rely on mechanical safety devices and documented procedures to prevent hazardous occurrences. This approach is largely based on estimating the probability and consequences of a hazard (risk analysis) and is focussed on reducing that probability, or minimizing the subsequent hazards, rather than eliminating the hazard. It is well documented that over 60% of all accidents at chemical manufacturing plants are caused by either mechanical failure or operational error. The concept of inherently safer design (ISD) seeks to avoid accidents and incidents by asking 'can the hazard be eliminated by redesigning the process'? At the heart of ISD is the avoidance of reliance on mechanical safety devices and/or procedures, both of which are ultimately fallible, and the adoption of the principle of in the words of ISD guru Trevor Kletz 'what you don't have can't harm you'.

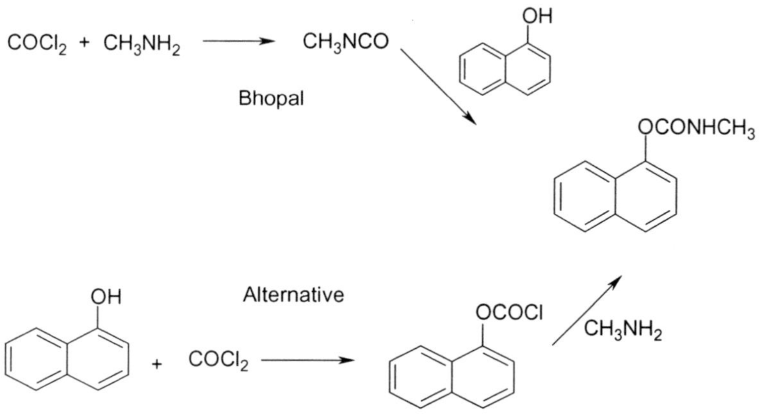

Scheme 8.1 Routes to carbaryl.

Chemical plants are designed to be safe but most are not designed to be inherently safe.

The concept of ISD arose as a consequence of the incidents at Flixborough and Bhopal; the latter has become the classic case study of what could have been prevented with ISD. The product being made at Bhopal was the carbamate insecticide, carbaryl. The synthetic procedure (Scheme 8.1) involved the reaction of methylamine and phosgene to give methyl isocyanate (MIC). MIC was then subsequently reacted with α-naphthol to give the product. None of these chemicals are particularly benign but MIC and phosgene pose the greatest threat. The accident at Bhopal resulted from the ingress of water into a large storage tank of MIC; this caused pressure build up. The subsequent explosion covered the nearby town with toxic gasses. The awful consequences were that at least 3000 people lost their lives and an estimated 200 000 more were seriously injured. How and why water got into the storage tank, and the fact that certain safety devices such as refrigeration plant and the flare were not working, could be considered to be somewhat irrelevant. The fact is that if MIC was stored then this accident was always a possibility, no matter how remote. Using the concept of ISD several questions would have been asked, such as:

- Do we need to use phosgene?
- Do we need to use MIC?
- Do we need to store MIC or phosgene?

The answer to the first question may have been yes but the answer to the second and third question would have been no, since an alternative safer route was already being used that did not involve MIC (Scheme 8.1).

The concepts and methodology of ISD can conveniently be sub-divided as follows.

8.3.1 Minimization

By minimizing inventories of hazardous material, *e.g.* through just-in-time production, the consequences of any accident will inevitably be reduced. As discussed above, storage of MIC was responsible for Bhopal and its use could have been avoided by using an alternative synthetic path. Phosgene, however, is required in both routes. Sometimes it is impossible (with current knowledge) to eliminate the use of a highly hazardous chemical altogether. However, increasingly, the chemical industry is avoiding storage of highly hazardous materials like phosgene and hydrogen cyanide through the development of small portable generators (as discussed in Chapter 7 for hydrogen cyanide).

Minimization goes much further than storage. For many processes the largest inventory of hazardous materials is in the reactor. If, through radical reactor design, inventories and equipment size can be reduced while throughput is maintained, then this presents opportunities for improved safety and possibly reduced capital costs. This is the concept behind process intensification, which is discussed more fully below.

Large reactor volumes are often employed because of slow reaction rates. Essentially there are two causes of this, either an inherently slow rate or, more commonly, poor heat and mass transfer. It is the latter that good reactor design can often resolve, even without going down the process intensification route.

The Flixborough disaster in 1974, in which a cyclohexane oxidation plant exploded, is a good example of what can happen when large inventories of hazardous material are used. The Flixborough process involved the oxidation of cyclohexane with air using a boric acid catalyst, the reaction proceeding *via* production of an intermediate hydroperoxide. Formation of the hydroperoxide was slow due to poor mixing of air and hydrocarbon. In addition, because of the possibility of local high concentrations of oxygen leading to unwanted higher oxidation products, reactor conversion was limited to under 10% per pass. Hence, to obtain the required throughput, six reactors were used in series, resulting in very high hydrocarbon inventories. The actual cause of the explosion was a simple broken pipe flange, but the seriousness of the consequences was a result of the large inventory.

At Flixborough air was added to the reactor through a simple sparge tube fitted at the base (similar to those used in laboratories). Mixing was with a conventional paddle stirrer. Although this was standard technology at the time better gas–liquid mixing devices are now available. For example, impeller blade designs (down pumping) are available that efficiently suck and mix headspace gasses; alternatively, intense mixing could be carried out inside a cyclone and the mix injected into the reactor. Options such as these could significantly reduce either the size or the number of reactors required, thereby reducing hydrocarbon inventory.

8.3.2 Simplification

Minimization and intensification often result in a simplified plant. Simplification will generally result in less mechanical equipment and fewer joints, both of which may fail and lead to an accident. Simplification therefore reduces the opportunity for error and malfunction. Although plants are not deliberately made overly complex, various cost reduction exercises usually avoid this; the standard design process often leads to complexity late in the day. Most safety studies are carried out late in the process and any risks identified are normally dealt with by the addition of further safety devices rather than by redesigning the process. By carrying out a detailed safety study early in the design process this inefficiency may be avoided. Frequent modification is another significant cause of complexity. Some modifications are carried out to make the plant more versatile but others are to overcome basic faults, like having valves in inaccessible places.

Some possible examples of over complexity include:

- running long lengths of pipe with many flanges, from say a reactor to a filter, due to lack of attention to layout;
- using solvent as a heat sink in a batch reaction rather than having a more efficient cooling system or using a tubular reactor;
- installing an excess of analytical equipment and sample ports, for example, an online near-infrared analyser may be preferable to a gas and liquid sample port and associated GC equipment.

8.3.3 Substitution

The substitution of hazardous materials by more benign ones is a core principle of green chemistry, and a key feature in ISD. Obvious examples would be the substitution of a flammable solvent by a non-flammable one and replacement of a harmful material by a safer one, as in the case of scCO$_2$ decaffeination of coffee. An area receiving much attention is the use of carbon dioxide as a replacement for

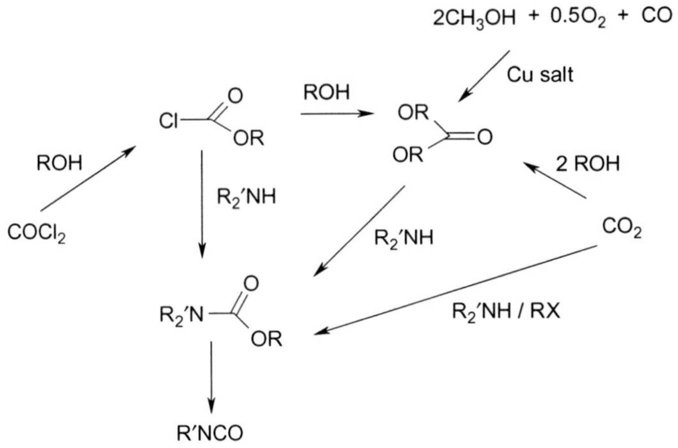

$$2CH_3OH + 0.5O_2 + CO$$

Scheme 8.2   Replacement of phosgene by CO_2.

phosgene. Safety is the main driver for doing this research but if some large processes used CO_2 as a feedstock it would contribute in some small way to reducing atmospheric CO_2 levels.

Approximately 8 million tpa of phosgene are used in the synthesis of isocyanates, urethanes, and carbonates; potentially CO_2 can be used in many of these reactions. Although the reactions using CO_2 depicted in Scheme 8.2 do work, they are generally slow, despite favourable thermodynamics, and hence are not currently commercially viable. As can be seen, dimethyl carbonate is also a safer alternative to phosgene. Dimethyl carbonate has traditionally been made from the reaction of methanol with phosgene but a commercial process involving reaction of methanol with oxygen and CO over a copper catalyst has been developed. Although possibly less eco-efficient than utilizing CO_2 directly, use of dimethyl carbonate in place of phosgene is growing, *e.g.* for production of diphenyl carbonate.

8.3.4   Moderation

In many cases it will not be possible to substitute a hazardous material, in which cases the onus should be on using the substance in a less hazardous form or under more benign conditions. Examples of this have been discussed in previous chapters, *e.g.* the unpredictable and potentially hazardous Grignard reaction can be moderated using ultrasound to avoid sudden exothermic reactions, and the hazards associated with using hydrogen in fuel cells can be moderated using Powerball technology. Concern over the storage of hazardous materials has resulted in significant moderation. Chlorine was frequently stored in pressurized containers on many chemical sites but, because of the potential consequences of an accident, it is now usually refrigerated at atmospheric pressure. Similarly, chlorine from cylinders was once used to disinfect swimming pools, now chlorine in the form of sodium hypochlorite is employed.

8.3.5   Limitation

Despite all the measures discussed above, chemicals manufacture will always involve some risk and will always rely to some extent on equipment integrity and appropriate operation. Limitation is the process of minimizing the effects of failure (equipment or people) or an incident, by design. One important aspect of the design process should be to limit the available energy to an appropriate level.

The Seveso accident in Italy serves as an appropriate reminder of the potential consequences of supplying too much energy to a reaction. The process involved the production of 2,4,5-trichlorophenol through the base hydrolysis of 1,2,4,5-tetrachlorobenzene. A partially completed batch was shut down at the weekend at a safe temperature of 158 °C. Although this was a safe temperature the vessel was heated with steam, from a turbine, capable of reaching a temperature of 300 °C. The reactor was only part full and, due to the high steam temperature, the wall above the reaction liquid reached a much higher temperature. Once the agitator had been turned off the top of the reaction liquid became hot enough (estimates vary between about 190 and 230 °C) to cause a runaway reaction and an explosion involving the formation of the carcinogenic 2,3,7,8-tetrachlorodibenzo-*p*-dioxin (Scheme 8.3). A wide area of land around the plant became contaminated with dioxins and around 2000 people had to receive medical treatment; fortunately, the long-term effects have not proved serious. Despite the lack of fail-safe devices on the plant this is a clear case in which it was known that a runaway

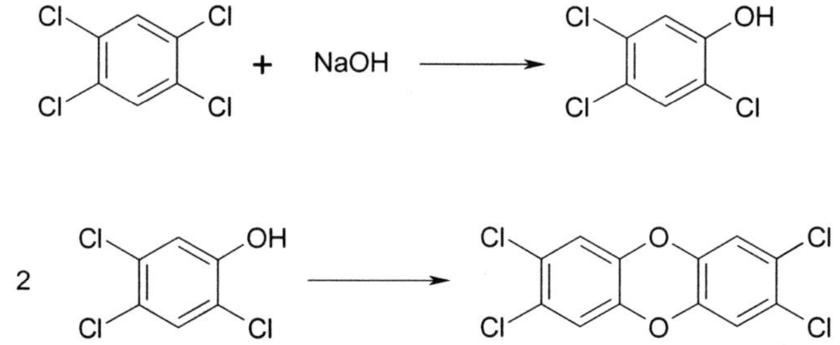

Scheme 8.3   Seveso chemistry.

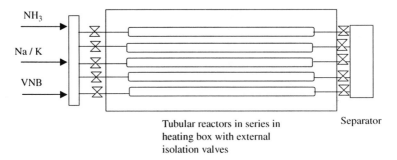

Figure 8.3 Schematic of an ENB reactor.

reaction was possible and the only inherently safe way to prevent it was to avoid the possibility of the reaction reaching runaway conditions. This could have been achieved by using more appropriate heating.

Reactor design may also have a large influence on the overall effect of an incident. In general, leaks from tubular, and especially banks of small tubular reactors in series, will have less adverse consequences than those from a batch reactor. This is due to the inventory in any one tube or part of the reactor being much lower than in a batch reactor. In the case of a leak the other areas of the unit can be isolated, limiting the total volume of lost material. Ethylidene norbornene manufacture (Chapter 9) involving the use of a hazardous sodium/potassium amalgam in liquid ammonia used such a reactor configuration (Figure 8.3).

In assessing COMAH sites (Control of Major Accidents and Hazards – sites with higher risk often due to large inventories of hazardous substances) the HSE states 'Major accident hazards should be avoided or reduced at source through the application of principles of inherent safety'. Similar statements are embedded in regulations in most parts of the world.

8.4 PROCESS INTENSIFICATION

Process intensification (PI) is commonly defined as:

Technologies and strategies that enable the physical sizes of conventional process engineering unit operations to be significantly reduced.This is achieved through:

- improving mass transfer rates to match that of the reaction;
- improving heat transfer rates to match the exothermicity of a reaction;
- having an appropriate residence time for the reaction.

Originally devised as a cost reduction concept, as a result of the development of novel smaller reactors and ancillary equipment, PI is now recognized as a way of providing safety improvements, greater throughput, and improved product quality through better control. All these features are important in the development of more sustainable processes. As a general rule, major equipment such as reactors and distillation columns account for only 20% of the price of a manufacturing plant, the remainder being pipework, instrumentation, labour, and engineering charges, *etc*. The concept behind PI was that even though novel pieces of key equipment may be a little more expensive the overall plant cost would be reduced as a result of simplification and size reduction. The scope of PI is now extensive, extending well beyond basic equipment design (Figure 8.4) into actual process methodology.

With PI, traditional process design criteria (particularly those focused around stirred batch reactors) are thrown out and the equipment is designed to match the chemistry. It is not unexpected therefore to find that PI has been applied successfully to reactions that are very fast and exothermic, where the process is being limited by poor design. Traditionally these processes have been handled through the use of large amounts of solvent, as a heat sink, or careful control.

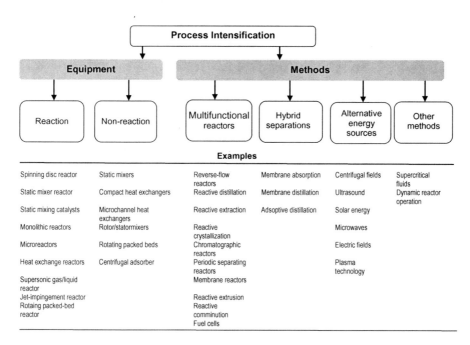

Figure 8.4 Scope of process intensification.

8.4.1 Some PI Equipment

Ensuring the efficient mixing of viscous or non-miscible liquids, or of gases and liquids, is a common problem that, if not solved, can lead to mass transfer limited reactions, with possible safety implications as noted above. Some options for improving gas–liquid mixing have already been discussed; the PI approach considers more radical, but often simpler, options. Mechanical mixers, as well as sometimes being inefficient, are prone to breakdown and the sealing arrangements on pressurized reactors can be complex and prone to leaking; a better option would be an efficient device without moving parts, *i.e.* let the liquid move not the reactor! Many such devices are now available.

The radial jet mixer (Figure 8.5) is perhaps the simplest devise for efficient liquid–liquid mixing; when the liquid mix hits the opposite tube wall fluid flow patterns are established that cause rapid mixing. This is a particularly good method of mixing in a tubular reactor with multiple injection points. Another type of static mixer frequently used is one containing structured packing, often referred to as a Sulzer mixer, after the company that initially developed the packings for distillation columns. Several different arrangements of structured packing are available; simply the mixers can be viewed as a column packed with a high surface area, honeycomb-like structure that disrupts liquid flow. In certain systems these can be prone to fouling, and may therefore be unsuitable. In other systems the honeycomb surface can be impregnated with a catalyst to produce a small efficient catalytic reactor. The three-way catalytic converter discussed in Chapter 4 is an example of this.

As noted above, stirred batch reactors are the most common type of reactor used for the production of fine and pharmaceutical chemicals; this is partly to do with tradition as well as the need to produce a flexible reactor design in which a whole range of products can be made. Spinning disc reactors (SDR), developed at Newcastle University, have been proposed as an efficient alternative for fast reactions (Figure 8.6). As its name suggests the SDR consists of a disc rotating at speeds up to

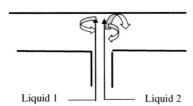

Liquid 1   Liquid 2

Figure 8.5   Jet mixer operating *via* liquid impact on wall.

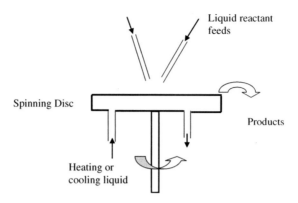

Figure 8.6 Schematic of a spinning disc reactor.

5000 rpm or more. The disc may be smooth or contain ridges to aid mixing; it may simply be a stainless steel plate acting as source of heat or have a catalytic surface on which the reaction is carried out. The reactant liquids are pumped onto the centre of the disc, the resulting flow patterns causing intense mixing as the liquids move towards the edge of the disc, where the products are collected. Because the liquid forms a thin film on the disc surface heat transfer is rapid (heating or cooling); this, together with the intense mixing overcomes any heat and mass transfer limitations, allowing the reaction to run under kinetic control. Typically, SDR reactors have residence times of between <1 s and 5 s, which translates to them being useful for reactions with half-lives of around 0.1–1 s.

Another form of related reactor is the rotating packed bed, initially called a HIGEE reactor, because of the high centrifugal force generated, by ICI who developed the technology. This reactor consists of a rotating bed containing packing, often metal gauze, but structured packing similar to those used in static mixers can be employed. These reactors are particularly efficient at gas–liquid mixing, the liquid being fed to the centre of the reactor and the gas coming in from the side (Figure 8.7). Although rotating packed beds provide exceptionally good mass transfer, heat transfer is not as efficient as in the SDR.

The production of membranes with specific pore sizes is now relatively easy, as membrane separation processes have become increasingly common. Catalytic membrane reactors (Figure 8.8) are now being developed in which the reaction and separation are carried out in a single process, thereby greatly intensifying the process. In addition to equipment reduction, since reaction and separation are being carried out together, membrane reactors offer potential for improved yield and

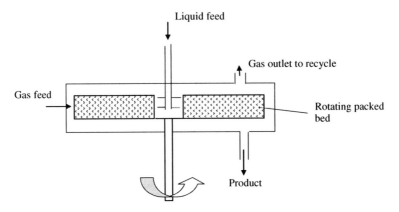

Figure 8.7   Schematic of a rotating fixed bed reactor.

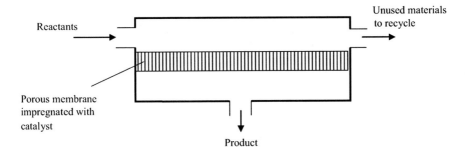

Figure 8.8   Schematic of a catalytic membrane reactor.

selectivity and increased overall rate, due to the driving of equilibrium reactions through product removal. This often has the additional advantage of preventing by-product formation from further reaction of the product. Catalytic membrane reactors offer several other advantages compared to fixed bed catalytic reactors. These include increased control through having membranes of specific pore size, improved mass and heat transfer, and a high surface area to volume ratio. There are some practical disadvantages to be overcome, including the current high cost of manufacture, the design of flow rates through the membrane to match reaction rates, and the selection of a suitable material capable of performing the separation at the reaction conditions. Polymer membranes, *e.g.* based on dihydroxydiphenyl sulfone, offer good separation properties but are frequently unstable at the reaction temperature. A wide variety of inorganic membranes are now commercially available, many being designed for fuel cell applications. Examples include membranes based on alumina, zirconium oxide, thorium oxide, silicon carbide, glass, and even stainless steel.

Microreactors are the ultimate in process intensive reactor design. These reactors usually consist of small diameter channels (sometimes less than 100 microns) sandwiched together in layers. Individual layers or individual channels may act as reactors, heat exchangers, or mixers. In many respects microreactors resemble printed circuit boards and much of the technology used in their fabrication comes from the electronics industry. One of the biggest advantages afforded by microreactor technology is that laboratory reactors become the full commercial reactors, simply by adding additional banks. This is often termed 'scale-out' as opposed to conventional 'scale-up'. Scale-out is seen as an important concept in minimizing the time taken from discovery to commercial production, particularly for pharmaceuticals.

There are many designs of microreactor for gas or liquid systems, some incorporating catalytic surfaces, which benefit from the high surface to volume ratio. Whilst heat transfer is usually very efficient, mass transfer can be more problematic. Owing to the small volumes there is usually laminar flow, resulting in the mixing being under diffusion control. Mass transfer is often improved by having a series of Y junctions in which two liquid streams enter a single channel, which subsequently divides into two before recombining, thereby producing good mixing.

8.4.2 Some Example of Intensified Processes

Perhaps the first example of a somewhat intensified process was the manufacture of nitroglycerine. The reaction, originally carried out in large stirred batch reactors, involves the nitration of propylene glycol (glycerine) with a mixture of concentrated nitric and sulfuric acid, and is highly exothermic. If heat is not removed quickly enough the nitroglycerine can explosively decompose. In the original process, external cooling was provided manually by the operator opening cold water circulating valves when the reactor had reached a certain temperature. In an attempt to ensure safe operator practice, one-legged stools were provided to prevent the operator falling asleep – definitely not an inherently safe design! The potential consequences of any explosion were severe due to the large size of the reactor. Eventually it was realized that large reactors were required to obtain throughput not because the reaction rate was slow but because mixing was poor. There were by now many ways in which mixing could be improved; the method developed in the 1950s involved having a rapid flow of acid into a small reactor, this created a partial vacuum (*cf.* water pump) that sucked glycerine into the acid stream, ensuring good mixing. Using this method residence time in the reactor was reduced from 2 h to 2 min. This enabled the reactor inventory to be reduced to 1 kg, a size at which the severity of an explosion could be mitigated by building a blast wall around the reactor.

Dow has commercialized a process for preparing hypochlorous acid using a packed rotating bed. The process (8.3) involves the reaction of gaseous chlorine with sodium hydroxide solution; however, in the presence of the reaction by-product NaCl the acid reacts to give sodium chlorate. Hypochlorous acid is formed very rapidly, and also reacts rapidly with NaCl; the rate of formation of the acid being limited by mass transfer of chlorine into the sodium hydroxide solution. After formation the acid, which is a gas, must be rapidly removed from the liquid to prevent further reaction; this removal is also thought to be a mass transfer limited process. Using conventional equipment yields are below 80%; however, by using a rotating packed bed yields of over 90% have been achieved:

$$Cl_2 + NaOH \rightarrow HOCl + NaCl \qquad (8.3)$$

Spinning disc technology is being explored for a range of mass transfer limited reactions. The synthesis of polyesters from a dibasic acid and a diol (8.4) is one, commercially important, example. This synthesis is normally carried out in large batch reactors, with the reaction being driven by water removal. Typical reaction times are over 12 hours due to low water removal rates, which in turn are attributed to mass transfer limitations that result from the increase in viscosity brought about by the formation of high molecular weight polyester:

$$HO(CH_2)_n OH + HO_2 C(CH_2)_m CO_2 H \rightarrow$$
$$- [O(CH_2)_n OC(O)(CH_2)_m C(O)]_x - + H_2 O \qquad (8.4)$$

In the early part of the reaction the SDR offers no benefit since the reaction is not mass transfer limited. However, at higher viscosities it was found that a single pass through the SDR, at 200 °C, gave a similar reduction in acid value (a measure of the degree of polymerization) to a batch reaction time of between 40 and 50 min. By recycling the polymer several times through the SDR the required degree of polymerization could be achieved, potentially reducing reaction time by several hours.

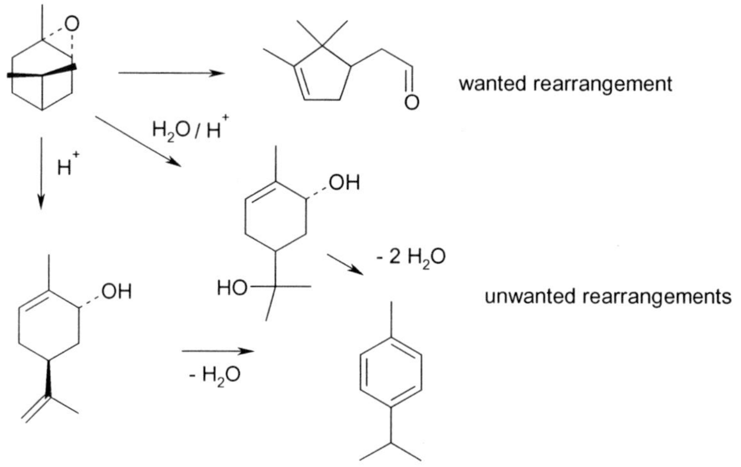

Scheme 8.4 α-Pinene oxide rearrangement.

A catalytic SDR reactor has also been used for the rearrangement of α-pinene oxide to campholenic aldehyde (Scheme 8.4), an important fragrance intermediate. Industrially the rearrangement is carried out in a batch reactor using a Lewis acid catalyst such as zinc bromide. The waste associated with the removal of catalyst using a water wash is one unattractive aspect of this process and, in addition, the reaction has relatively poor selectivity resulting from a series of other rearrangements. In this example the main aim of using the SDR was not mass transfer enhancement but the avoidance of waste, through the use of a supported catalyst and an improvement in selectivity through improved heat transfer.

The catalyst used was zinc triflate supported on silica, which was glued onto the surface of the SDR. Although total conversion could be achieved in a single pass, selectivity was highly dependent on residence time, increasing with shorter times. In a direct comparison with a batch process using the same catalyst at the same conversion and residence time a 200-fold increase in throughput could be obtained.

Several reactions have been demonstrated using microreactors. One of the potentially more important is the direct synthesis of MIC from oxygen and methyl formamide over a silver catalyst. Dupont have demonstrated this process using a microreactor cell similar to that described above in which the two reactants are mixed, then heated to 300 °C in a separate layer and subsequently passed through another tube coated with the silver catalyst. The estimated capacity of a single cell with tube diameters of a few millimetres is 18 tpa.

8.5   IN-PROCESS MONITORING

Traditionally process analysis has lagged behind the sophisticated analytical methods available in many research laboratories, with techniques such as NMR, mass spectrometry, and gel-permeation chromatography, for example, not being widely used to monitor the routine production of chemicals. Analysis in production environments has concentrated on fast and simple methods, which can be run on a 24/7 basis, without the requirement for Ph.D. level technicians. Hence methods such as pH, viscosity, melting point, and moisture content (Karl Fischer) are widely used, even though they provide little information on the chemical make-up of the sample. More recently, more informative analytical methods have become relatively routine, including GC, HPLC, and IR. For most batch reactions analysis is off-line, requiring an operator to take a sample, often becoming exposed to the reactor contents. The sample then needs to be taken to the analysis laboratory, often some distance away. Once results are obtained and communicated back to the plant an hour or more may have passed since it was decided to take the sample. During this time the batch may have gone out of specification and it may be too late to take simple remedial action.

In-line or on-line process monitoring improves overall efficiency. For batch processes this is through increased throughput (less waiting for results), reduced energy use (batch not on hold), and less rework of out of specification material, leading to less waste. For continuous processes the technique simply allows the reactor to be kept under the required conditions for a greater proportion of the time, thus optimizing production rate and quality and minimizing downstream disturbances. An additional benefit of in- or on-line-process analysis is that, once reliably established, it may be readily combined with automated control techniques. In these cases signals from the analyzer are fed directly to the process control unit (Figure 8.9) to make automatic adjustments, such as; increase or decrease temperature, add more reagent, or even in extreme cases initiate emergency shut down procedures, all without the operator needing to go near the plant. Although there are examples of this completely integrated approach there is often a lack of confidence in removing human control to this extent. Other barriers to the widespread adoption of in- or on-line process analysis including up front costs and a general lack of knowledge of the possibilities and the reliability of equipment. Maintenance and calibration of equipment is another issue since manual cleaning/intervention between each analysis is not desirable or perhaps even possible. In principle, most laboratory based analysis procedures could be adapted to run on-line. As with other safe

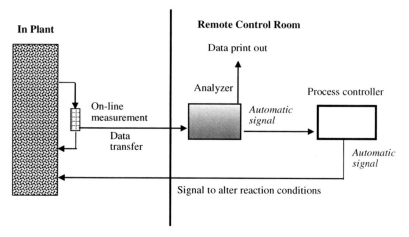

Figure 8.9 In-process analysis concept.

design tools it is much more cost-effective if process analysis methods are incorporated early into the design process.

It is worth briefly pointing out the difference between in-line and on-line analysis. In-line analysis does not involve removal of the sample from the process, *e.g.* in determining water or oxygen content. On-line analysis does involve removing a sample, usually as a side stream, adding to the complexity of the plant since the sample off-take equipment will often need to be built to the same integrity as the plant. There are four common techniques employed:

1. *Titration*: This method obviously requires physical removal of a sample from the plant and results in the sample being thrown away. In modern equipment samples can be taken at specified intervals from a flowing sample stream using flow injection analysis, thereby enabling the analysis time to be shortened.
2. *Chromatography*: GC is the most common analytical method used but liquid and supercritical fluid chromatographic methods are being increasingly developed. Like titration the sample is destroyed in the analysis process. The ideal situation depicted in Figure 8.9 can not normally be applied for titration or chromatographic analysis since the analysis equipment needs to be close to the sampling device; this is often termed at-line analysis.
3. *Spectroscopic*: A whole variety of spectroscopic methods is available, including IR, microwave, Raman, and X-ray spectroscopy. In all these cases real time analysis gives almost instantaneous feedback of results.

4. *Sensor based*: Many methods use sensors, the simplest being temperature measurement; this terminology is often used to cover viscosity, pH, oxygen, and humidity determination, *etc.* These are true in-line techniques and offer rapid, inexpensive real-time analysis. Humidity determination in drying ovens is a common example.

The above principles will be exemplified through a brief discussion of near-infrared spectroscopy.

8.5.1   Near-infrared Spectroscopy

When molecules are subjected to electromagnetic radiation, transition from the ground state to the first excited state may occur. This transition results in the absorption of a quantum of energy that can be measured, enabling a characteristic absorption spectrum to be produced. Other transitions sometimes occur, *e.g.* from the ground state to the second or third excited level; these are called overtones and have band frequencies in the near-infrared (NIR) region of the spectrum (780–2500 nm). The most intense bands are produced from light atoms; hence NIR spectroscopy is usually used to study O–H, C–H, and N–H bonds. The spectrum becomes more complex due to combination bands that arise from two or more overtone transitions occurring at the same time; these produce further bands at multiples of the original band.

NIR spectroscopy may seem an odd choice as an in-process monitoring technique compared to the more widely used laboratory analytical method of IR spectroscopy. There are two important reasons why it is so attractive. First and most importantly NIR signals can be transmitted over long distances through fibre optic cables, thus enabling the sensitive detection and analysis equipment to be remote from the light source; this has practical, cost, and safety benefits, since detectors are ignition sources. Special fibre optic cable made from silica with very few O–H groups is required to prevent interference. The second reason is that the sample path for NIR can be much larger than for IR, in many cases centimetres rather than microns; this makes equipment design and operation much simpler. However, NIR spectra are usually complex, and broad, hence very sophisticated and expensive data analysers are required to interpret the basic data and convert it into meaningful spectra.

There are several types of NIR instruments and equipment capable of analyzing the whole NIR spectrum can be very expensive, up to £75 000, which is well beyond most in-process analysis budgets! The cost can be significantly reduced by having an instrument that only looks at a small region of the spectrum, say a 20 nm band. This is perfectly adequate for most analyses since it is often only a singe bond that is important, *e.g.* observation of an O–H bond at 1900 nm to monitor water content. Alternatively, the cost may be justified by simultaneously using the NIR analyser for a wide range of analyses across a complex unit. In the simplest instruments the sample continuously flows past a fibre optic cable, sometimes in a small cell within or outside the reactor. The absorption data is then passed back up the cable to the remote analyser. Even with this lower-cost option NIR analysis is often only considered for the larger dedicated process plants.

This technique has been applied to the production of biodiesel, where methanol and free glycerol concentrations in the final product are important. An at-line laser based Fourier-transform NIR analyser was able to get results in 30 seconds compared to almost 1 hour for the traditionally used flash point and GC laboratory based analysis. Effective use of on-line NIR has also been demonstrated for continuous control of concentrations of sodium hydroxide solution fed into caustic scrubbing systems for the removal of acidic gases. Although the technique can minimize the amount of base used the initial cost is hard to justify.

8.6 PROCESS SAFETY

Even by adopting an approach based on the principles of green chemistry and engineering it is not possible to design out all risk associated with the manufacture of chemicals. A considerable part of the safety of any process is down to the training and approach taken by people and the culture of the organisation set by directors. It is now widely accepted that good leadership can have a significant impact and several critical aspects identified:

- have a Board Champion for process safety, with the board reviewing process safety indicators and performance on a frequent basis;
- have a clear and visible policy and expectations that are communicated throughout the workforce;
- senior managers must be visible in inspections and show interest in process safety;
- a set of relevant leading (*e.g.* up to date maintenance inspections) and lagging (*e.g.* number of lost time accidents) indicators are in place and used to monitor performance;

- a continuous improvement plan, based on the indicators should be in place and driven by senior management;
- incidents should be reviewed from across the industry as well as other sectors and programmes to implement lessons learned put in place.

REVIEW QUESTIONS

1. Describe the advantages and disadvantages of the following reactor types with reference to heat and mass transfer. For each reactor discuss one reaction for which it may be appropriate to use that reactor. (a) Fluidized bed reactor; (b) a continuous counter-current flow reactor; and (c) a monolith reactor.
2. Review two commercial processes for producing phenol from the point of view of 'Inherently Safer Design'.
3. All chemical processes have some inherent risk. Discuss the concept of risk related to process design, including how it is evaluated and how the cost–benefit is calculated.
4. What are the advantages and possible drawbacks of on-line GC monitoring compared to off-line monitoring? Discuss the differences required in the equipment used in the two situations.

FURTHER READING

J. M. Chalmers (ed.), *Spectroscopy in Process Analysis*, Sheffield Academic Press, Sheffield, 2000.

M. Gough (ed.) *Better Processes for Better Products*, BHR Publishing, Cranfield, 2001.

T. Kletz, *Process Plants: A Handbook for Inherently Safe Design*, Taylor & Francis, London, 1998.

D. Reay, C. Ramshaw, and A. Harvey, *Process Intensification*, Butterworth-Heinemann, Oxford, 2008.

CHAPTER 9

Industrial Case Studies

9.1 INTRODUCTION

This chapter discusses some examples of green chemical technology that have met with commercial success. The examples have been chosen to reflect a cross section of both the industry and the wide variety of applications in which chemicals are used.

9.2 METHYL METHACRYLATE

Methyl methacrylate (MMA) is produced on the scale of 3 million tonnes per annum with its polymers used in applications such as acrylic baths, plastic windows (Perspex), car headlights, dentures, and replacement lenses for patients with cataracts.

Historically, there have been several commercial processes. The most widely used one is called the acetone cyanohydrin route since it is based on the reaction of acetone with hydrogen cyanide (Scheme 9.1). It was developed by ICI in the 1930s. The resulting cyanohydrin is reacted with excess concentrated sulfuric acid to produce methacrylamide sulfate that is then reacted with methanol to produce MMA and an equivalent amount of ammonium hydrogen sulfate. The major disadvantages to the process are the production of the salt waste and the use of hazardous hydrogen cyanide. Sulfuric acid can be recovered from the salt by heating, but this is energy intensive. Hydrogen cyanide can be produced by reaction of methane with ammonia and air, but is also a by-product

Green Chemistry: An Introductory Text, 2nd Edition
By Mike Lancaster
© Mike Lancaster 2010
Published by the Royal Society of Chemistry, www.rsc.org

271

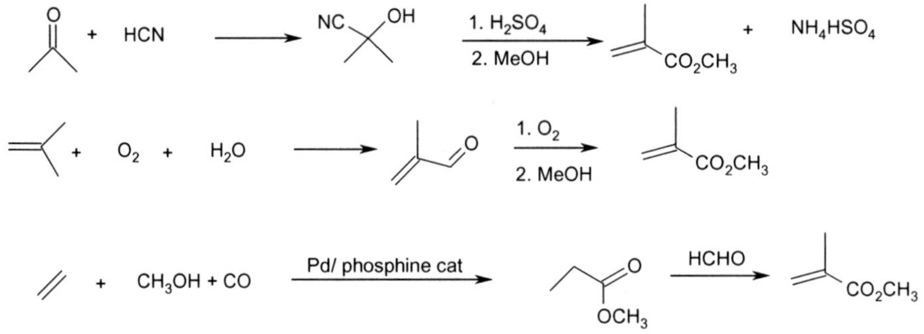

Scheme 9.1   Competing methyl methacrylate processes.

of propene ammoxidation used to produce acrylonitrile; hence MMA plants are usually co-located with acrylonitrile plants, reducing location flexibility and requiring the two processes to be in balance.

Various attempts have been made to overcome these drawbacks, including one multi-step and energy intensive process by Mitsubushi that involves hydrolysis of the cyanohydrins followed by reaction with methyl formate and dehydration to MMA. Formamide is a by-product of the process but is converted into cyanic acid that is then used in place of hydrogen cyanide.

An alternative process, largely only commercialized in Japan, is based on isobutene as a starting material and involves oxidation to methacrolein. The *t*-butyl alcohol is fed to a reactor, with air and steam containing a metal oxide catalyst system to make methacrolein, which is further reacted in the presence of a phosphorus-molybdenum catalyst to make methacrylic acid. Esterification with methanol produces crude MMA that is purified by distillation. A major disadvantage of the process is that considerable amounts of methacrolein need to be separated and recovered.

Eastman Chemical has developed a three-step syngas process. The first step of the new process is the production of propionic acid from ethylene and syngas – the source of carbon monoxide in the reaction – using a homogeneous iodine-promoted molybdenum-based hydrocarbonylation catalyst. The reaction takes place at moderate temperature (150–200 °C) and pressure (30–70 atm). In the second step, propionic acid is reacted with formaldehyde to produce methacrylic acid (MAA) using a heterogeneous (silica supported) acid–base niobium-based bifunctional catalyst. Finally, esterification with methanol takes place to produce methyl methacrylate.

Lucite International has developed an MMA process that uses readily available feedstocks ethylene, methanol, and carbon monoxide and only minor amounts of benign waste are produced. Called Alpha, the process is claimed to reduce the total cost of production by 40% compared to the cyanohydrin route and operates at mild conditions involving no toxic or corrosive chemicals. The first commercial plant began operation in Singapore in 2008 and was designed to produce 120 000 tonnes year^{-1}.

The Alpha process consists, of three steps: two separate catalytic reactions and a series of distillations in the final product separation stage. In the first stage, ethylene, carbon monoxide, and methanol are reacted in the liquid phase at very mild conditions (10 bar and 100 °C) over a homogeneous palladium-based phosphine ligand catalyst, to produce methyl propionate (MeP), with essentially no by-products. The MeP is then used in the second step to react with formaldehyde in the gaseous phase over a fixed bed proprietary heterogeneous catalyst in the presence of methanol, to produce MMA and water. The MMA is then separated from the other constituents using six distillation steps, ultimately producing a product stream, recycle stream, and waste stream.

In terms of selectivity from feedstocks the Alpha process is most efficient, with well over 90% selectivity, while the acetone cyanohydrin process is 87% and the C$_4$ process 70%.

9.3 GREENING OF ACETIC ACID MANUFACTURE

Acetic (ethanoic) acid is produced at a level of around 6.5 million tpa. Its main uses are in the production of vinyl acetate, which goes into emulsion paints, including those based on biodegradable poly(vinyl alcohol), and acetic anhydride used as an acylating agent and to produce cellulose acetate. Over the years there have been several industrial processes for acetic acid production, starting with fermentation of sugars to produce ethanol, which was subsequently oxidized by air. Although this may appear to be a very green process, being based on a renewable resource, it results in production of considerable amounts of (biodegradable) waste. In addition, as the process was carried out under dilute conditions, the recovery of pure acid by distillation was an expensive and energy intensive process. With the advent of a synthetic chemical industry based on coal the calcium carbide route gained prominence. This process involves the high temperature reaction of calcium with coal to give calcium carbide and subsequent reaction of the calcium chloride with water to give acetylene, which can be converted in two stages into acetic acid:

$$\text{Coal} + \text{Ca} \rightarrow \text{CaC} \rightarrow \text{HCCH} \rightarrow \text{CH}_3\text{CHO} \rightarrow \text{CH}_3\text{CO}_2\text{H} \quad (9.1)$$

With the growing prominence of the petrochemicals industry this technology was, in turn, replaced by direct air oxidation of naphtha or butane. Both these processes have low selectivities but the naphtha route is still used since it is a valuable source of the co-products, formic and propanoic acid. The Wacker process – which uses ethylene as a feedstock for palladium/copper chloride catalysed synthesis of acetaldehyde, for which it is still widely used (Box 9.1) – competed with the direct oxidation routes for several years. This process however produced undesirable amounts of chlorinated and oxychlorinated by-products, which require separation and disposal.

Box 9.1   Wacker oxidation process.

It has long been known that ethene can be oxidized to acetaldehyde in the presence of palladium chloride and water. This reaction was of no practical value since it requires molar amounts of precious metal:

$$C_2H_4 + PdCl_2 + H_2O \rightarrow CH_3CHO + Pd + 2HCl$$

The discovery of the oxidation of Pd^0 to Pd^{II} by oxygen mediated by a copper couple led to the development of the Wacker process during the early 1960s:

$$Pd + 2CuCl_2 \rightarrow PdCl_2 + 2CuCl$$
$$2CuCl + 2HCl + 0.5O_2 \rightarrow 2CuCl_2 + H_2O$$

Overall this equates to the direct oxidation of ethene in a 100% atom efficient process:

$$C_2H_4 + 0.5O_2 \rightarrow CH_3CHO$$

The process can be operated under moderate conditions (50–130 °C and 3–10 atm) in a single reactor. Regeneration of cupric chloride occurs in a separate oxidizer.

The free HCl and Cl⁻ generated in the catalytic cycle produce environmentally harmful chlorinated by-products to the extent that more that 3 kg of HCl need to be added to the reactor per tonne of acetaldehyde produced to keep the catalytic cycle going. Modified catalysts such as ones based on palladium/phosphomolybdovanadates have been suggested as a way of reducing by-product formation to less than 1% of that of the conventional Wacker process. However, these catalysts have yet to make an impact on commercial acetic acid production.

The most significant development both in terms of economic and environmental efficiency came with the development of the Monsanto methanol carbonylation process in 1970. In this process methanol and carbon monoxide are reacted together, under pressure (30 atm) at around 180 °C with a homogeneous rhodium/methyl iodide based catalyst. Yields of acetic acid (based on methanol) are virtually quantitative. The major advantage of this route is the very high yield and selectivity of the process but the major disadvantage is the requirement for exotic materials of construction (zirconium based) to prevent corrosion both in the reactor and other parts of the plant. This entails high capital expenditure. Since the 1970s until the late 1990s the Monsanto process held a dominant position.

It is hard to imagine that there would be a need for improvement in such high yielding, highly selective process based on inexpensive feedstocks. However, in the world of bulk chemicals, where profit margins are constantly under pressure, companies continually strive for improvements, especially ones that will enable existing plants to be debottlenecked or that will reduce capital costs associated with building new plants to meet growing world demand. In addition, for a bulk chemical company, licensing income can be a significant prize for the owner of the best technology. Through a careful study of the reaction mechanism (Scheme 9.2) key process limitations could be identified that offered opportunities for further efficiencies; three of the key limitations of the Monsanto process are:

1. *Less than perfect CO utilization:* The rate-determining step of the process is addition of methyl iodide to $[Rh(CO)_2I_2]^-$; HI generated elsewhere during the reaction cycle (Scheme 9.2) competes for this Rh species, generating hydrogen and subsequently carbon dioxide in a water-gas shift reaction summarized in Scheme 9.3. The H_2

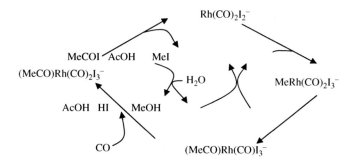

Scheme 9.2 Monsanto process carbonylation mechanism.

$$[Rh(CO)_2I_2]^- + 2HI \rightarrow [Rh(CO)_2I_4]^- + H_2$$
$$[Rh(CO)_2I_4]^- + H_2O + CO \rightarrow [Rh(CO)_2I_2]^- + CO_2 + 2HI$$

Scheme 9.3 Competing water-gas shift reaction.

and CO_2 need to be purged from the system to prevent build up; however, during the purging process CO is also inevitably lost, reducing overall CO utilization to less than 90%.

2. *Moderate reactor productivity:* The rhodium catalyst is continuously recycled; however, the catalyst is inherently unstable at low CO partial pressures, *e.g.* in the post reactor flash tank. Under these conditions the catalyst may lose CO and eventually form insoluble RhI_3, resulting in an unacceptable loss of expensive catalyst. This reaction is also more likely to occur at low water concentrations; hence to run the process satisfactorily catalyst concentrations are kept low and water concentrations relatively high. Hence through a combination of lower than optimum reaction rate (because of low catalyst concentrations) and water taking up valuable reactor volume the overall reactor utilization is less than optimum.

3. *Large energy requirement due extensive distillations:* Although in the normal sense of the word the Monsanto process is highly selective there is a very tight specification on the purity of the final product. In particular the propanoic acid concentrations must be reduced to below 400 ppm in the final product. To achieve this three distillation stages, one solely to separate propanoic acid from acetic acid, are required. The overall distillation stage is both energy and capital intensive. Propanoic acid is produced by the carbonylation of ethanol, which arises from hydrogenation of acetaldehyde, eliminated from the reaction of HI with $(MeCO)Rh(CO)I_3^-$. As mentioned above a high water content is required for catalyst stability, removal of which also entails a large energy requirement.

If the above limitations could be overcome then greater throughput could be obtained from the same sized reactor, energy usage could be reduced, and fewer raw materials would be consumed. This would obviously generate both economic and environmental benefits.

A process named CATIVA, by BP, uses an iridium catalyst promoted by ruthenium carbonyl. This catalyst results in significantly faster reaction rates than the Monsanto process, at low water concentrations. In-process high-pressure infrared spectroscopy indicates that the reason

$$Ru(CO)_3I_2 + HI \rightarrow [Ru(CO)_3I_3]^-[H]^+$$
$$[Ru(CO)_3I_3]^-[H]^+ + MeOH \rightarrow MeI + H_2O + Ru(CO)_3I_2$$

Scheme 9.4 CATIVA – possible role of ruthenium promoter.

for this is a higher relative concentration of the active catalyst species $[MeIr(CO)_2I_3]^-$. Since the water-gas shift reaction is not significant in the CATIVA process, the inactive species $[Ir(CO)_2I_4]^-$ that would be formed during this process is not present, leaving more Ir present in a catalytically active form compared to the Monsanto process. There are several possible explanations for the lack of the water-gas shift reaction, one being that HI, which is required, is removed by the promoter according to Scheme 9.4.

As well as increasing the reaction rate and catalyst stability, at all important low water concentrations, and low CO partial pressures, the iridium system also produces lower levels of by-products. These improvements combine to give the CATIVA process the following advantages:

- improved CO utilization to well over 90%;
- due to reduced requirements for water and propanoic acid removal there is only a need for two distillation stages – this results in lower energy consumption, which contributes to the overall reduction in CO_2 emissions to 0.31 te per te product compared to 0.48 te per te product with the Monsanto process;
- enhanced reaction rate, which can be used to increase plant capacity by up to 75%;
- significant reductions in the cost of a new plant, largely due to reduced distillation requirements.

9.4 EPDM RUBBERS

Ethene/propene/diene monomer rubbers (EPDM) are elastomeric terpolymers used in the production of sealants, tubing, gaskets, and, in the USA, is used in roofing applications. As the name suggests they are prepared by the polymerization of mixtures of ethene, propene, and diene monomers to form crosslinks. By far the most common diene used is 5-ethylidene-2-norbornene (ENB).

ENB is normally prepared in a two-step process (Scheme 9.5). The first step involves *in situ* cracking of dicyclopentadiene to cyclopentadiene, and subsequent Diels–Alder reaction with 1,3-butadiene to give

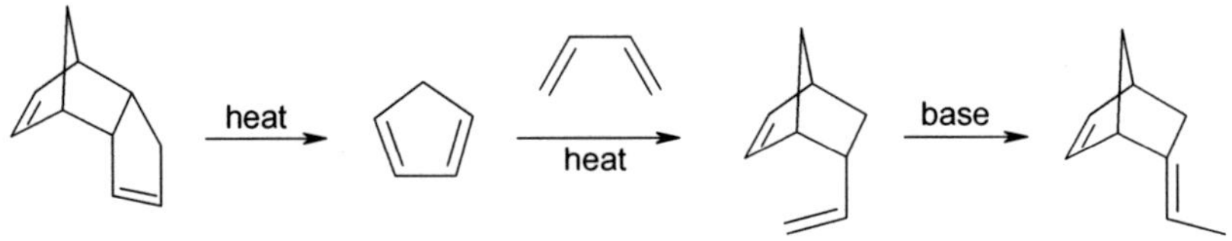

Scheme 9.5 Synthesis of ENB (5-ethylidene-2-norbornene).

Table 9.1 Efficiency issues associated with EPDM manufacture.

Stage[a]	Inefficiency
VNB	By-products from competing Diels–Alder reactions
VNB	Reactor fouling – heat transfer & product loss issues
VNB	Distillation required – energy intensive
ENB	Poor reactor volume utilization
ENB	Hazardous catalyst
ENB	Distillation required – energy intensive
EPDM	Low reactor utilization
EPDM	Energy intensive diluent recovery

[a]VNB = 5-vinyl-2-norbornene; ENB = 5-ethylidene-2-norbornene;
EPDM = ethene/propene/diene monomer rubbers.

5-vinyl-2-norbornene (VNB). The second step involves isomerization of VNB to ENB.

As both cyclopentadiene and butadiene can act as diene and dienophile several by-products are formed. VNB is separated through several energy intensive distillation steps. The purified VNB is then isomerized to ENB using a base. Very strong bases are required, industrially a sodium/potassium alloy (liquid at temperatures close to ambient), in liquid ammonia is the most commonly used. Although, due to its low boiling point, ammonia can be easily separated and recycled, recovery and reuse of the metals is more problematic and hazardous. The production of the EPDM rubber itself is usually carried out in batch reactors using a Ziegler–Natta type catalyst and an inert hydrocarbon solvent to render all reactants soluble. This hydrocarbon solvent significantly increases the energy requirements of the process and reduces reactor utilization. Table 9.1 summarizes the cost and environmental issues.

Several different companies have 'greened' various steps of the process. In the VNB production by-products come from competing Diels–Alder reactions and polymerization, largely of cyclopentadiene. The reaction is usually carried out in a continuous tube reactor, but this results in fouling, due to polymerization, at the front end where the dicyclopentadiene is cracked to cyclopentadiene at temperatures over

175 °C. Although fouling does not have a very significant effect on yield, over time it builds up, hence increasing energy requirements. Increased pressure drop also results from fouling; once a limiting pressure drop value is reached the plant must be shut down for cleaning. Several different ways have been identified for reducing the fouling, including carrying out the reaction in an inert hydrocarbon diluent and the addition of organonickel compounds. The first approach significantly reduces reactor utilization, increases energy consumption, and makes product separation more difficult; the second approach involves the use of a highly hazardous material. Neither method has been used commercially. Sumitomo has found that fouling can be minimized by the addition of amine radical inhibitors such as *N,N*-diethylhydroxyl-amine or 4-oxo-2,2,6,6-tetramethylpiperidine-1-oxyl at levels of around 1000 pm. Using these inhibitors, fouling can be minimized and the time between reactor shutdowns extended.

Although there are several Diels–Alder reactions occurring that lead to unwanted by-products, one of the main ones, and the most trouble-some to remove by distillation, is 4-vinylcyclohexene (VCH), formed from two butadiene molecules. Sumitomo discovered that most indus-trial processes actually ran as two-phase systems, with the gas phase containing a high proportion of the butadiene. In this region of high butadiene concentration VCH formation is exacerbated (Scheme 9.6). By increasing the reaction pressure the system becomes liquid phase only, thereby eliminating the region of high butadiene concentration. This shifts the selectivity of the reaction so as to favour cyclopentadiene/ butadiene reactions and hence VNB production. Under such conditions VCH formation can be reduced around by 60%.

Several alternatives have also been suggested for the second stage of the reaction to improve the efficiency of VNB isomerization to ENB and overcome the hazards associated with handling Na/K and liquid am-monia. Lewis bases such as triethylaluminium used in conjunction with titanium tetra-alkoxides have been successful in increasing reactor effi-ciency but can not be recovered in a reusable form. Although solid base catalysts such as sodium on carbon have been known for many years their performance has always been poor for this reaction. Sumitomo has

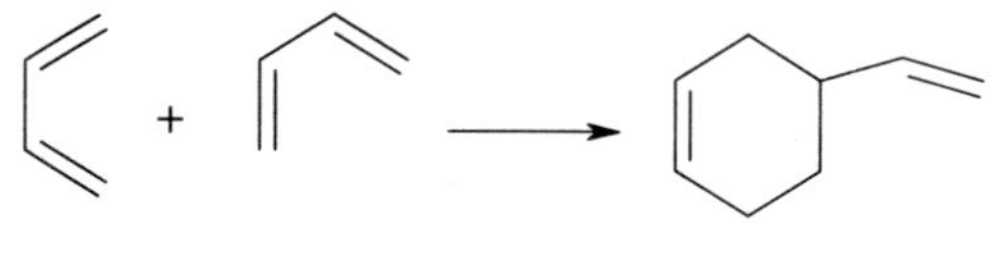

Scheme 9.6 Synthesis of VCH (4-vinylcyclohexene).

now developed one of the first heterogeneous base catalysts to be used commercially. This catalyst consists of sodium and sodium hydroxide on alumina, the resulting high activity being a consequence of the catalyst preparation method. To obtain the high catalytic activity a high surface area alumina such as γ-alumina is used and impregnation with the hydroxide and metal is carried out at temperatures in excess of 250 °C. Catalyst activity is such that the isomerization takes place at ambient temperature. If highly pure VNB is used the ENB produced may be used without further purification, since the conversion and selectivity are almost 100%.

As noted earlier, EPDM is often prepared in a batch process using an inert hydrocarbon diluent and Ziegler–Natta type catalysts. Complex and costly solvent recovery adds significantly to the energy requirement of the process. Union Carbide developed a fluidized bed process that reduces overall energy requirements, removes the need for diluent, avoids catalyst residues, and reduces overall costs. Again the secret behind the success of this process is the catalyst. In this case the support is an anhydrous silica in which the silanol groups have been removed by treatment with triethylaluminium. This support is then impregnated with a magnesium chloride/titanium chloride complex in tetrahydrofuran. The fluidized bed is run at temperatures of around 50 °C, under pressure; a nitrogen stream is used to prevent polymer blockages. Regulating the various partial pressures of the reactants controls EPDM structure. As well as producing EPDM to specifications required by current commercial grades this technology can be used to extend the range of materials that can be produced, thereby opening up other commercial opportunities.

9.5   VITAMIN C

Vitamin C (L-ascorbic acid) is found in many fruits and vegetables, being naturally produced from D-glucose. Although a small amount of vitamin C is extracted from fruits and rose hips this is not a commercially viable route for the quantities currently demanded (over 50 000 tpa). The basic synthetic route was developed in the 1930s and, with some improvements (Scheme 9.7), is still operated today. Being based on glucose the synthesis does have some green credentials; however, the overall synthesis is complex, producing considerable effluent, and requires sophisticated separation technology.

Initially the process used potassium permanganate as the oxidant as well as strong acids and bases at other stages of the process. The overall yield from glucose was only around 20%, with hazardous waste being

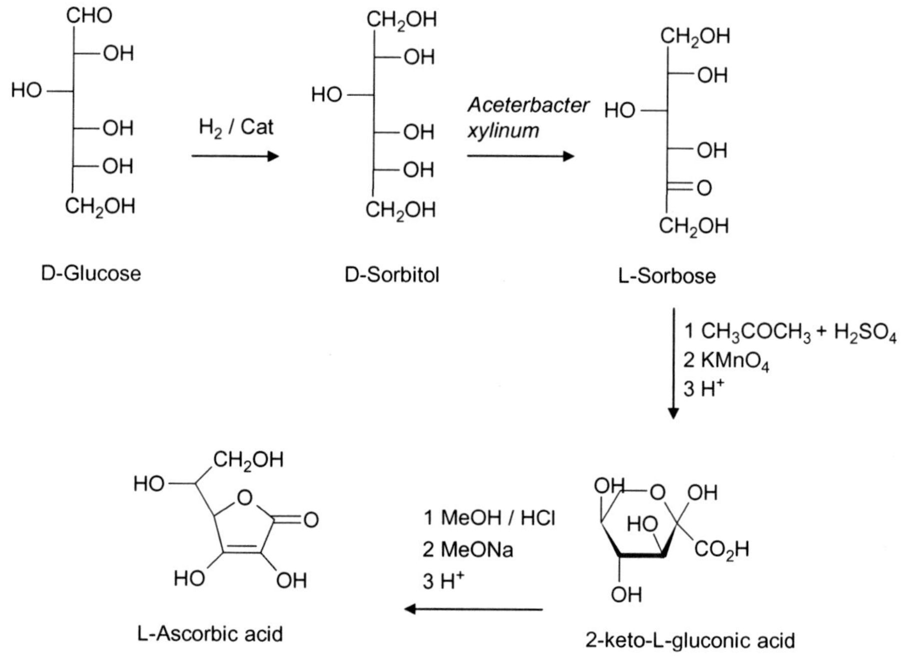

Scheme 9.7 Original route to vitamin C.

produced in relatively large amounts. Several process improvements have been made during the last 50 years but the basic steps are still the same. The bacterial oxidation process is now carried out using *Acetobacter suboxydans* as a source of the active enzyme sorbitol dehydrogenase. This bacterium is more resistant to trace amounts of nickel catalyst carried over from the hydrogenation step and can operate at relatively high sorbitol concentrations. This biotransformation inverts the sugars from the D- to the L-isomer, a step that, until recently, was considered to be very difficult to achieve chemically. Although the potassium permanganate step proceeds in over 90% yield environmental and cost concerns over the production of stoichiometric amounts of manganese waste have resulted in its replacement. One common alternative uses a basic solution of sodium hypochlorite in the presence of a nickel chloride catalyst. Although used in small amounts the catalyst must be totally removed from the product and environmental concerns over its toxicity remain. A more benign alternative uses a catalyst of palladium on carbon with air being used as the oxidant.

Hoffmann La Roche has developed a procedure using a second enzyme system for the direct conversion of 2-keto-D-gluconic acid into L-ascorbic acid. In this case the enzyme is a lactonase, *e.g.* gluconolactonase from

Zymomonas mobilis. This route avoids the use of toxic methanol and reduces the requirement for strong acids. A one-step synthesis from glucose has been patented. The process uses a hydrous oxide of cobalt ($CoO_2 \cdot xH_2O$), formed by treatment of a cobalt salt with hypochlorous acid. Although the mechanism of the reaction has not been disclosed it is claimed that the direct conversion of D-glucose into L-ascorbic acid at room temperature and pH 5.5 can be achieved with a yield of 50%. Furthermore, the by-products of the reaction, *e.g.* sorbose can be isolated and recycled.

9.6 LEATHER MANUFACTURE

Because it comes from a natural resource, leather is often considered to be a more environmentally benign alternative to man-made textiles and fabrics. Modern leather manufacturing processes, however, consume large volumes of chemicals, including many environmentally unsound ones such as chromium and formaldehyde. For every kilogram of finished leather product approximately seven kilograms of chemicals (excluding water) are used. In many parts of the world the large effluent volume coming from the manufacturing process is regulated and many more eco-friendly alternative chemicals are being developed. Figure 9.1 shows a simple outline of the processes involved together with the major chemical types used.

After removal from the animal the hides start to deteriorate rapidly through bacterial action. To minimize this, the hides are usually salted either in concentrated brine or by covering with dry salt. In some instances small amounts of organic materials, *e.g.* naphthalene, are added to enhance the preservation. All the salt used in the process ends up in an aqueous waste stream at the tannery, with approximately 5-L of saturated brine being produced per kilogram of hide. In Australia cold storage and refrigerated transportation to the tannery is becoming increasingly common; although this reduces chemical waste it is a more energy intensive approach. Another alternative, which has achieved little commercial success, is irradiation. Irradiation with either electron beams (10^6 V) or γ-rays sterilizes the hides, which, if packed appropriately, can then be transported long distances without deterioration. This technique is widely used for some medical products but is currently considered too costly for leather.

At the tannery the hides are soaked in an aqueous solution of detergents, biocides, and salt to clean and re-hydrate the hides following preservation. Several changes of soaking solution may be required to achieve this, resulting in significant effluent. The next stage is called

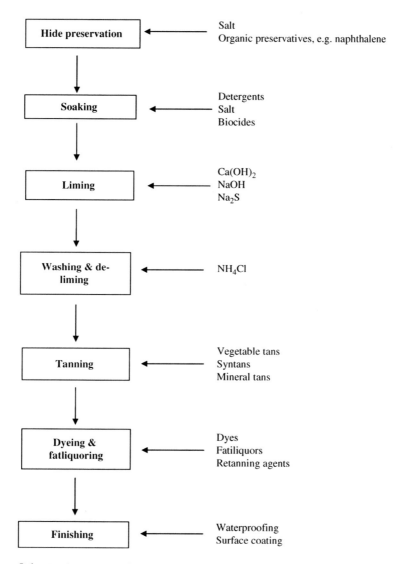

Figure 9.1 Main aspects of traditional leather manufacturing process.

liming, which, as the name suggests, involves treatment with calcium hydroxide. The process also involves treatment with sodium sulfide and/or sodium hydrogen sulfide as well as sodium hydroxide. This is done to remove hair but has the added advantage of plumping the hides, hence improving the efficiency of the subsequent tanning process. The effluent from this process not only contains the chemical residues but also has very high COD and BOD levels; these result from dissolution and decomposition of the hair by the lime. In more recent processes introduced by BASF and others the process conditions in the early stages are controlled such that the hair is loosened from the hide before significant decomposition has occurred. The hair is then removed by filtration before the final stages of the liming process are undertaken; the hair may be successfully composted. Such processes produce effluent with less than 50% of the oxygen demand of those in which the hair is dissolved. De-liming is the final stage before tanning and involves the neutralizing of the hide and the removal of the liming chemicals. Historically, the deliming process has been carried out using ammonium chloride or sulfate. There are several significant environmental concerns surrounding this step. First the process results in the production of ammonia; this must be dealt with as a gaseous effluent, but this is not always an easy task for small tanneries in developing countries. Second, if the pH becomes too low there is the possibility of H_2S evolution. A number of incidents have occurred that have resulted in H_2S levels of well over 100 ppm inside the tanneries. Several approaches that partially tackle these problems have met with some commercial success, including neutralization with organic mono- and di-carboxylic acids and propane diol esters of sulfurous acid. All these alternatives are both expensive and add significant quantities of organic material to the effluent. Neutralization with CO_2 is usually slow and leaves the internal structure in the hide highly basic. This has been identified as a mass transfer problem, for which Sol SPA has engineered a solution. This solution involves use of multiple CO_2 injection points to generate a high CO_2 concentration throughout the process, which is performed in rotating drums. With this modified process ammonia evolution, possible H_2S evolution, and organic effluent production are all avoided.

9.6.1 Tanning

The unique properties of leather are due to collagen, a protein consisting of three amino acid chains held in a helical formation. In the presence of water the collagen fibres are relatively soft and flexible but, unfortunately, are prone to decay by bacterial action. When dry they are quite stable but the leather becomes hard and inflexible due to inter-fibre hydrogen bonding replacing collagen–water hydrogen bonding. During the tanning process the hydrogen-bonded water is replaced by a material that forms stronger bonds, preventing decay yet maintaining the attractive properties of the leather.

To make the hide ready for tanning it is usually 'pickled' in a solution of sulfuric acid and sodium chloride for 2–3 h, a process that produces a large salt-laden effluent stream. The main purpose of the salt is to prevent swelling of the hide. Although it is more expensive, some tanneries now pickle in a salt-free solution of phenol sulfonic acid and obtain the same results. The comparative environmental implications of these competing processes need to be assessed locally. In addition to the pickling process, hides containing a high fat content, *e.g.* sheepskin, may require further treatment to remove some of the fat prior to tanning. There are two traditional processes with differing environmental consequences. The first involves the use of hydrocarbon solvents, giving rise to VOCs and solvent waste. The second involves washing with water and non-ionic surfactants, giving an aqueous waste with a high COD content. Recently, it has been found that the defatting process can be successfully carried out with $scCO_2$ (Chapter 5), thus avoiding the production of a waste stream. For this process the hides must be relatively dry, which adds to the time, cost, and energy consumed in the process. However, fat extraction is efficient and the fat is recovered in a form that can be used in other applications. Although the process has not yet been commercialized it has many similarities with $scCO_2$ processes used in other areas of the textile industry.

The most widely used tanning agent is chromium(III) sulfate, applied as sodium dichromate. Chrome tanning imparts properties difficult to achieve by other means, this is due to the ability of chromium sulfate to link into large complexes providing crosslinks between collagen fibres through the sulfate group. In contrast to organic tanning agents the chromium compounds occupy relatively small amounts of space within the gaps between the fibres, giving a soft, flexible leather. In addition, due to the strong nature of the crosslinks, chrome tanned leather can withstand boiling water. Although Cr(III) is non-toxic there is considerable concern over residues entering the environment, in case it is converted into highly toxic Cr(VI). Several eco-friendly alternatives are now used industrially. These materials reduce the amount of chromium used and therefore reduce the amount entering the environment as waste. Most tanneries now use increasing amounts of mineral tans based on aluminium, iron, and zirconium. Aluminium is particularly useful for white leather. Although the feel and robustness of the leather is not generally as good as with chrome tans, it is sufficient for most applications. One of the largest sources of chrome waste is leather shavings, which are normally disposed of in landfill. Increasingly industry is finding value in this waste. By treating the waste with protease enzyme the collagen can be broken down into a protein suitable for use as an animal feed, releasing the chromium for recycle into the tannery.

The oldest form of tanning is vegetable tanning using extracts from plants, in particular extracts containing poly-phenolic compounds based on catechol or pyrogallol. Some of the most useful tannins are found in mimosa and the bark of the sumac tree. The tannins are extracted with water and alcohol. The tanning action has many similarities with wine clarification using egg whites, and involves the hydroxyl groups on the tannins reacting with the protein to form crosslinked insoluble products. Whereas with chrome tanning only around 3% by weight uptake of chrome is required to form a stable leather, for vegetable tanning the figure in around 50%. As a result of this, much of the inter-fibre space becomes filled, giving much fuller and firmer leather. Fullness is a valuable property in many applications but the firmness produced by pure vegetable tanning limits it usefulness to applications such as shoe soles. Whilst vegetable tanning may be the most sustainable form of tanning it can not produce all the properties society demands of leather ware.

Synthetic tanning agents (syntans), which where developed to improve upon the properties of vegetable tans, are widely used, often in conjunction with mineral tans. As such, many syntans are based on phenolic compounds. Scheme 9.8 shows a typical type of syntan. Sulfonated phenol is used for two main reasons: first to impart water solubility to the resin and second to limit the degree of polymerization both during syntan manufacture and in subsequent tanning, with the sulfonic acid group blocking an active site on the aromatic ring. High

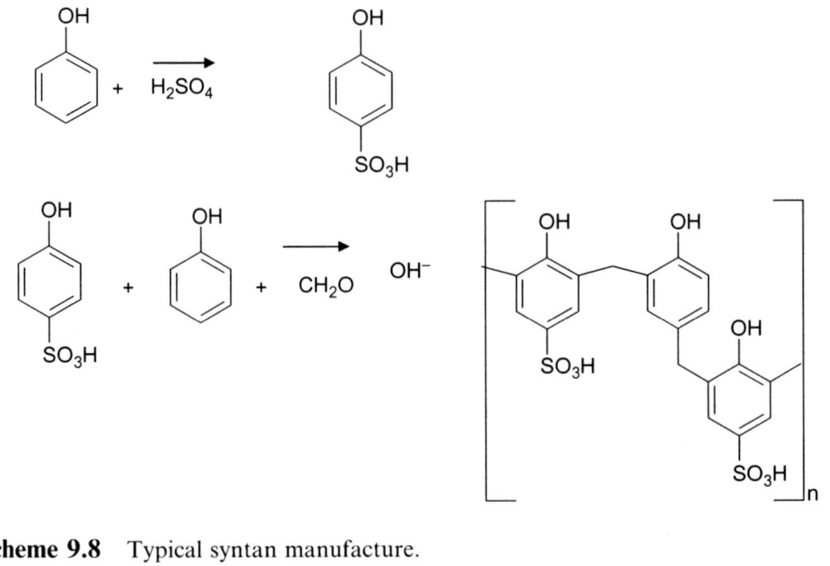

Scheme 9.8   Typical syntan manufacture.

molecular weight syntans do not penetrate into the leather sufficiently, giving brittle products. Depending on the application, phenolic syntans usually have molecular weights in the range of several hundred to a few thousand. There are a huge variety of syntans produced commercially, varying in molecular weight degree of sulfonation and degree of incorporation of other materials such as ammonia or urea.

Although hazardous, phenol and formaldehyde can be handled safely during syntan manufacture. The situation is often different in tanneries, where workers are often exposed to the chemicals used. Partially due to residual levels of phenol in the syntan and to regeneration of some formaldehyde during the tanning process (the process is carried out in drums, often at pH 1) the presence of these chemicals in the workplace and in effluent gives rise to environmental concern. There is also a real concern that small amounts of these materials may remain in the finished article. Syntan manufacturing methods have been improved in recent years and low–free phenol products are now widely available; these are frequently produced by vacuum stripping the product to remove residual phenol as an azeotrope with water. Stabilizing the product with amines and optimizing tanning conditions reduces subsequent release of formaldehyde.

An increasing variety of phenol and formaldehyde free syntans are becoming available, many of which are based on acrylate or acrylamide copolymers. One interesting alternative has been developed by Rhodia. Tetrakis(hydroxymethyl)phosphonium (THP) salts are known to interact strongly with proteins, but as tanning materials they can be considered as being too efficient, forming many crosslinks, giving hard brittle products. By adding natural products such as sucrose or sorbitol, the tanning efficiency can be modified such that leathers with a firm but soft feel can be produced.

9.6.1.1 Reverse Tanning. A new 'greener' and cleaner chemical process called reverse tanning could dramatically reduce the amount of chemicals used. The process essentially works backward from the point where conventional tanning ends and saves time, money, and energy while cutting water use and pollution. The new approach flips the process around and eliminates some of the steps, which results in multiple and substantial production efficiencies.

In the new process, prior to tanning the skins are treated with chemicals normally used after tanning is completed. The reverse process produces leather that is comparable to conventional tanning, but requires 42% less time, 54% fewer chemicals, 42% less energy, 65% less water, and cuts emissions of key pollutants by up to 79%. The results were achieved without changing chemicals or using new ones. In addition to costing less and being 'greener' than conventional tanning, the reverse process is said to be easy-to-adopt and could help the global industry overcome emerging environmental and economic concerns.

9.6.2 Fatliquoring

Most of the natural oils and fats present in the hide have been removed during the above process steps; one of the final stages in leather manufacture is to replace these, to give leather a supple feel and prevent the fibres sticking together when dry. This process is called fatliquoring. In general fatliquoring uses environmentally benign products from renewable resources. The most common fatliquors are vegetable oil based fatty acids. To ensure penetration they are applied as an emulsion in water, often under mildly basic conditions with subsequent fixing at lower pH. The level of unsaturation in the acid has an important bearing on the properties. High levels of unsaturation give poor lightfastness and oxidative stability due to UV-initiated oxidation. Often, highly unsaturated oils are sulfited or sulfated to both overcome this and provide additional water solubility and compatibility. Mineral oils, although relatively environmentally unfriendly, are still used in some applications largely due to their good fastness and waterproofing properties.

Like any complex industrial process leather manufacturing produces waste and uses some non-environmentally benign materials. During the last few years all aspects of the process have undergone some greening. Perhaps the most significant development has been the co-operation between chemical manufacturers in the developed world and the tanneries in the developing countries. This has seen development of more benign products as well as transfer of greener technology to the tanners.

9.7 DYEING TO BE GREEN

The synthetic dye industry is well over 100 years old with manufacture and application of dyes being carried out globally. Environmentally, the industry has had and continues to have problems from both manufacturing and use aspects. Many dyes are based on aromatic amine chemistry, often involving multi-step processes; as such, waste generation from manufacturing is high. More importantly many of the early products containing specific groups (*e.g.* toxic dye intermediate groups **9.1–9.3**) were recognized in the 1960s as leading to higher than normal instances of bladder cancer in industry workers. Dyes containing toxic groups such as these are now banned and the whole industry is covered by strict legislation in most countries. As with the leather industry, much of the actual dyeing process is carried out in third world countries in small, often unregulated, factories. Traditional dying processes use 100 kg water per kg of textile. Effluent from these factories is causing real concern in places like India and China, and end of pipe remediation technology is being increasingly employed. Developments in membrane technology and electrochemical oxidation processes have proved useful in cleaning up effluent from dyehouses. One inherent cause of this effluent is that the dyeing process is less than perfect. With many dyes, reaction with, or absorption onto, the fabric being dyed is relatively poor, leaving a considerable amount of dye, together with auxiliary chemicals, to be disposed of. This is obviously not good environmental or economic practice. In recent years new, more efficient, but more expensive, dyes have been introduced to help solve this problem. As old, traditional dye manufacturing processes have followed the end users to third world countries, the chemical industry in the West has responded by introducing these more 'high tech' dyes that command higher margins, but which offer environmental and overall process cost advantages.

9.1   9.2   9.3

9.7.1 Some Manufacturing Improvements

Since many dyes contain aromatic amines one of the most important reactions carried out in the synthesis of dye intermediates is aromatic nitration, involving the use of stoichiometric amounts of nitric and sulfuric acid as discussed previously. Many cleaner nitration processes have been proposed; here discussion will be limited to use of lanthanide triflate catalysts. Lanthanide triflates were discussed as 'green' Lewis acids for Friedel–Crafts and aldol reactions in Chapter 4; the same properties of water stability and ease of recovery also make them valuable in nitration. Work with lanthanide triflates led to the conclusion that increasing activity was related to the increase in size to charge ratio of the metal as well as the nature of the counter-ion. This has led to the development of active catalysts based on readily available metals

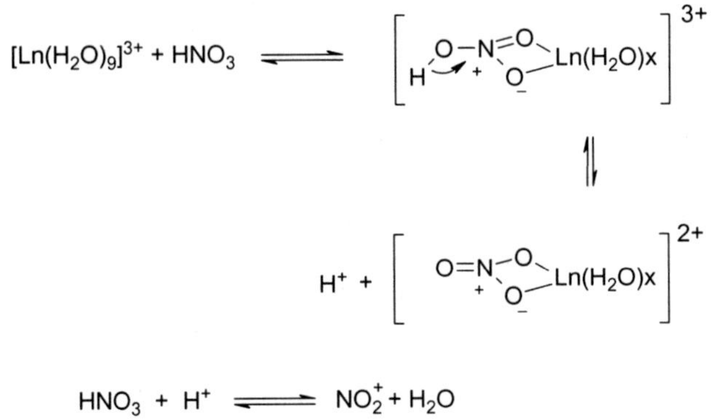

$$[Ln(H_2O)_9]^{3+} + HNO_3 \rightleftharpoons \left[\begin{array}{c} \text{O}-\text{N}\overset{\text{O}}{=} \\ H\overset{+}{\diagdown} \quad \text{O}^- \end{array} Ln(H_2O)x \right]^{3+}$$

$$\left[\begin{array}{c} \text{O}=\text{N}-\text{O} \\ \overset{+}{}\quad \text{O}^- \end{array} Ln(H_2O)x \right]^{2+} + H^+$$

$$HNO_3 + H^+ \rightleftharpoons NO_2^+ + H_2O$$

Scheme 9.9   Lanthanide-catalysed generation of nitronium ions.

such as zirconium, which has a large size to charge ratio. The counter-ion effectively is a phase transfer agent facilitating transport of the nitronium ion from the aqueous phase were it is generated (by inter-action with the metal, Scheme 9.9) to the organic phase were the reaction occurs.

Much research has been carried out into direct amination of aromatic substrates, typified by the direct conversion of benzene into aniline using ammonia and a catalyst. Although there have been many patented routes conversions are normally low, making them uneconomic. Modern catalysts based on rhodium and iridium, together with nickel oxide (which becomes reduced), have proved more active; such is the research activity in this area that it is only a matter of time before such process become widely used.

In terms of manufacturing the most important advances have been made in the synthesis of new more potent dyes, resulting in less dye being used and wasted. Owing to their ease of application and brilliance of colour, reactive dyes for cotton are generally regarded to be the most significant advance in the dye industry during the last 50 years, with production totalling some 120 000 tpa. There are three main types of reactive dye, the most common being based on chloro-*s*-triazines; other types include chloropyrimidines and vinylsulfones. Manufacturing methods have been improved through traditional process development but serve as a useful reminder of the issues connected with using water as a solvent. Since, for application reasons, the dyes need to be relatively water soluble it is difficult to have an aqueous based manufacturing process that does not result in a contaminated aqueous effluent; for some procedures dye losses due to hydrolysis and directly to water have been reported to be as high as 50%! Taking the dye Reactive Blue as an

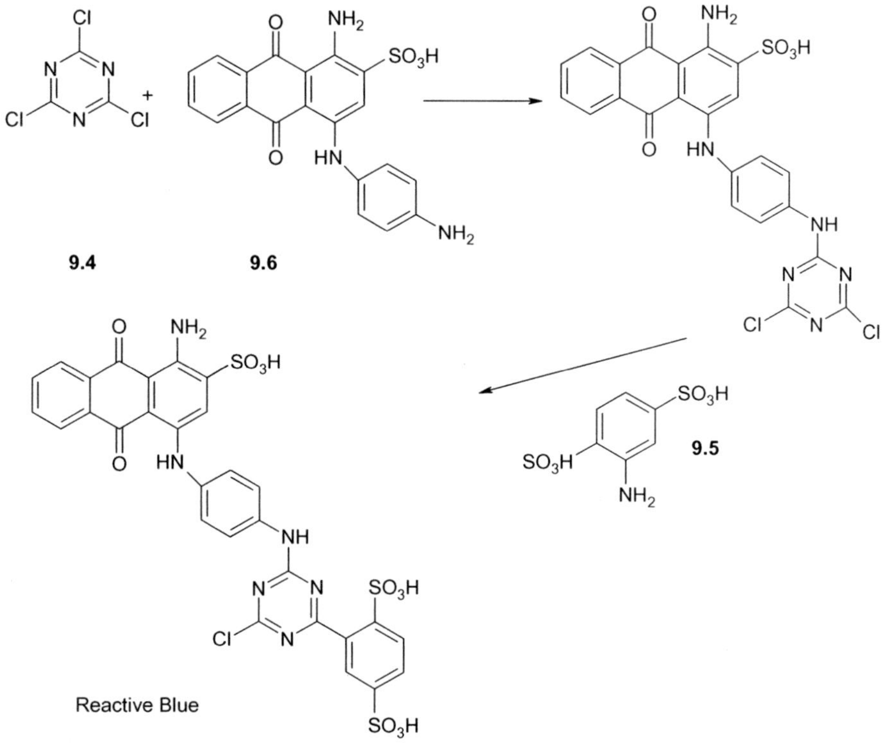

Scheme 9.10 Optimized synthesis of Reactive Blue.

example (Scheme 9.10), much can be done to improve yields by altering the order in which the substrates are reacted, using an excess of one reagent, and careful control of pH. Synthesis involves coupling three reagents, cyanuric chloride (**9.4**), 2-aminobenzene-1,4-disulfonic acid (**9.5**) and 1-amino-4-(4-aminoanilino)anthraquinone-2-sulfonic acid (**9.6**). The main problems are associated with hydrolysis of **9.4** and generation of amine hydrochloride salts, inhibiting further reaction. By reacting in the order shown in Scheme 9.10, using an excess of **9.6** and adding sodium bicarbonate to control pH at 6.7 throughout the reaction yields of 89% can be obtained.

Reactive dyes are so called because they react with the surface of the cotton, forming much stronger bonds and being more resistant to washing out than absorbed dyes. Chlorazine dyes attach themselves to the cotton *via* the residual Cl reacting with hydroxyl groups on cellulose. Competing hydrolysis reactions, however, result in considerable dye wastage. Typically, for every tonne of reactive dye applied 0.3 tonnes end up in the aqueous effluent, which also contains 35 tonnes of salt (added to force the dye onto the cotton). To minimize the effect of hydrolysis DyStar has developed a range of reactive dyes containing multi-chlorazine groups (**9.7**), such that if one is hydrolysed attachment to the fabric can still take place. Overall dye wastage is reduced, and because of an increased affinity for cotton less salt can be used, resulting in a lower effluent volume. Because of the higher molecular weight water solubility is lower; the dying temperature therefore needs to be raised but overall, due to much faster throughput and reduced losses, energy consumption per tonne of cotton dyed has been reduced by 50%. The energy savings have been important in persuading dyehouses to use these more environmentally benign but more expensive dyes. This work, which has provided commercial benefits for the dye producer and user as well as reducing the environmental burden of the process won DyStar the first Industrial UK Green Chemistry Award in 2000.

9.7 (L = linking group)

9.7.2 Dye Application

Owing to significant dye losses in the water effluent there has been considerable interest in use of scCO$_2$ as a solvent, since at the end of the process the solvent could be vented, leaving residual dye for reuse. There are many technical and cost challenges to be overcome before there is widespread commercial acceptance. Cotton, for example, is dehydrated by scCO$_2$, resulting in poor dyeing; this can be largely overcome by adding a plasticizer such as propylene glycol, but this must be subsequently removed by washing. Most of the work has been done with relatively non-polar disperse dyes since these are likely to be more soluble in scCO$_2$; however, many of the commercially important dyes have very low solubilities. Recent findings suggest that this may not be a significant problem since it is the partition coefficient between the fabric and the solvent that is important. PET fibres, for example, have high coefficients and can be dyed with substances that only have a solubility in scCO$_2$ of a few ppm. Dyes containing fluorotriazine groups, in the presence of small amounts of acetic acid, have been found to be particularly good for dying cotton. Commercially, dying using scCO$_2$ is still in its infancy but is being used on a fairly small scale for dying buttons, zips, *etc.*

9.8 POLYETHYLENE

Polyolefins are often regarded as environmentally unfriendly due to their persistence in the environment. In recent years the manufacture of polyethene, which accounts for almost 25% of all polymer production, has undergone significant change, which provides an excellent example of two important facets of green chemistry. First, getting more product out of the same reactor and, second, producing a superior quality product so that less material is required to perform the same function.

9.8.1 Radical Process

Low-density polythene (LDPE) was accidentally discovered in 1935 by ICI when ethene was heated at high temperature and pressure (presumably with a trace of oxygen). Since then this free radical chain growth polymerization process has gained widespread commercial use, producing material for use in films, extrusion coating, and wire and cable. The basic process involves heating ethene in the presence of a small amount of initiator (typically oxygen or organic peroxide) at temperatures of 200–300 °C and pressures of 1500–3000 bar. The mechanism (Scheme 9.11) involves thermal breakdown of the initiator, propagation by reaction of the growing chain with ethene and termination *via* hydrogen abstraction, recombination of two chains, or disproportionation. An important aspect of the process is branching; these

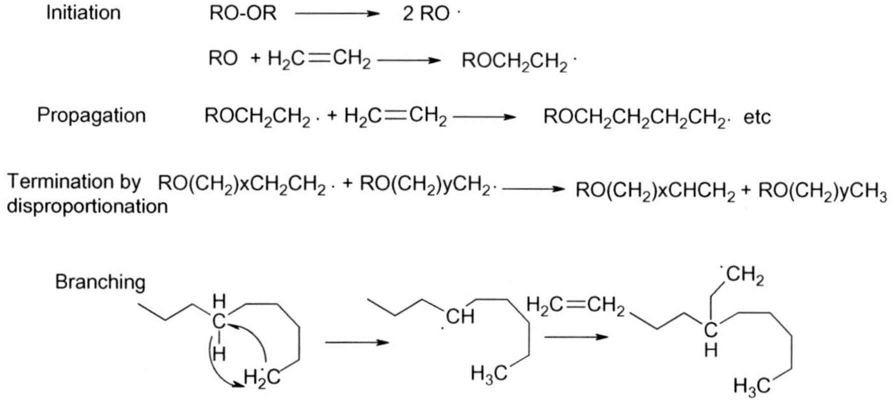

Scheme 9.11 Mechanism of LDPE (low-density polythene) synthesis.

branches are formed by internal chain transfer or backbiting. Typically there are around two such branches per 100 chain carbon atoms. The degree of branching affects the degree of crystallinity and hence the packing and density of the polymer; branching is controlled by temperature, the degree increasing with increasing temperature.

The 'green' credentials of the process could be improved in several respects, the most obvious being the high-energy requirement to operate under very high pressure and relatively high temperature. Ethene polymerization is a highly exothermic process ($-120\,kJ\,mol^{-1}$) and above 300 °C ethene decomposition may occur. Hence the process is operating in a critical regime where slight variations in temperature, *e.g.* through poor heat transfer, may lead to explosion. To run safely, provide adequate heat transfer, and keep reactant viscosity low the process is operated at low conversion, often around 20% per pass. This large recycle stream adds to the energy use as well as the cost of capital employed.

9.8.2 Ziegler–Natta Catalysis

The development of Ziegler–Natta catalysts (*e.g.* titanium chloride and triethylaluminium) in the 1950s improved the process economics and safety of polyethene production by enabling the reaction to be carried out below 100 °C and 50 bar. The product, however, does not contain any branches and therefore has a higher density (HDPE); because of this it is more suitable to blow moulding and injection moulding applications. It was subsequently discovered that small amounts of higher alkenes (hex-1-ene or but-1-ene) could be copolymerized to give branches and produce a material of similar density to LDPE. This material is known as linear low-density polyethene (LLDPE), the actual density being controlled by the type and quantity of co-monomer (typically up to 3 wt%).

To obtain good mixing of ethene with the catalyst, the original Ziegler–Natta processes used hexane as a solvent. Despite being almost completely recovered the use of solvent, particularly a hazardous material like hexane, detracts from the greenness of the process. Since the catalyst is highly moisture sensitive it needed to be deactivated at the end of the process by addition of water or alcohol; this produced a small waste stream. Despite these two environmental negatives the energy, safety, and cost advantages of the process ensured its commercial success.

The next important break through was the development of fluidized bed technology and supported catalysts such as the Phillips chromium on silica catalyst. A typical fluidized bed process operates at around 80 °C and 20 bar, with a feed of ethene, butene, and hydrogen (to control

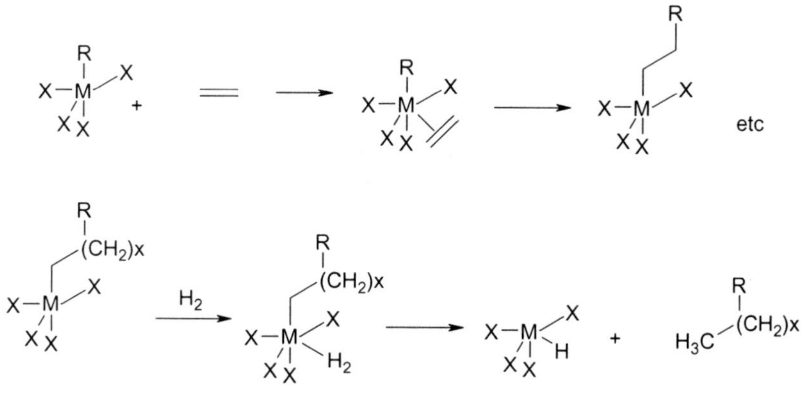

Scheme 9.12 Outline mechanism of LLDPE (linear low-density polyethene) process.

molecular weight). This gas stream passes through a gauze and fluidizes the small catalyst particles onto which the polymer grows. The catalyst activity is such (up to 1000 kg polymer per g metal) that its removal is not required. Unused gases are recirculated and polymer is withdrawn from the base of the reactor, with each particle having an average residence time of around 4 h. An important feature of the process is the cyclone through which the unused gasses pass; it removes fine polymer particle, preventing equipment fouling. These fines are returned to the reactor. High reaction rates are maintained throughout the polymerization process due to the polymers growing by replication. In essence this means that the shape of the catalyst particle is maintained but that the polymerization process breaks down the catalyst particle into smaller ones held together by polymer; this helps prevent diffusion becoming rate limiting. Overall, the mechanism of these heterogeneous processes is similar (Scheme 9.12). Such processes have the same energy and safety benefits of the solution process but avoid the need for solvent and aqueous quench. The cost advantage of such a process is large, with production costs being over 10% less than the solution method and some 15% less than the LDPE process.

9.8.3 Metallocene Catalysis

The latest processes involve the use of metallocene catalysts, which were first identified in the 1970s. There are now many different types of metallocene catalyst available but they essentially consist of a transition metal cyclopentadiene or dicyclopentadiene complex (**9.8**) and an alumoxane (**9.9**), usually the metal will be Ti, Zr, or Hf. There are several advantages to using metallocene catalysts connected to improved efficiency, ease of work up, and product quality:

- *Improved co-monomer incorporation:* Metallocene catalysts are very efficient at co-monomer incorporation; this means that co-monomer use can be reduced by a factor of ten or more. This has some cost advantages but, more importantly, there is less un-incorporated co-monomer in the final product, improving efficiency and mass balance and reducing VOCs.
- *Homogeneous polymer production:* Ziegler–Natta catalysts produce some chains with many branches and others with few – in contrast the distribution of branches when using metallocenes is much more uniform. This leads to less wasted co-monomer and a more homogeneous polymer, which has beneficial effects on polymer strength and permeability. Importantly, reduced co-monomer levels reduce product stickiness, which enables the process to operate at a higher temperature, significantly increasing product throughput.
- *Improved hydrogen chain transfer:* Because the Ziegler process is fairly inefficient in hydrogen transfer as much as 5 mol% may be present in the gas stream. Owing to the poor heat capacity of hydrogen this reduces the overall efficiency of heat removal. Metallocene catalysts are more efficient at hydrogen transfer, enabling lower amounts to be used (down to 0.01 mol%); hence the proportion of high heat capacity monomer is relatively increased. The effect of this is an estimated 10% increase in production capacity. In addition it is much easier to recover unused gasses.
- *Condensed mode operation:* To avoid accumulation of liquid in the reactor the dew point of the gas must be kept above reactor temperature. When operating in condensed mode higher boiling components in the recirculating gas are allowed to condense in the heat exchanger; this liquid is fed back to the reactor, where it evaporates, efficiently removing heat. Owing to lower co-monomer levels and the improved homogeneity of the product the amount of condensate can be increased, when using metallocene catalysts, before resin stickiness occurs. This extra heat removing capacity can improve reactor throughput by up to 30%.

9.8   9.9

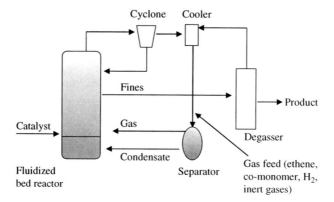

Figure 9.2   Outline metallocene process.

As will be evident from the above discussion metallocene catalysis have brought about huge gains in reactor productivity and energy efficiency. Similar to our discussion of acetic acid production, improved reactor efficiency results in fewer chemical plants and a reduced environmental burden. Owing to the improved homogeneity and narrower molecular weight, polymer properties such as strength are superior when using metallocenes; this is reflected in thinner polymer films being able to be used in some packaging applications, again reducing the overall environmental burden. Figure 9.2 depicts an outline process.

9.8.4   Post Metallocene Catalysts

Metallocene catalysts are homogeneous single-site systems, implying that there is a single, uniform type of catalyst present in the system. This is in contrast to the Ziegler–Natta catalysts that are heterogeneous catalysts and contain a range of catalytic sites. The catalytic properties of single-site catalysts can be controlled by modification of the structure of the catalyst. Because of the oxophilicity of metals used in metallocene catalysts they are not useful for copolymerizations involving polar monomers. In an effort to overcome this, metallocene-type catalysts based on palladium and nickel have been studied – frequently called post-metallocene catalysts. These catalysts have now been commercialized for polyethylene production. The catalysts can homopolymerize ethylene to various structures that range from high density through hydrocarbon to elastomers by a mechanism referred to as 'chain-walking'. By reducing the bulk of the α-diimine used (**9.10**) the product distribution of these systems can be 'tuned' to consist of hydrocarbon oils (α-olefins), similar to those produced by more traditional nickel(II) oligo/polymerization catalysts.

9.10 (M = Ni, Pd)

9.9 ECO-FRIENDLY PESTICIDES

Pesticides, which includes insecticides, fungicides, and herbicides, are widely used throughout the world to help maximize food and other crop yields, the total market being worth around $30 000 million. Of course it may be argued that no-pesticide is 'green' and that the organic option is the only eco-friendly one. The opposing argument is that use of pesticides has contributed enormously to the worldwide availability of affordable good quality food. Historically, pesticides have been highly toxic chemicals, being poisonous to humans as well as the insect or fungus. Early examples include organomercurials used as antifungal seed dressings and calcium arsenate used to control boll weevil on cotton. Since the 1930s various organic pesticides have been used that have gradually become more target specific, less toxic to mammals and birds, and more biodegradable. Producing highly selective pesticides that degrade at a suitable rate is obviously difficult. The DDT story discussed previously (Chapter 2) is testimony to these difficulties.

9.9.1 Insecticides

Even with modern insecticides it is estimated that some 30% of food crops are consumed by insects, with the vast destructive power of swarms of locusts accounting for a significant proportion of this. One of the earliest insecticides used to control aphids was the natural product nicotine. Despite being 'natural' it is highly toxic to all mammals, perhaps surprisingly (given its presence in tobacco) a single dose of 40 mg is enough to kill a human being! Nicotine is rarely used outside the Far East these days. A much less toxic natural insecticide is Derris, an extract from plant roots. Whilst this is quite toxic to caterpillars it has a fairly low mammalian toxicity and is quickly broken down in the soil. The most widely used natural insecticide is pyrethrum, which is extracted from certain chrysanthemum flowers and is particularly active against flying insects but again has very low mammalian toxicity. The main problem (which is also an environmental benefit) is that it is very rapidly destroyed by sunlight. A range of synthetic pyrethroids (**9.11** and **9.12**) has therefore been produced that aim to combine low mammalian toxicity with sufficient stability to environmental conditions and high activity.

9.11 Permethrin, lasts for 7 days on folliage

9.12 Tefluthrin, active against soil pests

Since the problems with DDT, organochlorine insecticides have largely been replaced by organophosphates. Early organophosphates developed in the 1940s were highly toxic, being closely related to nerve gases! One of the first active but relatively non-toxic (to mammals) organophosphates with acceptable persistence was malathion, although its synthesis (Scheme 9.13) leaves something to be desired. This material was not highly selective, and could kill beneficial insects. A significant advance was made by ICI with the introduction of menazon (Scheme 9.13) since this material controls aphids but in addition to low mammalian toxicity it is harmless to many beneficial insects such as bees.

Organophosphates and the other common class of insecticides, carbamates, act by inhibiting acetylcholinesterase in the central nervous system. This enzyme is used to destroy the neurotransmitter acetylcholine, which is responsible for transmitting messages between nerve cells. By reacting with this enzyme in an irreversible way the nerve cells are continuously stimulated causing death. Most insects and mammals have similar nerve functions using acetylcholine esterase; hence development of a species-specific insecticide is very difficult with this class of insecticide. Recently, new insecticides have been developed that have different modes of action, interrupting vital processes that are specific to the target insect.

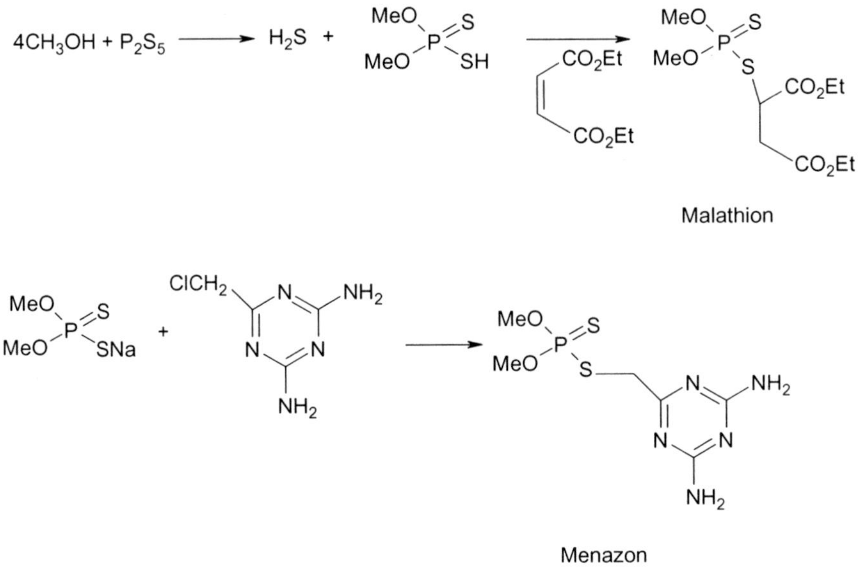

Malathion

Menazon

Scheme 9.13   Synthesis of organophosphates.

Caterpillars and other moulting insects excrete the hormone α-ecdysone, which at moulting time becomes hydroxylated to 20-hydroxyecdysone (20-E), which in turn triggers the moulting process, enabling the insect to shed its exoskeleton and resume feeding. Rohm and Hass have developed a novel insecticide, tebufenozide (**9.13**), which mimics 20-hydroxyecdysone, binding to the same site. The consequence of this is that the insect stops feeding and starts to shed its exoskeleton. At this stage in the normal moulting process levels of 20-E would drop, enabling a new exoskeleton to be grown and feeding to begin; tebufenozide, however, binds more permanently, preventing resumption of feeding causing death. Since this process is specific to certain insects, tebufenozide has very low toxicity to mammals, birds, fish, and other insects. This material has been classed by the EPA as a reduced risk pesticide and won its developers a coveted Presidential Green Challenge Award.

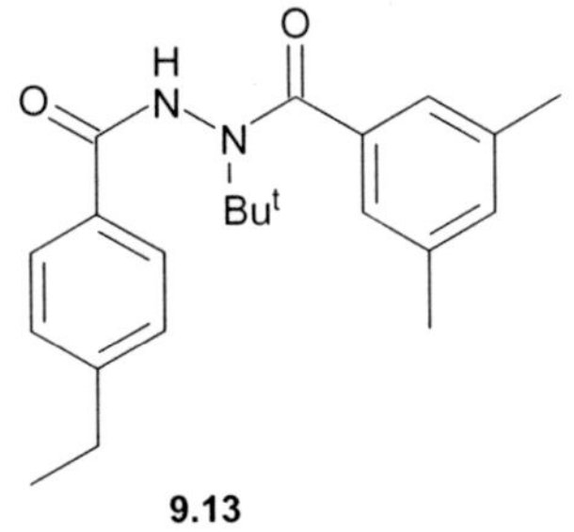

9.13

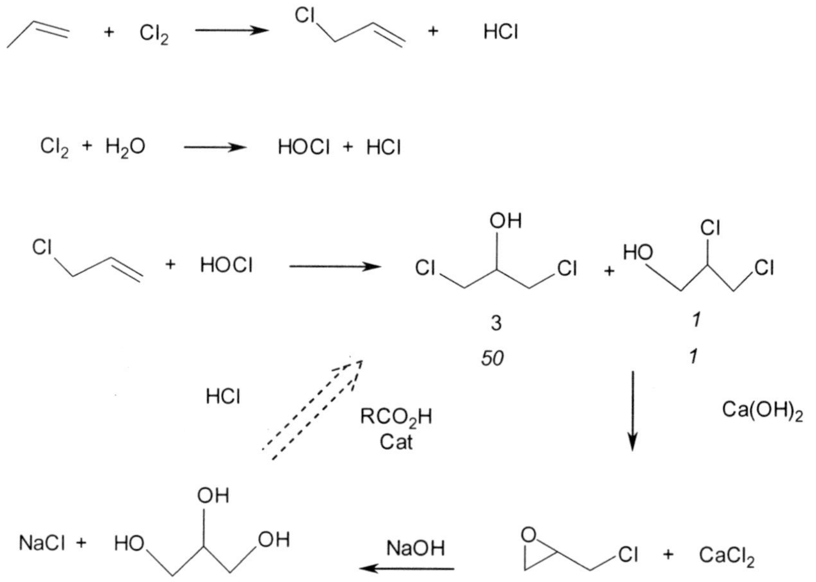

Scheme 9.14 Traditional manufacture of epichlorohydrin and glycerol.

9.10 EPICHLOROHYDRIN

Epichlorohydrin is a key raw material for the manufacture of epoxy resins that are used in the automotive, electronics, and packaging industries as well as in the production of many sports goods and wind turbines.

Until recently epichlorohydrin has been manufactured by the high-temperature radical initiated chlorination of propene with chlorine to give allyl chloride (Scheme 9.14). Although substitution of allylic hydrogen is quite selective several unwanted chlorinated by-products are produced. The overall yield of allyl chloride is around 85%. Allyl chloride is then reacted with hypochlorous acid, at slightly above ambient, to give a mixture of dichlorohydrins. These are separated and treated with lime to give epichlorohydrin, in an overall yield from propene of about 73%.

Other than epoxy resins one of the main markets for epichlorohydrin has been as a raw material for glycerol manufacture, made by hydrolysis of epichlorohydrin with dilute sodium hydroxide. Pure glycerine was obtained through a series of energy intensive distillations.

Overall the process uses hazardous material (chlorine), produces troublesome chlorinated by-products, as well as a considerable amount of salt waste, and because of the distillations and high-temperature first stage is relatively energy intensive. In terms of atom economy only 50% of the chlorine used ends up in the epichlorohydrin and none in the glycerol.

With the advent of biodiesel and the co-production of copious amounts of inexpensive glycerol there is no longer a need for a process converting epichlorohydrin into glycerol. On the contrary, researchers are actively looking for new processes that can use bio-glycerol as a raw material. The basic reaction of glycerol with hydrogen chloride to produce a mixture of dichlorohydrins has been known for many years (dotted arrow in Scheme 9.14). Several companies, including Dow and Solvay, re-examined this work with a view to making it a commercial success; the main problem being driving the reaction to completion to get viable conversions without the need for repeated distillations and recycle of the glycerol stream.

One solution to this problem is to use a small overpressure of HCl to drive the equilibrium forward at a reaction temperature of about 120 °C. Addition of about 2 wt% carboxylic acids speeds up the reaction – although acetic acid works well, the use of a higher molecular weight acid makes recovery easier. An added advantage of this process is that the ratio of 1,3-dichlorohydrin to the 1,2-isomer is much higher; the former is much more readily converted into epichlorohydrin, making the second part of the process faster. Overall advantages of the new process, commercialized by Solvay in 2007, include a switch of chlorine source from elemental chlorine to hydrogen chloride, which is a often a by-product on a large petrochemical site, unwanted chlorinated by-products have been reduced by a factor of 8, water consumption by a factor of 3, and of course it is now based on a renewable feedstock.

REVIEW QUESTIONS

1. Discuss the application of catalytic carbonylation processes in the development of green chemical technology. Highlight the application challenges that need still need to be met and review current research in the area to meet these challenges.
2. Assess the Diels–Alder reaction between cyclopentadiene and 1,3-butadiene, drawing structures for all likely products from the reaction. Suggest ways in which the selectivity of the reaction may be improved.

3. Review the effect that legislation has had on the manufacture of dyes during the last 25 years. Highlight two dyes that are no longer produced due to legislation and show how they have been effectively replaced by more environmentally friendly alternatives.
4. For a bulk chemical of your choice (not reviewed in this chapter) discuss how the process has been developed during the last 50 years to improve the process economics and its environmental impact.

CHAPTER 10

The Future's Green: An Integrated Approach to a Greener Chemical Industry

10.1 SOCIETY AND SUSTAINABILITY

Historically, society in general, and industry in particular, developed with more or less complete disregard for the environmental consequences. It can be argued that this pioneering, risk taking approach moved society forward in many respects in the nineteenth and early twentieth centuries. With global industrialization and a growing impact of civilization on the earth these attitudes are no longer appropriate. Many environmental organizations argue that the unsustainable use of resources are a direct consequence of the profit driven, capitalist, competitive society in which many of us now live; although evidence from former communist states suggests that this is a too simplistic an argument. In reality the environmental problems we have today and predict for the future are, at least in part, due to societies' collective pursuit of short-term economic growth. Even though chemical development has gradually reduced the amount of waste and overall environmental burden per tonne of product there is still much to do before we have a sustainable chemical industry, let alone a sustainable society. Clearly, for a sustainable future we need a different framework in which to operate. The chemical industry is, however, a key solution provider on the road to sustainability. An independent study has shown that, on average, for every tonne of greenhouse gas emitted during chemical

Green Chemistry: An Introductory Text, 2nd Edition
By Mike Lancaster
© Mike Lancaster 2010
Published by the Royal Society of Chemistry, www.rsc.org

production three tones are saved by society by using products of the chemical industry. Chief amongst these products are insulation, advanced materials for reducing weight and hence fuel consumption of cars and planes, specialist coating for reducing drag of ships, and fuel additives.

One of the conclusions from the 1992 United Nations Conference on Environment and Development in Rio de Janeiro (the Earth Summit) was the urgent need to find a more sustainable way of life, based on careful use of resources and a reduction in environmental emissions. There was also a call to move towards a model in which environmental enhancement is fully integrated with economic development. The consequences of this summit have been far-reaching, not least by the fact that, in Europe and elsewhere, environmental protection requirements are now integrated into many European policies rather than being separate pieces of legislation. Indeed Article 2 of the EC treaty states that the Community shall . . .

'promote throughout the Community harmonious, balanced and sustainable development of economic activities.'

This may be viewed as the first step towards creating a framework for a win–win situation in which economic growth goes hand-in-hand with environmental protection. This chapter discusses some aspects of recent thinking that will help establish a framework from which it will become easier to develop more sustainable products and processes.

10.2 BARRIERS & DRIVERS

It is agreed that sustainable development has not been achieved; perhaps the two words will prove mutually exclusive. What is clear is that all sectors of society must make greater effort to achieve sustainable development if the world as we know it is not to become irreparably damaged by the end of the twenty-first century. Although many chemists now generally believe this, and significant progress is being made, there are barriers that are hindering greater adoption of greener technologies (Table 10.1).

The barriers can be broadly divided into three categories: knowledge, legal, and economic. To break down the knowledge barriers we must start to have school and university chemistry courses underpinned by green chemistry such that when young chemists enter industry they are concerned with issues of waste, energy efficiency, and safer design, *etc.* When developing synthetic procedures chemists will automatically turn

Table 10.1 Adoption of greener technologies: some barriers & drivers.

Barriers	*Drivers*
Lack of global harmonization on regulation/ environmental policy	Legislation – cost of end of pipe treatment
Lack of sophisticated accounting practices focussed on individual processes	Competition
Difficult to obtain R&D funding	Public pressure
Notification processes hinder new product & process development	Potential for reducing costs
Short term view by industry and investors	Less hassle from HSE/ Environment Agency
Difficult to obtain information on best practice	Licensing opportunities
Lack of clean, sustainable chemistry examples & topics taught in schools & universities	Chemical debottlenecking
Culture geared to looking at chemistry not the overall process/life cycle of materials	Good PR – fewer problems with neighbours
Lack of communication/understanding between chemists & engineers	
Lack of technically acceptable 'green' substitute products and processes	

to a benign solvent system as a first choice and simplicity not complexity will be become the trade-mark of a great organic synthetic procedure. Adequate knowledge is also lacking in the industrial arena; there are communication gaps between academia and industry and between scientists and engineers. Green chemistry may provide the common language to help bridge this gap and aid technology translation.

Relevant legislation on a global basis creates an uneven playing field and may encourage adoption of less environmentally friendly practices in some parts of the world. In addition, regulation and the associated costs regarding introduction of (often more environmentally friendly) new products can act as a real deterrent. Finally, and perhaps most importantly, there are economic barriers, both real and perceived. Increasingly stakeholders demand short-term profits from industry; this, with the possible exception of the pharmaceutical sector, has had an adverse effect on long-term R&D and development of new technology. There is also the perception that new green technology must be more expensive; this coupled with the lack of 'demonstration' facilities and case studies has hindered adoption of some developments coming from the research base. This will improve with time but can be fast-tracked by more widespread industry co-operation, aided by government incentives.

To overcome these barriers and to make the drivers more appealing a more unified approach from industry, government, and society in general is required. In the short-term greater co-operation and a culture that values the sharing of best practice combined with a legislative and tax framework that encourages green technology is needed. In the longer term the chemist's answer to sustainable development lies with the development of novel technology that is energy efficient, produces little waste, but does produce a benign recyclable and economic product that helps other sectors of society become more sustainable.

10.3 ROLE OF LEGISLATION

Aspects of legislation, particularly those concerned with waste minimization and development of new processes, have been discussed in Chapters 2 and 3, and only a brief summary of the role it can play is required here. Legislation also has a major impact on the development of new products and this aspect is discussed more fully in Chapter 3 (Section 3.1). Traditionally introduction of environmental legislation has been deemed necessary to control environmental pollution and as the mechanism society uses to control and change the behaviour of industry towards environmental, health, and safety issues. Not surprisingly industry has largely viewed this negatively as the imposition of extra costs. The argument is sometimes used that the costs to industry of environmental legislation has had a negative effect on R&D spending and have forced industry to be less risk taking. The consequence of this is that introduction of new 'greener' process technologies have been delayed in favour of quick end of pipe 'fixes'.

Recently there has been a shift in many countries to develop regulatory systems that provide incentives for industry to change whilst maintaining environmental standards. Here the emphasis is on co-operation between industry and the regulating body, with the regulatory body providing help and guidance and only using legislation as a last resort. Social pressures also play a role here in encouraging industry to adopt sound environmental policies. In general it is unlikely that social pressure can have the same effect as legislation and it is more likely that social pressures will influence government policies rather than industry directly. By improving its reputation through communication of the benefits the chemical industry brings to society as well as proving a good track record on health, safety, and environment issues the industry can gain the trust of society and thereby create less legislative pressure.

The introduction of the Alkali Act in 1863 to curb the adverse health effects produced from emissions of HCl from the Leblanc sodium carbonate process was discussed in Chapter 2. This act stated the particular steps companies had to take to reduce emissions. Although this approach, if used wisely, could ensure the adoption of the latest best practice it tends to stifle innovative solutions to problems. Also, whilst helping to ensure an even playing field, it may not be necessary or even appropriate to adopt the same technical solution to different processes in different locations.

This prescriptive approach to legislation has today largely been superseded by one of compliance standards. Under this approach legal limits are set for the discharge of material to water air and land, how these limits are actually met is left to the discretion of the company, hence encouraging an innovative low cost approach. A variation on the imposition of absolute limits is the Polluter Pays Principle, which operates through a system of increasing charges as base emission levels are exceeded. Whilst regulations of this type leave it to the individual company to decide on the relative merits of various solutions (*e.g.* end of pipe *versus* integrated process pollution prevention) the reality is that end of pipe solutions are often adopted. The reasons for this include:

- the need to reduce emission levels quickly;
- 'off the shelf' availability of many products;
- low risk;
- minimal disruption to production.

Although this approach enables companies to comply with legislation and harmful emissions will be prevented from entering the environment it does not provide the best framework for sustainable development. A much better approach would be to develop legislation that encourages pollution prevention at source. In this respect, the old prescriptive approach has something to offer; however, it is generally accepted that with the number of diverse operations we have today this would be unworkable.

One ongoing debate concerns the role that adoption of the 'Precautionary Principle' can have in future legislation. The Precautionary Principle advocates that where there is a potential serious risk to human health or the environment then a decision should be made to eliminate the cause of that risk even though there may be no definite scientific proof of hazard or risk of causing harm. A wide ranging example of the precautionary principle is an EU directive that restricts the deliberate release of genetically modified organisms despite there being no conclusive scientific evidence that they would cause harm.

Although the principle has been incorporated into several international directives and conventions, *e.g.* Convention on Climate Change, it has found only limited judicial support. For example, the EU failed in a case to ban US beef on the grounds that it contained growth hormone that may have a detrimental effect on human health.

In general, industry is against more regulation and legislation, particularly if it creates an uneven global playing field. The alternative is voluntary regulation in which industry sectors agree to limit emission of certain substances; for example, European chlorine manufactures have volunteered strict limits on mercury discharges into the North Sea. In this case manufacturers are usually more willing to share best practice to ensure compliance with the code, avoiding possible legislation. This approach has worked particularly well in Holland, but here breaking a voluntary code can result in similar consequences to breaking a law. Although difficult to obtain, a global legislative framework is required that encourages innovation and development of more sustainable products and processes. Current legislation can be viewed as occupying the middle ground between what industry wants for increased profitability and what some NGOs want to reduce chemicals in the environment. Perhaps more contrasting legislation designed to reward sustainable development and 'punish' those who ignore societies changing requirements would serve us better.

10.4 GREEN CHEMICAL SUPPLY STRATEGIES

Most of this book has been concerned with greening of chemical synthesis and production processes, which is the core business of the chemical and pharmaceutical industries. The chemical industry has traditionally thrived on selling ever increasing amounts of chemicals to end users who then produce a product for sale to the customer (ultimately the general public); this supply chain is depicted in Figure 10.1.

In short, to increase profits (and hence shareholder value) the chemical company has had to increase sales of chemicals. The consequence of this has been increasing amounts of chemicals requiring treatment and disposal, at added cost to the end user. The relationship between the supplier and end user can be viewed as a competitive one, the supplier wishing to aggressively increase sales of chemicals and the end user wishing to minimize their use per unit of finished product, increased

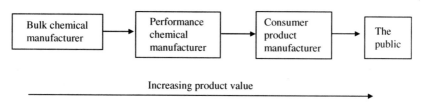

Figure 10.1   Example of chemical supply chain.

profits for one often coming at the expense of the other. The environmental implications of this kind of relationship are significant. A much better relationship, at least from an environmental point of view, would be one in which both suppler and end user could gain from a reduction in the amount of chemicals used.

The chemicals sold to end users such as car manufacturers, white goods manufacturers, electroplating firms, and paper manufacturers, *etc.* are termed performance chemicals. These chemicals are sold to produce an effect or performance and are often complex formulations. The end user is really only interested in buying a performance and often knows little about how the chemicals used produce the desired effect. With the current trend of focussing on core business, the car producer, for example, will wish to concentrate on the overall manufacturing and marketing of the car rather than spending time optimizing spray paint finishing if someone else can do it at an acceptable cost. This kind of approach opens up the way for improved supply relationships to be negotiated. Broadly there are three kinds of relationship, as identified below:

1. *Traditional* – Chemicals are sold on a price per kilogram basis, with the supplier not being involved in the users process – unless specific problems related to the chemical occur. Disposal of unused or spent chemicals is the end users problem. Suppliers are chosen mainly on price, quality, and ability to supply. The supplier has an incentive to sell more chemicals.
2. *Service* – Chemicals are still purchased on a price per kilogram basis. The supplier offers additional services such as just-in-time delivery, direct electronic ordering *via* an electronic data interface, routine analysis of process streams and advice on chemicals usage, *etc.* The supplier still has an incentive to sell more chemicals, but does have more to lose should the end user change supplier.
3. *Full chemical management service or chemical leasing* – This relationship is structured around an agreed management fee per month or per unit of production to the suppler covering all aspects of chemical supply, performance, monitoring, problem solving and, importantly, including recovery or safe disposal of spent chemicals. In such a relationship the supplier has an incentive to use fewer chemicals since he is receiving a management fee for a service not the supply of chemicals. This type of a relationship is sometimes called 'Shared Saving' since it often includes a clause agreeing to share any savings in reduced chemical use or improved performance between the supplier and the end user.

Table 10.2 Some supplier and user benefits of a full chemical management agreement.

Supplier benefits	User benefits
More secure relationship	Known charges for budgeting purposes
Increased profits	Reduced chemical handling problems
Improved customer & sector knowledge	Easier access to technical experts
Potential to benefit from process improvements	Reduced expenditure on analysis/ monitoring
Incentive to develop more cost effective eco-friendly alternatives	Potential to benefit from process improvements
	Reduced waste
	Can concentrate on core business

From an environmental aspect there are obvious advantages to be gained from a full chemical management relationship. Increasingly both supplier and end user are seeing the mutual benefits (Table 10.2) of a full chemical management agreement.

A small but growing number of these relationships are being reported. Obvious areas to start are those peripheral to the main business of an organization such as waste treatment. For example, a glue and detergent manufacturer who went into partnership with a company that treats waste made a 50% reduction in use of iron chloride and a 15% reduction in sodium hydroxide due to the expertise of the waste management company.

10.5 GREENER ENERGY

A considerable portion of this book has been devoted to greener forms of energy such as biofuels, fuel cells, and solar energy. It is a fact, however, that fossil fuels will still form a large part of the societies' fuel needs for the foreseeable future. Many countries, for example, are now looking at new coal fuelled power stations as forming an important part of the energy mix for the next 50 years. With CO_2 emissions now being a major cause of climate change it is clear that new technology will be needed to prevent these emissions. Carbon capture and storage (CCS) is often seen as the answer.

In essence, CCS involves the capture of CO_2 from large single point sources, transport (usually by pipeline), and storage in underground geological formations. Estimates suggest that this will cut CO_2 emissions from power stations by over 80%; however, capture and compression is energy intensive and may increase the fuel requirements of the power station by an extra 25%. Several demonstration CCS projects based around new coal fired power stations are planned across Europe in the next 10 years.

A possible alternative to storage in geological structures is mineral storage. Here, naturally occurring Ca and Mg minerals are reacted to give the carbonates. These are extremely stable and avoid the possibility of the CO_2 leaking out of underground formations. The idea would be to carry out these reactions at power stations, avoiding the need for compression and long CO_2 pipelines. To be commercially viable reaction rates and energy consumption need to be improved.

10.6 CONCLUSIONS

Hopefully the reader will have concluded that Green Chemistry is not a new branch of science but more a new philosophical approach that underpins all of chemistry and has technological, environmental, and societal goals. Through applying and extending the Principles of Green Chemistry chemists can contribute to sustainable development. There are those who suggest that science and technology is responsible for the current poor state of the environment, climate change, *etc.*, and equally there are those who suggest that science and technology has all the answers. The truth probably lies somewhere in between these extreme views. Science and chemistry has a key role to play in sustainable development but we must not become isolated from other professions and society in general. Just as we discussed the importance of team work in waste minimization so scientists and engineers must work with social scientists, economists, and politicians to develop the appropriate culture, infrastructure, and society as well as the technology developments needed on the journey towards sustainability. Chemists from different disciplines as well as chemists and chemical engineers also need to work more closely together. Increasingly it is being shown that the greatest opportunities for step-change developments come from collaborative projects, *e.g.* from using a new catalyst and a new benign solvent in a new intensified reactor.

REVIEW QUESTIONS

1. With reference to Europe, America, and a typical developing country discuss how legislation affects the chemical industry, with particular emphasis on building new chemical plants.
2. What are the likely effects of a full chemical management supply strategy on the bulk chemicals industry? How may the industry respond to mitigate these effects?

3. Discuss the role chemists can play in sustainable development.
4. For an industry sector of your choice (not chemistry related) describe the environmental impact it has had in the last 25 years and discuss some of the initiatives being taken to make it more sustainable. What lessons can chemistry-based industries learn from this?

FURTHER READING

T. J. Bierma and F. L. Waterstraat Jr., *Chemical Management: Reducing Waste and Cost through Innovative Supply Strategies*, John Wiley & Sons, Inc., New York, 2000.

R. Hofer (ed.), *Sustainable Solutions for Modern Economies*, Royal Society of Chemistry, Cambridge, 2009.

B. Lomberg, *The Sceptical Environmentalist – Measuring the Real State of the World*, Cambridge University Press, Cambridge, 2001.

Subject Index

Breinigsville, PA USA
28 January 2011

254312BV00004B/15/P